SOLUTIONS MANUAL—STUDENT'S VERSION

Chemistry: The Science in Context

SOLUTIONS MANUAL—STUDENT'S VERSION

Chemistry: The Science in Context

Thomas R. Gilbert, Rein V. Kirss, Geoffrey Davies

Daniel A. Durfey

NAVAL ACADEMY PREPARATORY SCHOOL

Edward Witten

NORTHEASTERN UNIVERSITY

W • W • NORTON & COMPANY • NEW YORK • LONDON

Printed in the United States of America

ISBN 0-393-97986-5 (pbk.)

W. W. Norton & Company, Inc., 500 Fifth Avenue, New York, NY 10110
www.wwnorton.com

W. W. Norton & Company, Ltd., Castle House, 75/76 Wells Street, London W1T 3QT

1 2 3 4 5 6 7 8 9 0

CONTENTS

CHAPTER 1 Matter and Its Origins

1. a. Matter is anything that has mass and occupies space. An example of matter is the textbook that you are reading.
   b. An element is matter made up of only a single type of atoms. There are 92 known elements that occur naturally and another 20 or so that are man-made. Some examples are carbon, hydrogen, and iron.
   c. A compound is a pure substance made up of two or more elements chemically combined. Some examples of compounds are water (H_2O), glucose ($C_6H_{12}O_6$), and sodium chloride (NaCl).
   d. A metal is an element that is lustrous, malleable, ductile, and a good conductor of electricity. Metals are found on the left-hand side of the periodic table. About three-quarters of the known elements are metals. Examples of metals are sodium, calcium, nickel, and uranium.
   e. A nonmetal is an element that is neither malleable nor ductile and is a poor conductor of heat and electricity. Examples are chlorine, oxygen, sulfur, and nitrogen.
   f. A cation is a positively charged particle. Some examples are Na^+, Fe^{2+}, and Al^{3+}.
   g. An anion is a negatively charged particle. Some examples are Cl^-, O^{2-}, and N^{3-}.

2. A substance (or pure substance) is a particular kind of matter with well-defined properties and a fixed chemical composition. The two types of pure substances are elements and compounds.

   An element is the simplest form of a pure substance. It is made up of atoms of only a single type.

   A compound is a pure substance made up of two or more elements chemically combined. The masses of the elements that make up a compound have fixed proportions.

3. Physical properties of metals are:
   i) Luster (reflects light)
   ii) Malleability (can be hammered into sheets)
   iii) Ductility (can be drawn into wires)
   iv) Conductivity (good conductor of heat and electricity)

4. Salt is soluble in water. If you add water to the mixture of salt and sand, the salt will dissolve and the sand can be separated by filtration. The salt can then be recovered by evaporating the water from the salt solution.

5. Salt has a very high boiling point (1413°C) compared with water (100°C). Seawater can be desalinated by distillation in an apparatus like that shown in Figure 1.7. When heat is applied to the boiling flask, the water is vaporized to steam while the salt remains in the flask. The pure steam is then cooled by the condenser to give pure liquid water.

6. a. Distillation involves physical changes. In distillation, a mixture is separated into the components that make it up. There are no changes in the identities of the chemicals present before and after distillation.
   b. Combustion is a chemical change. In the combustion process a substance reacts with oxygen to form at least one new substance.
   c. Filtration is a physical process. Again, a mixture is separated into the components that make it up. When a mixture of sand and water is filtered, the sand remains in the filter paper while the water passes through it. In this operation, we have separated the components of this mixture but have not changed the identity of either component.
   d. Condensation is a physical change. It is a change from the vapor (gaseous) state to the liquid state. Changing

water from steam to liquid does not alter its chemical makeup. It is still H_2O.

7. Iron (Fe) and sodium chloride (NaCl) are pure substances.
   a. Fe (iron) is a pure substance (a metallic element).
   b. Gasoline is a mixture of several compounds. It is not a pure substance.
   c. Cow's milk is a mixture of several components. It is not a pure substance.
   d. Sodium chloride (NaCl) is a pure substance (an ionic compound).

9. A physical property is one that can be observed without changing the identity of a substance. A chemical property can only be observed when a substance undergoes a reaction and becomes changed from its original identity.
   a. Density is a physical property. It can be determined by measuring mass and volume and these operations do not bring about any change in the substance measured.
   b. Melting point is a physical property. The state of a substance changes from solid to liquid when melting occurs but it still has the same chemical composition.
   c. Conduction is a physical property. When heat or electricity pass through a metal, the chemical makeup of the metal is unchanged.
   d. The softness of sodium metal is a physical property. If the metal is cut with a knife, it still remains the same substance. Only its shape is changed.
   e. The tarnishing of sodium in contact with air is a chemical property. The change in the surface from shiny to powdery shows that metallic sodium has reacted with oxygen in the air to form a new substance, sodium oxide.
   f. The reactivity of sodium with water is a chemical property. In the observed reaction sodium is changed to sodium hydroxide.

11. Cow's milk and orange juice are heterogeneous mixtures.
   a. Iron (Fe) is an element. It is a pure substance.
   b. Gasoline is a homogeneous mixture. It consists of a number of liquid compounds that are miscible with (soluble in) each other.
   c. The milk produced by a cow is a heterogeneous mixture. It contains large globules of butterfat, which separate out as cream on standing. Most of the milk sold in supermarkets is homogenized milk, which has been processed to break up the globules of butterfat into particles so small that they do not separate out and remain suspended within the body of the liquid milk.
   d. Sodium chloride (NaCl) is a compound. It is a pure substance.
   e. Sodium chloride (NaCl) in water is a homogeneous mixture.
   f. Orange juice is a heterogeneous mixture. It contains particles that settle out to the bottom. Shaking orange juice distributes these particles more evenly throughout the juice.
   g. Nitrogen gas (N_2) is an element. It is a pure substance.
   h. Ammonia gas (NH_3) is a compound. It is a pure substance.

13. You can separate water and ethanol by using the technique of distillation. The lower-boiling ethanol will vaporize more readily and can be collected first. The process can be monitored by constantly reading the temperature. When the temperature rises above the boiling point of ethanol, the vapor being condensed is a mixture of water and ethanol.

    You can use the apparatus pictured in Figure 1.7.

15. A hypothesis is a tentative explanation for an observation or a series of observations. One would need observations or data to make a hypothesis.

16. A hypothesis becomes a theory if it stands up to repeated tests.

17. Yes. Scientific theories cannot be proven correct, but they can be shown to be incorrect or in need of modifications.

18. The theory that matter consists of atoms is universally accepted because it explains a vast array of observations and chemical laws such as the laws of conservation of mass and definite proportions. Predictions based on this theory have led to innumerable advances in science and technology. Moreover, despite the passage of 200 years since the modern atomic theory was formulated, no experimental evidence has been found that is inconsistent with this theory.

19. Hubble discovered galaxies outside our Milky Way and found that:
    i) These other galaxies are moving away from our own.
    ii) The speeds with which these galaxies are receding are proportional to their distance from us. That is, the galaxies farthest away are moving away the fastest.

    If all the galaxies appear to be moving away from the Earth, a reasonable explanation for this observation is that the universe is expanding in all directions, that is, getting bigger.

20. Today one could convince someone that Earth was round with pictures taken from space. All of these pictures show the Earth as round, so it must be a sphere. Of course, people knew the Earth was round long before this. Here is some of the evidence that led to this conclusion:
    i) When ships sailed out to sea, they "sank"; that is, the bottom of the ship disappeared until only the top of the mast was visible. The reverse happened when ships were observed sailing directly toward land.

ii) During lunar eclipse, when the Earth passes between the sun and the moon, the Earth's shadow is always curved.
iii) Polaris (the Pole star) appears higher in the sky as one travels North and lower in the sky as one travels South.
iv) The length of a shadow cast by a measured rod at noon was shorter in places farther south and longer in places farther north.
v) Magellan circumnavigated the globe by continually sailing westward in 1519–1522.

21. X-rays are able to penetrate skin and other tissues and can cause damage to the human body. The lead shielding prevents X-rays from penetrating your skin other than in the area of interest.

22. If two wave trains with different wavelengths are moving at the same speed, the number of waves passing a reference point in a unit of time (such as a second) is smaller for the wave train with the longer wavelength. That is, the longer the wavelength the lower the frequency. Figure 1.14 shows this concept graphically. If both of the wave trains shown move at the same speed, the crests of the upper series of waves, which have longer wavelength, will pass a reference point less often than the crests of the lower series of waves, which have shorter wavelengths.

23. Ultraviolet radiation has a shorter wavelength than infrared radiation, so ultraviolet radiation has a greater frequency and greater energy than does infrared radiation. The higher energy associated with ultraviolet radiation is able to cause changes in the molecules that make up skin tissue.

24. In diffraction, rays of light appear to be deflected from a straight line path and to produce fringes of light and dark or colored bands (interference patterns). Both the deflection of light rays and the interference patterns produced are characteristic of the behavior of waves when they pass through a narrow opening (narrow compared to the wavelength) or around an obstacle in their path. Deflection and interference can easily be observed in water waves. Particles do not behave in the same fashion at all.

25. The frequency can be calculated from the formula that relates wavelength and frequency

$$c = \lambda\nu$$

where

c is the speed of light (3.00×10^8 m/s)
λ is the wavelength of the radiation in meters
ν is the frequency of the radiation in 1/s (s^{-1} or Hertz)

The speed of light has units of m/s; the wavelength in this problem (616 nm) is given in nanometers. We need to convert the wavelength measurement from nanometers to meters before we can use the formula. The conversion factor needed to do this is

$$1 \text{ nm} = 1 \times 10^{-9} \text{ m}$$

so the wavelength in nanometers is

$$616 \text{ nm} \times \frac{1 \times 10^{-9} \text{ m}}{1\text{nm}} = 6.16 \times 10^{-7} \text{ m}$$

Now we are ready to calculate the frequency.

$$\nu = c/\lambda$$

$$\nu = \frac{c}{\lambda} = \frac{3 \times 10^8 \text{ m/s}}{6.16 \times 10^{-7} \text{ m}} = 4.87 \times 10^{14} \text{ s}^{-1}$$

Note: Let's see how we end up with these units.

$$\frac{\frac{\text{m}}{\text{s}}}{\text{m}} = \frac{\text{m}}{\text{s}} \times \frac{1}{\text{m}} = \frac{1}{\text{s}} \text{ or s}^{-1}$$

When a unit appears on the bottom of a fraction, as in 1/s, it can be rewritten with the unit on the top of the fraction, as long as the sign of the power of the unit is reversed. For example, 1/s can be written as s^{-1}, and $1/s^2$ can be rewritten as s^{-2}. This is the same rule that applies to exponents, $1/10$ is the same as 10^{-1}, and $1/10^5$ has the same value as 10^{-5}.

27. The formula needed is:

$$c = \lambda\nu$$

a. KKNB 104.1 MHz (or megahertz)

$$\text{mega} = 1{,}000{,}000 \text{ or } 1 \times 10^6$$

To do this problem, first convert the frequencies from megahertz to hertz (1 cycle/s or 1/s). Then calculate the wavelength using $c = \lambda\nu$.

$$104.1 \text{ MHz} \times \frac{1 \times 10^6 \text{ Hz}}{1 \text{ MHz}}$$
$$= 1.041 \times 10^8 \text{ Hz or } 1.041 \times 10^8 \text{ s}^{-1}$$

Next, determine the wavelength.

$$\lambda = \frac{c}{\nu} = \frac{3 \times 10^8 \text{ m/s}}{1.041 \times 10^8 \text{ s}^{-1}}$$
$$= 2.882 \text{ m (to four significant figures)}$$

b. WFNX 101.7 MHz

$$101.7 \text{ MHz} \times \frac{1 \times 10^6 \text{ Hz}}{1 \text{ MHz}} = 1.017 \times 10^8 \text{ Hz or s}^{-1}$$

$$\lambda = \frac{c}{\nu} = \frac{3 \times 10^8 \text{ m/s}}{1.017 \times 10^8 \text{ s}^{-1}} = 2.950 \text{ m}$$

c. KRTX = 100.7 MHz

$$100.7\ \text{MHz} \times \frac{1 \times 10^6\ \text{Hz}}{1\ \text{MHz}} = 1.007 \times 10^8\ \text{Hz or s}^{-1}$$

$$\lambda = \frac{c}{\nu} = \frac{3 \times 10^8\ \text{m/s}}{1.007 \times 10^8\ \text{s}^{-1}} = 2.979\ \text{m}$$

29. Visible light has much shorter wavelength than any radio station waves. Since wavelength and frequency are inversely proportional, the longer the wavelength the lower the frequency. Since the radio station has the longer waves, its radiation has a lower frequency.

We can also solve this problem by calculation if we calculate the frequency for green light with a wavelength of 550 nm and compare this with the frequency for the radio station, 1090 kHz (1.090×10^3 Hz).

$1\ \text{nm} = 1 \times 10^{-9}\ \text{m}$ $\qquad$ $550\ \text{nm} = 5.50 \times 10^{-7}\ \text{m}$

Frequency of green light

$$\nu = \frac{c}{\lambda} = \frac{3.00 \times 10^8\ \text{m/s}}{5.50 \times 10^{-7}\ \text{m}} = 5.45 \times 10^{14}\ \text{s}^{-1}$$

The radio transmission at 1090 Hz has a much lower frequency than green light with a frequency of 5.45×10^{14} Hz (s^{-1}).

31. The car would need to have been traveling fast enough to cause a change in wavelength of about 150 nm (the difference between 700 nm and 550 nm).

The magnitude of the Doppler effect is determined by how fast the source and observer are approaching or receding from one another. For example, if the car were approaching the stop light at one tenth the speed of light, the observed wavelength of the light would be increased by ten percent. The percentage change in the velocity is equal to the percentage change in the observed wavelength.

% change in wavelength

$$= \frac{\text{change in wavelength}}{\text{wavelength of the source}} \times 100$$

$$= \frac{150\ \text{nm}}{700\ \text{nm}} \times 100 = 21.4\%$$

The car would need to have traveled at 21.4% of the speed of light or

$$\frac{3 \times 10^8\ \text{m}}{1\ \text{s}} \times 0.214$$

$$= 6.42 \times 10^7\ \text{m/s (or about } 1.43 \times 10^8\ \text{mph)}$$

Obviously the physicist's defense is not tenable.

33. See Table 1.2

Mass: kilogram	Length: meter
Temperature: kelvin	Time: second
Amount of substance: mole	

34. One can compare 110 μg and 1.00 mg by converting both to grams.

$1\ \mu\text{g} = 1 \times 10^{-6}\ \text{g}$ $\qquad$ $1\ \text{mg} = 1 \times 10^{-3}\ \text{g}$

$$110\ \mu\text{g} \times \frac{1 \times 10^{-6}\ \text{g}}{1\ \mu\text{g}} = 1.1 \times 10^{-4}\ \text{g}$$

$$1\ \text{mg} \times \frac{1.0 \times 10^{-3}\ \text{g}}{1\ \text{mg}} = 1.0 \times 10^{-3}\ \text{g}$$

1.0×10^{-3} g is larger than 1.1×10^{-4} g (or 1 mg is a larger quantity than 110 μg).

35. The speed of light as given is 2.9979×10^8 m/s. We need to convert this velocity from the units of m/s to miles/hr. To do this we need to use the following conversion factors:

1 min = 60 s $\qquad$ 1 hr = 60 min
1 mile = 1.61 km $\qquad$ 1 km = 1000 m

Now we are ready to do the conversions. Make sure that units of miles are in the numerator (the top) and units of hours are in the denominator (the bottom). When you do the conversions, make sure the units cancel out and when you are finished, be sure that you have units of miles in the numerator and hours in the denominator.

$$\frac{2.9979 \times 10^8\ \text{m}}{1\ \text{s}} \times \frac{60\ \text{s}}{1\ \text{min}} \times \frac{60\ \text{min}}{1\ \text{hr}} \times \frac{1\ \text{km}}{1000\ \text{m}} \times \frac{1\ \text{mi}}{1.61\ \text{km}} = 6.7034 \times 10^8\ \text{mi/hr (mph)}$$

Note: We shall assume that all conversion factors have an infinite number of significant figures. Thus the only measured quantity is the speed of light. This quantity contains five significant figures, so the answer should have five significant figures.

37. We need to convert 1500 m to miles, or 1 mile to meters, in order to compare the lengths 1500 m and 1 mile. In the problem, we are given the conversion factors 1 mile = 5280 ft and 1 in = 2.54 cm. We can use these factors to convert from miles to kilometers as follows:

$$1\ \text{mile} = 5280\ \text{ft} \times \frac{12\ \text{in}}{1\ \text{ft}} \times \frac{2.54\ \text{cm}}{1\ \text{in}} \times \frac{10^{-2}\ \text{m}}{1\ \text{cm}} = 1609\ \text{m}$$

(Or we could look this value up in Table 1.2.)

The percentage can be calculated as follows:

$$\%\ \text{of mile} = \frac{\text{km in Olympic mile}}{\text{km in standard mile}} \times 100$$

$$= \frac{1500\ \text{km}}{1610\ \text{km}} \times 100 = 93.2\%$$

An Olympic (metric) mile is 93.2% of the distance of a standard mile.

Alternatively, we can convert 1500 km to miles

$$1500 \text{ m} \times \frac{1 \text{ km}}{1000 \text{ m}} \times \frac{1 \text{ mi}}{1.61 \text{ km}} = 0.932 \text{ mile}$$

The decimal fraction 0.932 can be converted to a percentage by multiplying by 100. Thus an Olympic mile is 93.2% of a standard mile.

39. We are given the price of gasoline in terms of different volume units. To compare them, we can convert $1.25/gal to dollars per liter, or $0.35/L to dollars per gallon. Let's convert the price of gasoline in Canada ($0.35/L) to dollars per gallon.

The needed conversion factors are

$$1.000 \text{ L} = 1.057 \text{ qt} \qquad 1 \text{ gal} = 4 \text{ qt}$$

The price of gasoline in Canada is

$$\frac{\$0.35}{\text{L}} \times \frac{1 \text{ L}}{1.057 \text{ qt}} \times \frac{4 \text{ qt}}{1 \text{ gal}} = \$1.32/\text{gal}$$

Since the unit we are converting (liters) is on the bottom, our conversion factor must have liters on the top (1 L/1.057 qt) in order to cancel out this unit.

According to this calculation, gas is a little cheaper in the United States ($1.25/gal) than in Canada ($1.32/gal).

41. From the question, we are given three pieces of information:

A wheelchair-marathon racer is moving 13.1 mi/hr
The rate that he/she is expending energy is 665 Cal/hr
The distance the racer is traveling is 26.2 miles.

To determine the total amount of energy required in completing a marathon race (26.2 miles) we can use the rates 13.1 mi/hr and 665 Cal/hr as conversion factors.

$$13.1 \text{ mi} = 1 \text{ hr} \qquad 665 \text{ Cal} = 1 \text{ hr}$$

We can start with the 26.2 miles and convert this to hours using the travel rate (13.1 mi/hr). Next we convert hours to Calories using the rate of energy use (665 Cal/hr).

Energy required for a marathon

$$26.2 \text{ mi} \times \frac{1 \text{ hr}}{13.1 \text{ mi}} \times \frac{665 \text{ Cal}}{1 \text{ hr}} = 1330 \text{ Calories}$$

Note: The answer should have three significant figures because all of the measured quantities have three significant figures. The answer 1330 Calories has only three significant figures. The zero on the right is not considered to be significant because there is no decimal point after it.

It would probably be best to report the answer in scientific notation (1.33×10^3 Calories; three significant figures) to reduce any confusion as to the number of significant figures in the answer.

When numbers are expressed in standard scientific notation (a non-zero digit and then a decimal point), all numbers shown are considered significant.

43. This is a simple conversion factor problem.

To convert from meters to miles use

$$1 \text{ mile} = 1.609 \text{ km} \qquad 1 \text{ km} = 1000 \text{ m}$$

Number of miles in 4.0×10^3 m of silk

$$4.0 \times 10^3 \text{ m} \times \frac{1 \text{ km}}{1000 \text{ m}} \times \frac{1 \text{ mi}}{1.609 \text{ km}}$$
$$= 2.5 \text{ miles (to two significant figures)}$$

Then convert from miles to feet. To do this use

$$1\text{mile} = 5280 \text{ ft}$$

Number of feet in 4.0×10^3 m of silk

$$2.486 \text{ mi} \times \frac{5280 \text{ ft}}{1 \text{ mi}}$$
$$= 1.3 \times 10^4 \text{ ft (to two significant figures)}$$

45. We are given the density of magnesium metal and the dimensions of the magnesium block. From the dimensions, we can calculate the volume of the block. If we know both the density and the volume, the mass can be calculated because $d = m/v$.

Volume of the magnesium block $= l \times w \times h$

$$v = (2.5 \text{ cm})(3.5 \text{ cm})(1.5 \text{ cm}) = 13.1 \text{ cm}^3$$
$$d = 1.74 \text{ g/cm}^3$$
$$m = d \times v = (1.74 \text{ g/cm}^3)(13.1 \text{ cm}^3) = 23 \text{ g}$$

Note: The answer should be rounded to two significant figures because the volume value should have had only two significant figures. We carried an extra significant figure through to the mass calculation and rounded off after finishing the calculation.

47. We are given the following information:

Density of Earth = 5.5 g/cm^3
Mass of Venus is 81.5% of the mass of the Earth
Volume of Venus is 88% of the volume of the Earth

The information given does not tell us the mass or volume of either the Earth or Venus. However, we can still determine the density of Venus from the density of the Earth and the relative mass and volume of Venus compared to the Earth. The mass of Venus is only 81.5% of the Earth's value, while the volume is 88% of the Earth's. This tells us that, compared to the Earth, the mass of

Venus is decreased a greater amount than its volume. We should expect the density of Venus to be less than the density of Earth. We can calculate the actual density as follows:

$$\text{Density of Earth} = \frac{\text{mass of Earth}}{\text{volume of Earth}} = 5.5\ \text{g/cm}^3$$

$$\text{Density of Venus} = \frac{\text{mass of Venus}}{\text{volume of Venus}}$$

$$\text{Mass of Venus} = 0.815 \times \text{mass of Earth}$$

$$\text{Volume of Venus} = 0.88 \times \text{volume of Earth}$$

Density of Venus

$$\frac{0.815 \times \text{mass of Earth}}{0.88 \times \text{volume of Earth}}$$

$$= \frac{0.815}{0.88} \times \frac{\text{mass of Earth}}{\text{volume of Earth}}$$

$$= \frac{0.815 \times 5.5\ \text{g/cm}^3}{0.88}$$

$$= 5.1\ \text{g/cm}^3$$

As expected the density of Venus is slightly lower than the density of Earth.

49. If the plastic cube is to float on water, it needs to be less dense than water. To see if the cube will float, we need to compare the density of the plastic cube to the density of water. The density of water is about 1 g/cm^3.

Let us examine the data given to see if we can calculate the density of the cube.

$$\text{Mass of the cube } m = 1.10 \times 10^{-3}\ \text{kg}$$

$$\text{Length of the cube} = 1.2 \times 10^{-5}\ \text{km}$$

We know the mass but we do not know the volume of the cube. However, the volume can be easily calculated.

$$\text{volume of a cube} = \text{length} \times \text{width} \times \text{height}$$

Or

$$V = l \times w \times h = l^3\ (\text{since } l = w = h)$$

$$V = l^3 = (1.2 \times 10^{-5}\ \text{km})^3$$

Before actually doing this calculation we should check to see if the data has the proper units. Since the density of water is usually reported as 1 g/cm^3, the units of mass and volume that we want are grams and cube centimeters. In the problem, the mass is given in kilograms, and our previous calculation would yeild a volume in units of km^3. We need to do some conversions before determining the density of the cube.

First let's find the volume of the cube in cm^3. To do this we first need to convert the length of the cube from kilometers to centimeters, using the conversion factors

$$1\ \text{km} = 1000\ \text{m} \quad \text{and} \quad 1\ \text{m} = 100\ \text{cm}$$

$$1.2 \times 10^{-5}\ \text{km} \times \frac{1000\ \text{m}}{1\ \text{km}} \times \frac{100\ \text{cm}}{1\ \text{m}} = 1.2\ \text{cm}$$

The length of the cube in centimeters is 1.2 cm.

Now we can calculate the volume of the cube.

$$V = l^3 = (1.2\ \text{cm})(1.2\ \text{cm})(1.2\ \text{cm})$$

$$= (1.2\ \text{cm})^3 = 1.728\ \text{cm}^3$$

Next we need to convert the mass unit from kilograms to grams. The needed conversion factor is

$$1\ \text{kg} = 1000\ \text{g}$$

Mass of the cube in kilograms

$$1.10 \times 10^{-3}\ \text{kg} \times \frac{1000\ \text{g}}{1\ \text{kg}} = 1.10\ \text{g}$$

Density of the plastic cube

$$d = \frac{1.10\ \text{g}}{1.728\ \text{cm}^3} = 0.64\ \text{g/cm}^3 (\text{to two significant figures})$$

Even though the mass measurement has three significant figures the answer should have two significant figures because the length measurement has only two significant figures.

The plastic cube will float on water because it is less dense than water.

51. Let's start this problem by listing the information given.
Density of diamond = 3.51 g/cm^3
Mass of the diamond = 5.0 carat
1 carat = 0.200 g (a conversion factor)

From the information given, we might expect that the density formula ($d = \frac{m}{V}$) could be used to find the volume of the diamond.

Inspecting the data, we should realize that carats are a unit of mass. We can convert from carats to grams using 1 carat = 0.200 g.

Mass of the diamond in grams

$$5.0\ \text{carats} \times \frac{0.200\ \text{g}}{1\ \text{carat}} = 1.0\ \text{g}$$

We know the mass of the diamond and its density. Since we know two of the three values in the density formula, we can solve for the unknown quantity, the volume.

$$d = \frac{m}{V} \quad \text{or} \quad V = \frac{m}{d}$$

Volume of the diamond

$$V = \frac{m}{d} = \frac{1.00\ \text{g}}{3.51\ \text{g/cm}^3} = 0.28\ \text{cm}^3$$

The answer has two significant figures because the mass of the diamond is given to only two significant figures (5.0 carats).

Note: The units cm^3 in the answer are a volume unit and can be determined as follows:

$$\frac{g}{g/cm^3} = \frac{g \times cm^3}{g} = cm^3$$

Remember to treat units in calculations just like you would numbers.

53. a. Accuracy is a measure of how close a measurement (or an average value) is to the known value.

    Precision measures how close measured values are to each other. The smaller the range the more precise the value.

    Note: Data can be precise but not accurate. For example, if a person throws darts at a target and all of the darts hit the target in the same general area, the person's aim would be precise. If this shooter was aiming for the bull's-eye and the darts were not close to it, we would say the shooter's aim was precise but not accurate. On the other hand, data can be accurate but not precise, as when the data shows a large range but the average is close to the known value. Lastly, data can be both accurate and precise. In this case, the range is small and the average is close to the correct one.

    b. The lawyer uses the term "precisely accurate." He appears to be using it in the sense of "very accurate," since he qualifies his remark with "down to the ounce." His argument is that if the measurement of ounces is not accurate, then the measurement of pounds is not accurate either. His language is technically imprecise. The cartoon shows the lawyer standing on the scale and taking a single measurement. Precision refers to how closely a number of measurements cluster together, so the lawyer would have to take a number of measurements to estimate the precision of the scale. Since bodyweight is commonly noted in pounds, rather than ounces, the lawyer is incorrectly rejecting the accuracy of the scale on the basis of a measurement (ounces) that is insignificant in the context. Note that the sign on the wall reads "PRECISE WEIGHT," although what it intends to convey is "accurate weight."

    c. The term "precisely accurate" is used in common speech to mean "very accurate," rather than "precise and accurate." For this reason the term "precisely accurate" is confusing. Since precision and accuracy have different meanings, it is necessary to address them separately. For example, a champion darts player who can consistently hit the bull's-eye, is better described as "both precise and accurate" rather than "precisely accurate."

    d. "Accurate Weight to the nearest pound" means the data is only accurate to the nearest pound. The uncertainty in this measurement would be ± 0.5 pounds. Depending on what you are measuring, this may or not meet the precision desired. An uncertainty of ± 0.5 lb would be acceptable in determining the weight of a grown man (150–200 lb), but not in determining the weight of a newborn (approximately 7 lb).

55. *b*, *c*, and *d* have four significant figures.

    Remember, zeros before the first non-zero digit are not significant.

    a. The number of eggs in a dozen is 12. This is an exact number. It has an infinite number of significant figures.
    b. 0.08206 has four significant figures (8 is the first non-zero digit)
    c. 8.314 has four significant figures
    d. 5,400. has four significant figures
    e. 5.4×10^3 has two significant figures
    f. 5,400.0 has five significant figures

57. a. $0.6274 \times 1.00 \times 10^3/[2.205 \times (2.54)^3] =$

    In this question the only mathematical operations present are multiplication and division. The number that has the fewest significant figures determines the number of significant figures that we should report in the answer.

    The number of significant figures in each measured quantity is

    0.6274 (four) $\quad$ 1.00×10^3 (three)
    2.205 (four) $\quad$ 2.54 (three)

    The measurements with the fewest significant figures are 2.54 and 1.00×10^3. They both contain three significant digits. The answer should have three significant figures.

    $$0.6274 \times 1.00 \times 10^3/[2.205 \times (2.54)^3] = 17.3634002$$

    Rounded to three significant figures, the answer is 17.4

    b. $6 \times 10^{-18} \times (1.00 \times 10^3) \times 17.4 =$

    Only multiplication and division are involved in this calculation, so the answer should have the same number of significant figures as the measured quantity with the fewest significant figures. The number of significant figures in each measured quantity is

    6×10^{-18} (one) $\quad$ 1.00×10^3 (three)
    17.4 (three)

    This time the answer should have one significant figure.

    $$6 \times 10^{-18} \times (1.00 \times 10^3) \times 17.4 = 1.044 \times 10^{-13}$$

Rounded to one significant figure, the answer is 1×10^{-13}

c. $(4.00 \times 58.69)/(6.02 \times 10^{23} \times 6.84) =$
As in parts *a* and *b*, the only mathematical operations present are multiplication and division. The number of significant figures in each measured quantity is:

4.00 (three)	58.69 (four)
6.02×10^{23} (three)	6.84 (four)

The answer should be reported to three significant figures.

$$(4.00 \times 58.69)/(6.02 \times 10^{23} \times 6.84)$$

$$= 5.701268676 \times 10^{-23}$$

The answer should be rounded to 5.70×10^{-23}

d. $[(26.0 \times 60.0) + 43.53]/(1.000 \times 10^{4})$
The mathematical operations involved in this problem are multiplication, division, and addition. When adding or subtracting numbers, count the number of decimal places in each value, and report your answer to the lowest number of decimal places given. When applying this rule, the numbers being compared must be of the same magnitude. For example,

$$4.15 + 5.332 = 9.47$$

but

$$4.14 \times 10^{-2} + 5.332 = 0.0414 + 5.332 = 5.373$$

Because (26.0×60.0) is in parenthesis, we need to perform this operation before we can do the addition.

26.0 (three significant figures)	60.0 (three significant figures)

$$26.0 \times 60.0 = 1560$$

Note: We will round the answer to three significant figures at the end of the calculation.

Next, we should perform the addition.

$$1560 + 43.53 = 1603.53$$

The numbers 1560 and 43.53 have zero and two decimal places respectively. The answer to the addition should be reported to zero decimal places, the smallest number of decimal places of any of the added quantities. Rounded to zero places, the answer is 1604.

Finally, perform the division. Remember, your answer needs to be rounded to three significant figures at the end of the calculation.

$$1604/1.000 \times 10^{4} = 1.604 \times 10^{-1}$$

Rounded to three significant figures, the answer is 0.160 or 1.60×10^{-1}.

59. a. ${}^{0}_{-1}e$ An electron has a mass of 0 and a charge of -1.
b. ${}^{1}_{1}p$ An proton has a mass of 1 and a charge of $+1$.
c. ${}^{1}_{0}n$ An neutron has a mass of 1 and a charge of 0.
d. ${}^{2}_{1}d$ An deuteron has a mass of 2 and a charge of $+1$.
e. ${}^{4}_{2}\alpha$ An alpha particle has a mass of 4 and a charge of $+2$.

60. The half-life is the time it takes for a half of the radioactive material to decay. Originally, there is 100 g of the radioactive element. This means that it always takes the same amount of time for half of a sample to decay. Once half of a sample has decayed, it takes the same length of time for half of the remaining material to decay.

After 1 half-life 50 g of the radioactive material has decayed, so 50 g remain.
After 2 half-lives 25 g (half of 50) remain.
After 3 half-lives 12.5 g (half of 25) remain.
After 4 half-lives 6.25 g (half of 12.5) remain.
After 5 half-lives 3.125 g (half of 6.25) of the radioactive material remain.

61. To solve this question, we need to use formula

$$\frac{A_t}{A_0} = 0.5^n$$

Where

A_t is the amount of radioactive remaining after a given time interval
A_0 is the initial amount of radioactive present
n is the number of half-lives that have occurred in the given time period

Also, we can define the ratio $\frac{A_t}{A_0}$ to be the fraction of radioactive material present after the given time period.

In this problem, we are asked to find the fraction (or percentage) of radioactivity present after 151 days. Since we are never told how much of the radioactive sample we started with, we cannot find the actual amount left. However, if we know how many half-lives have occurred, then we can solve for the ratio A_t/A_0.

To find the fraction A_t/A_0, we need to determine n the number of half-lives in 151 days.

We are told that the half-life of ^{60}Co is 5.26 yr.

We need to convert 151 days to years and then divide by 5.26 yr to find out how many half-lives have elapsed.

Time elapsed in years

$$\frac{151 \text{ days}}{365 \text{ days/yr}} = 0.414 \text{ yr}$$

Number of half-lives (n)

$$\frac{\text{time elapsed}}{\text{half-life}} = \frac{0.414 \text{ yr}}{5.26 \text{ yr/half-life}} = 0.0787 \text{ half-lives}$$

Now we can use this information with the formula $\frac{A_t}{A_0} = 0.5^n$ and solve for the ratio A_t/A_0.

$$n = 0.0787$$

$$\frac{A_t}{A_0} = 0.5^n = 0.5^{0.0787} = 0.947$$

Of the original radioactivity the fraction 0.947 remains, or 94.7% of the radioactivity remains. Only 5.3% of the radioactive sample has decayed.

63. In a weighted average each quantity does not contribute equally in the calculation. For example, many course grades are calculated as a weighted average. If there are two mid-term exams that each count for 25 percent of the final grade and the final exam counts for 50 percent, the final grade is a weighted average calculated as follows:

    0.25 (midterm 1 grade) + 0.25 (midterm 2 grade) + 0.50 (final exam grade) = course grade

    A similar method is used for calculating the average atomic mass for elements that have several different isotopes.

64. A nuclide consists only of nuclear particles. An isotope consists of a nuclide plus the electrons that are outside the nucleus.

65. The average atomic mass is a weighted average that takes into account the mass of each isotope and its naturally occurring abundance. We multiply the mass of each isotope by its fractional abundance, and the sum of these values gives us the average atomic mass, which is a weighted average of the masses of the individual isotopes.

66. If there are only two isotopes of an element, then the one whose weight is closer to the average atomic weight must be present in the greater percentage.
    a. ^{11}B is more abundant because 10.81 amu is closer to 11 amu than to 10 amu.
    b. ^{7}Li is more abundant because 6.941 amu is closer to 7 amu than 6 amu.
    c. ^{14}N is more abundant because 14.01 amu is closer to 14 amu than 15 amu.
    d. ^{20}Ne is more abundant because 20.18 amu is closer to 20 amu than 22 amu.

    Only in *a* and *b* is the heavier element more abundant.

67. Average atomic weight of Cu = (mass ^{63}Cu)(abundance ^{63}Cu) + (mass ^{65}Cu)(abundance ^{65}Cu)

    To find an average atomic weight we need to multiply the mass of each isotope by its abundance (as a fraction not a percentage). The sum of these values is equal to the average atomic weight.

$$\text{avg. at. wt.} = (62.9298)(0.6909) + (64.9278)(0.3091)$$
$$= 43.48 + 20.07 = 63.55 \text{ amu}$$

    The average atomic weight of copper is 63.5 amu. This agrees with value found on the periodic table in the front cover of the text.

69. Average atomic weight of S = (31.97207)(0.950) + (32.97146)(0.0076) + (33.96786)(0.0422) + (35.96709)(0.00014)

$$\text{avg. at. wt.} = 30.373 + 0.25 + 1.43 + 0.0050$$
$$= 32.06 \text{ amu}$$

    The average atomic weight of sulfur is 32.06 amu (rounded to two decimal places).

71. Average atomic weight of Ar = (mass ^{36}Ar)(abundance ^{36}Ar) + (mass ^{38}Ar)(abundance ^{38}Ar) + (mass ^{40}Ar)(abundance ^{40}Ar)

    There are seven values in this expression, six of which are known. The only unknown quantity is the mass of ^{40}Ar. All we need to do is to solve for this unknown.

    Plugging in the appropriate values (remember to change % to fractions) yields

$$39.948 = (35.96755)(0.00337) + (37.96272)(0.00063) + (?)(0.9960)$$
$$39.948 = 0.121 + 0.024 + (?)(0.9960)$$
$$39.948 - 0.121 - 0.024 = (?)(0.9960)$$
$$39.803 = (?)(0.9960)$$

    ? = mass of ^{40}Ar = 39.803/0.9960 = 39.96 amu

    Note: The answer should have four significant figures.

73. The formula that relates the wavelength of a solid to its temperature on the Kelvin scale is

$$\lambda_{max} = \frac{2.897 \times 10^6 \text{ nm}}{T \text{ (in kelvins)}}$$

74. An absolute temperature scale is one whose lowest point is zero. On the Fahrenheit scale and the Celsius (centigrade) scales, the lowest possible temperatures are expressed as negative values. At the time these scales came into use, scientists did not know what was the lowest possible temperature so they could not set it at zero.

75. We can convert 4 K to °C using

$$°C = K - 273$$

    Note: We will typically use a value of absolute zero of −273°C instead of −273.15°C.

    Boiling point of helium on the centigrade scale

$$°C = 4 - 273 = -269°C$$

    Note: The answer should have three significant figures. You may ask why should it have three significant figures when the only measured quantity (4K) has only one significant figure. The measurement 4K has a digit in the units place. Because the mathematical operation is addition (rather than multiplication or division) and the other number involved has three significant figures and a digit in the units place, the number of significant figures in the sum remains three. The remaining calculations do not change the number of significant figures, so the answer −269°C has three significant figures.

Note: You can assume all numbers in the temperature conversion formulas are exact numbers and not measured quantities, so they have an infinite number of significant figures.

77. To convert from 102.5°F to °C use

$$°C = 5/9(°F - 32)$$

$$°C = 5/9(102.5 - 32)$$

$$= 5/9(70.5)$$

$$= 39.2°C$$

Note: The answer should have one decimal place because the only measured quantity 102.5°F has one decimal place and because the mathematical operation performed is a subtraction. Assume the 32 in the formula is an exact quantity and has an unlimited number of significant figures.

79. Convert −128.6°F to °C and then to K

$$°C = \frac{5}{9}(°F - 32)$$

$$= \frac{5}{9}(-128.6 - 32)$$

$$= \frac{5}{9}(-160.6)$$

$$= -89.2°C$$

$$K = °C + 273 = -89.2 + 273 = 183.8\ K$$

81. The temperatures to consider are 92K, −250°C, and −221.3°F.

Converting all the temperatures to Celsius gives For 92K

$$°C = 92 - 273 = -181°C$$

For −221.3°F

$$°C = 5/9(-221.3 - 32) = 5/9(-253.3) = -141°C$$

The three temperatures (in degrees Celsius) are

$$-181°C, -141°C, \text{and} - 250°C$$

The highest temperature is −141°C or −221.3°F, so the highest temperature superconductor is $HgBa_2CaCu_2O_6$.

83. The formula for calculating the maximum wavelength of blackbody radiation is

$$\lambda_{max} = \frac{2.897 \times 10^6\ nm}{T}$$

Note: λ_{max} in this equation is in nm (nanometers)
The temperature must be in kelvins (K = °C + 273).

a. $T = 37°C = 310\ K$

$$\lambda_{max} = \frac{2.897 \times 10^6}{310} = 9.34 \times 10^3\ nm$$

b. $T = 100°C = 373\ K$

$$\lambda_{max} = \frac{2.897 \times 10^6}{373} = 7.77 \times 10^3\ nm$$

c. $T = 2000\ K$

$$\lambda_{max} = \frac{2.897 \times 10^6}{2000} = 1.448 \times 10^3\ nm$$

d. $T = 1 \times 10^7\ K$

$$\lambda_{max} = \frac{2.897 \times 10^6}{1 \times 10^7} = 3 \times 10^{-1}\ nm$$

CHAPTER 2 Nuclear Chemistry and the Origins of the Elements

1. | | | | |
|---|---|---|---|
| Electron | ${}_{-1}^{0}e$ | Beta particle | ${}_{-1}^{0}e$ or ${}_{-1}^{0}\beta$ |
| Positron | ${}_{1}^{0}e$ | Proton | ${}_{1}^{1}H$ or ${}_{1}^{1}p$ |
| Neutron | ${}_{0}^{1}n$ | Alpha particle | ${}_{2}^{4}\alpha$ or ${}_{2}^{4}He$ |
| Deuteron | ${}_{1}^{2}H$ | | |

2. Negatively charged: electron, β particle
   Positively charged: positron, proton, α particle, and deuteron
   Neutral: neutron

3. Electrons = β particles = positrons < protons < neutrons < deuterons < α particles

   Electrons, β particles, and positrons have the same mass (they all have the mass of an electron, the lightest subatomic particle). The mass of each of these particles is 9.11×10^{-28} g or 0.00055 amu.

   A protons has a mass of 1.67262×10^{-24} g or 1.00728 amu.

   A neutron can be considered to consist of a proton and electron. The mass of a neutron is 1.6749×10^{-24} g or 1.00867 amu, slightly heavier than a proton.

   A deuteron is a particle that consists of one proton and one neutron. It has a mass of 3.3436×10^{-24} g or 2.01355 amu.

   An α particle consists of two protons and two neutrons. The mass of an α particle is 6.6446×24^{-24} g or 4.0015 amu.

4. The balanced nuclear equation for this process is:

$$ {}_{1}^{1}H + {}_{1}^{1}H \rightarrow {}_{1}^{2}H + {}_{1}^{0}e $$

   This process produces 1 positron. If the positron comes in contact with an electron, the two particles annihilate each other and 2 γ rays are produced. The energy associated with this process is in the gamma-ray region.

5. A hydrogen atom is made up of one proton and one electron. Anti-hydrogen is made up of an anti-proton (negatively charged proton) and a positron (a positively charged electron). An anti-proton is a particle that has the mass of a proton but the charge of an electron.

6. No. An anti-particle is a particle with the same mass but the opposite charge. Since a neutron has no charge, it is not possible for there to be an anti neutron.

7. The mass defect is the difference between the mass of the particles (nucleons) that make up a nuclide and the actual mass of the nuclide. Some mass is lost (and converted into energy) when fundamental particles come to together to form a nuclide.

   The binding energy is the energy released when neutrons and protons come together to form a nuclide, or the energy required to separate the particles in a nucleus into individual nucleons. The greater the binding energy, the more stable the nucleus.

8. When lighter particles and elements fuse, they do so with the release of energy. This results in a particle that is lower in energy (more stable) than the original one. To form a particle with a mass number greater than 26 by fusion requires energy (that is, energy must be supplied to this process). This results in a particle that is higher in energy (less stable) than the starting species. In general, processes that result in a loss of energy are more likely to occur.

9. The mass of a proton and an anti-proton are the same; they differ in that the proton has a positive charge and the anti-proton has a negative charge. The mass of a proton is 1.67262×10^{-24} g (or 1.67262×10^{-27} kg).

We can calculate the energy by using the equation $E = \Delta mc^2$.

$$m = 1.67262 \times 10^{-27}\ \text{kg}$$

$$c = 2.9979 \times 10^{8}\ \text{m/s}$$

Substituting into the formula yields

$$E = (1.67262 \times 10^{-27}\ \text{kg})(2.9979 \times 10^{8}\ \text{m/s})^2$$

$$= 1.5033 \times 10^{-10}\ \text{J}$$

Note: Since the units of a Joule (J) are kg · m^2/s^2, we need to use a mass unit of kilograms when using the equation $E = \Delta mc^2$.

The wavelength of these γ rays can be calculated using the equations

$$c = \text{wavelength} \times \text{frequency} = \lambda\nu$$

$$E = h\nu \quad \text{or} \quad E = hc/\lambda$$

Solving the second of these formulas for wavelength (λ)

$$\lambda = \frac{hc}{E}$$

$$= \frac{(6.626 \times 10^{-34}\ \text{J} \cdot \text{s})(2.9979 \times 10^{8}\ \text{m/s})}{1.5033 \times 10^{-10}\ \text{J}}$$

$$= 1.321 \times 10^{-15}\ \text{m}$$

As expected, the wavelength of the radiation is very short.

11. In this problem, we are given the mass of a ^{60}Ni atom. The mass of the nickel atom includes the masses of the 28 electrons present in this atom. To determine the nuclear mass of nickel, we need to subtract the mass of the electrons.

$$\text{mass of } ^{60}\text{Ni nucleus} = \text{mass of } ^{60}\text{Ni atom} - \text{mass of 28 electrons}$$

$$\text{mass of } ^{60}\text{Ni nucleus} = 59.9308\ \text{amu} - (28)(0.00055\ \text{amu})$$

$$= (59.9308 - 0.01540)\ \text{amu}$$

$$= 59.9154\ \text{amu}$$

Now we can calculate the mass defect, the mass lost when protons and neutrons combine to form a nucleus.

$$\text{mass defect} = (\text{mass protons} + \text{mass neutrons}) - (\text{mass nucleus})$$

The ^{60}Ni nucleus contains 28 protons and 32 neutrons.

$$\text{mass defect} = [(28)(1.00728) + (32)(1.00867)] - 59.9154$$

$$= (28.20384 + 32.27744) - 59.9154$$

$$= 60.4813 - 59.9154 = 0.5659\ \text{amu}$$

Now we are ready to use the formula $E = \Delta mc^2$. Before plugging in numbers, we need to deal with units. The mass unit that we currently have is amu. To get an answer in joules (kg · m^2/s^2), we need to convert the mass defect into kilograms using the conversion factor

$$1\ \text{amu} = 1.66056 \times 10^{-27}\ \text{kg}$$

$$\text{mass defect in kg} = 0.5659\ \text{amu} \times \frac{1.66056 \times 10^{-27}\ \text{kg}}{1\ \text{amu}}$$

$$= 9.3968 \times 10^{-28}\ \text{kg}$$

Next convert the mass defect into energy released, using $E = \Delta mc^2$.

$$E = \Delta mc^2 = (9.3968 \times 10^{-28}\ \text{kg})(3.00 \times 10^{8}\ \text{m/s})^2$$

$$= 8.46 \times 10^{-11}\ \text{J}$$

13. In this problem, the numbers of electrons on each side of the equations are equal, so their masses do not have to be considered when calculating Δm for the fusion reactions. For example in Part A there are 14 electrons on the left side of the arrow (two nitrogen atoms each containing 7 electrons) and 14 electrons on the right side of the arrow (1 silicon atom containing 14 electrons).
   a. Determine the change in mass when the substances on the left side of the arrow are converted into the specie(s) on the right side of the arrow.

      Let's define the following terms:

$$\text{mass of products} = \text{mass of substances on the right side of the arrow}$$

$$\text{mass of reactants} = \text{mass of substances on the left side of the arrow}$$

$$\Delta m = \text{mass of products} - \text{mass of reactants}$$

$$= \text{mass of } ^{28}\text{Si} - (\text{mass of } ^{14}\text{N} + \text{mass of } ^{14}\text{N})$$

$$= 27.97693 - (14.00307 + 14.00307)$$

$$= -0.0292\ \text{amu}$$

The negative sign indicates that mass has been converted into energy.

We can convert Δm into energy by using $E = \Delta mc^2$.

$$E = (-0.0292\ \text{amu}) \times \frac{1.66056 \times 10^{-27}\ \text{kg}}{1\ \text{amu}} \times (3.00 \times 10^{8}\ \text{m/s})^2$$

$$= -4.36 \times 10^{-12}\ \text{J}$$

b. Δm = mass of ^{28}Si − (mass of ^{10}B + mass of ^{16}O + mass of ^{2}H)

$= 27.97693 - (10.0129 + 15.99491 + 2.0146)$

$= -0.0455$ amu

$$E = -0.0455 \text{ amu} \times \frac{1.55056 \times 10^{-27}\text{ kg}}{1\text{ amu}} \times (3.00 \times 10^{8}\text{ m/s})^2$$

$= -6.80 \times 10^{-12}$ J

c. Δm = mass of ^{28}Si − (mass of ^{16}O + mass of ^{12}C)

$= 27.97693 - (15.99491 + 12.0000)$

$= -0.01798$ amu

$$E = -0.01798 \text{ amu} \times \frac{1.66056 \times 10^{-27}\text{ kg}}{1\text{ amu}} \times (3.00 \times 10^{8}\text{ m/s})^2$$

$= -2.687 \times 10^{-12}$ J

d. Δm = mass of ^{28}Si − (mass of ^{24}Mg + mass of ^{4}He)

$= 27.97693 - (23.98504 + 4.00260)$

$= -0.01071$ amu

$$E = -0.01071 \text{ amu} \times \frac{1.66056 \times 10^{-27}\text{ kg}}{1\text{ amu}} \times (3.00 \times 10^{8}\text{ m/s})^2$$

$= -1.601 \times 10^{-12}$ J

15. In this problem we need to calculate the nuclear binding energy per nucleon.

First, find the mass of a helium nucleus. The mass given for ^{4}He is the atomic mass in amu. It includes 2 protons, 2 neutrons, and 2 electrons. To determine the mass of the helium nucleus, we need to subtract the mass of two electrons (2 × 0.00055 amu) from the mass of the nucleus (4.0026 amu)

$$\text{mass helium nucleus} = \text{mass of helium atom} - \text{mass of 2 electrons}$$

$= 4.00260 - 2(0.00055)$

$= 4.00150$ amu

The mass defect is

$$\Delta m = (\text{mass 2 protons} + \text{mass 2 neutrons}) - (\text{mass }^{4}\text{He nucleus})$$

$= [(2 \times 1.00728) + (2 \times 1.00867)] - (4.00150)$

$= (2.01456 + 2.01734) - (4.00150)$

$= 4.03190 - 4.00150$

$= 0.0304$ amu

Now, calculate the binding energy for $^{4}_{2}$He

$$E = 0.03040 \text{ amu} \times \frac{1.66056 \times 10^{-27}\text{ kg}}{1\text{ amu}} \times (3.00 \times 10^{8}\text{ m/s})^2$$

$= 4.5433 \times 10^{-12}$ J

Lastly, determine the energy per nucleon by dividing the total energy by four

$$E = \frac{4.5433 \times 10^{-12}\text{ J}}{\text{nucleus}} \times \frac{1\text{ nucleus}}{4\text{ nucleons}}$$

$= 1.1358 \times 10^{-12}$ J/nucleon

17. High temperature and high particle velocity favor proton capture.

Low temperature and low particle velocity favor neutron capture.

18. In β-decay a particle with an atomic number of zero and a charge of −1 is produced. To balance this process, the additional nuclide formed must increase its atomic number by 1 from the starting one.

19. The belt of stability is a plot of neutrons versus protons in the nuclei of stable and radioactive nuclides. Stable nuclides are represented by green dots in Figure 2.9. Nuclides with neutron-proton ratios above the belt are neutron rich and tend to undergo β-decay. Nuclides below the belt of stability are proton rich and undergo positron emission or electron capture.

20. The net result for both electron capture and positron emission is that the neutron-to-proton ratio increases.

In electron capture, a proton and an electron combine to form a neutron. This increases the number of neutrons by one, while decreasing the number of protons by one. In positron emission a positively charged particle is lost from the nucleus; however, the mass remains the same. This means that the nucleus will contain one more neutron and one less proton, just as with electron capture.

21. This statement is false. It should read:

"Elements with atomic numbers between 27 and 83 form by β-decay because the neutron-to-proton ratio is too high."

22. To answer this question, let's write down the nuclear reactions that occur in this decay sequence.

$$^{137}_{53}\text{I} \rightarrow \, ^{137}_{54}\text{Xe} + \, ^{0}_{-1}\text{e}$$

$$^{137}_{54}\text{Xe} \rightarrow \, ^{137}_{55}\text{Cs} + \, ^{0}_{-1}\text{e}$$

Both processes involve β decay; therefore

a. false

b. false

c. true

23. Let's answer this question by describing some nuclear reactions starting with ^{116}Sn absorbing a neutron and then emitting a β particle.

Absorbing a neutron increases the mass by one but does not change the number of protons.

Beta decay converts a neutron into a proton and an electron (that is ejected from the nucleus). This increases the atomic number by one.

a. No, the nuclide ^{117}Sn cannot be produced by neutron capture because the ensuing β decay converts it into an atom of antimony (Sb).

b. No. The nuclide, $^{116}_{49}$In, cannot be made by neutron capture followed by β decay. This nuclide, $^{116}_{49}$In, has a nuclear charge of 49. Neutron capture does not change this quantity. Beta decay increases the nuclear charge by one, so this scheme cannot produce a nuclide with an atomic number less than 50.

c. Yes. The nuclude, $^{121}_{51}$Sb, could be made by absorbing five neutrons followed by β decay.

$$^{116}_{50}\text{Sn} + 5^{1}_{0}\text{n} \rightarrow {}^{121}_{50}\text{Sn}$$

$$^{121}_{50}\text{Sn} \rightarrow {}^{0}_{-1}\text{e} + {}^{121}_{51}\text{Sb}$$

d. No. If $^{121}_{51}$Sb in part C gives off another β particle then the nuclide, $^{121}_{52}$Te, is produced.

$$^{121}_{51}\text{Sb} \rightarrow {}^{0}_{-1}\text{e} + {}^{121}_{52}\text{Te}$$

25. The neutron-to-proton ratio is greater than one.

The mass number is the sum of the protons and neutrons, while the atomic number corresponds to the number of protons. If the mass number is exactly twice the atomic number, then the atom has an equal number of protons and neutrons, that is, a neutron-to-proton ratio of 1:1. If this ratio is greater than 2:1, the atom has more neutrons than protons making the neutron-to-proton ratio greater than 1.

27. Let's write a nuclear reaction showing all the information we know,

$$^{26}_{13}\text{Al} \rightarrow {}^{26}_{12}\text{Mg} + {}^{a}_{b}?$$

Let's solve for (a) and (b) so that we can determine the identity of the unknown element.

For the atomic mass (a)

$$26 = 26 + a$$

$$a = 0$$

For the atomic number (b)

$$13 = 12 + b$$

$$b = 1$$

The particle that has a mass number of 0 and an atomic number of 1 is the positron ($^{0}_{+1}$e).

The balanced nuclear reaction for the decay of aluminum-26 is

$$^{26}_{13}\text{Al} \rightarrow {}^{26}_{12}\text{Mg} + {}^{0}_{+1}\text{e}$$

29. a. ^{32}P, 17 neutrons, 15 protons

This isotope is above the belt of stability. The only stable isotope of P contains 16 neutrons and 15 protons. It should undergo β-decay.

b. ^{10}C, 4 neutrons, 6 protons

This isotope is below the belt of stability. There are no known stable isotopes with a neutron-to-proton ratio of less than 1 (electron capture or positron emission).

c. ^{50}Ti, 28 neutrons, 22 protons

This isotope is stable. It is shown as a green dot in Figure 2.9. It does not undergo decay.

d. ^{19}Ne, 9 neutrons, 10 protons

This isotope is below the belt of stability. There are no known stable isotopes with a neutron-to-proton ratio of less than 1 (electron capture or positron emission).

e. ^{116}Sb, 65 neutrons, 51 protons

This isotope is below the belt of stability. Its neutron-to-proton ratio is too low (electron capture).

31. From Figure 2.9, we see that both of these isotopes are below the belt of stability. That is, they have too few neutrons. These nuclides need to increase their neutron-to-proton ratio. This is accomplished by either positron emission or electron capture. Likely decay schemes to stable neutron-to-proton combinations are given below.

For ^{44}Ti $\quad ^{44}_{22}\text{Ti} + {}^{0}_{-1}\text{e} \rightarrow {}^{44}_{21}\text{Sc}$

It turns out that this nuclide is also radioactive; it will further decay.

$$^{44}_{21}\text{Sc} + {}^{0}_{-1}\text{e} \rightarrow {}^{44}_{20}\text{Ca (stable nuclide)}$$

For ^{56}Co $\quad ^{56}_{27}\text{Co} + {}^{0}_{-1}\text{e} \rightarrow {}^{56}_{26}\text{Fe (stable nuclide)}$

33. In fission a heavier atom is split into two lighter atoms. In fusion two lighter atoms are fused to make a heavier atom.

34. A fissionable nucleus is one that can be split into two or more nuclei when it absorbs a neutron in the course of neutron bombardment. Not all nuclides (or atoms) are fissionable. For example ^{238}U is not fissionable and cannot be split by neutron absorption; however, ^{235}U is fissionable and can be split into several new nuclides by neutron absorption.

35. Fission of ^{235}U is initiated by absorption of a neutron by the nucleus.

36. When ^{235}U absorbs a neutron and splitting of the nucleus occurs, 3 neutrons are created. The following

nuclear reactions is one of several that occur:

$$^{1}_{0}n + ^{235}_{92}U \rightarrow ^{142}_{56}Ba + ^{91}_{36}Kr + 3\,^{1}_{0}n$$

These neutrons can then split more nuclides, each event producing 3 neutrons. This chain will proceed as long as there are enough ^{235}U nuclei present to absorb the neutrons being produced. On average, at least one neutron from each fission decay must cause the fission of another nucleus for the chain reaction to be self-sustaining. When an uncontrolled chain reaction results, it is accompanied by an enormous release of energy.

37. The rate of energy released is controlled in a nuclear reactor by lowering control rods (made of cadmium or boron) into the reactor. The function of these rods is to absorb neutrons and therefore slow the reaction.

38. A breeder reactor is a nuclear reactor in which fissionable material (^{239}Pu) is produced during normal reactor operation.

In a breeder reactor ^{238}U (non-fissionable) absorbs a neutron, forming ^{239}Pu (fissionable product). The reactions that occur are:

$$^{1}_{0}n + ^{238}_{92}U \rightarrow ^{239}_{92}U$$

$$^{239}_{92}U \rightarrow ^{239}_{93}Np + ^{0}_{-1}e$$

$$^{239}_{93}Np \rightarrow ^{239}_{94}Pu + ^{0}_{-1}e$$

The fissionable ^{239}Pu nuclide absorbs a neutron, is split into other particles, and neutrons, and produces energy. One possible nuclear process is

$$^{239}_{94}Pu + ^{1}_{0}n \rightarrow ^{144}_{58}Ce + ^{90}_{38}Sr + 2\,^{0}_{-1}e + 6\,^{1}_{0}n$$

Some of these neutrons are then absorbed by ^{238}U and eventually form more ^{239}Pu. The amount of ^{239}Pu produced in this process is great enough to fuel this reactor and possibly another one.

39. Neutrons are always by-products of the fission of heavier elements because the heavier naturally-occurring nuclides are converted into several lighter nuclides which have lower neutron-to-proton ratios. The extra neutrons may collide with and split other nuclides.

40. Possible nuclear reactions are:

i) For ^{12}C

$$^{12}_{6}C + ^{1}_{1}H \rightarrow ^{13}_{7}N \rightarrow ^{10}_{5}B + 2\,^{1}_{1}H + ^{1}_{0}n$$

$$^{12}_{6}C + ^{1}_{1}H \rightarrow ^{13}_{7}N \rightarrow ^{10}_{5}B + 2\,^{0}_{1}e + 3\,^{1}_{0}n$$

ii) For ^{14}N

$$^{14}_{7}N + ^{1}_{1}H \rightarrow ^{15}_{8}O \rightarrow ^{10}_{5}B + ^{1}_{1}H + ^{4}_{2}\alpha$$

$$^{14}_{7}N + ^{1}_{1}H \rightarrow ^{15}_{8}O \rightarrow ^{10}_{5}B + ^{0}_{1}e + ^{4}_{2}\alpha + ^{1}_{0}n$$

41. To balance a nuclear equation, the sum of the superscript values must be equal on both sides of the arrow. The same is true of the subscript values.

a. $^{235}U + ^{1}n \rightarrow ^{96}Zr + ? + 2\,^{1}n$

The first thing we notice is that only superscript values have been given.

So let's go back and put in subscripts. The atomic number of uranium is 92, the atomic number of zirconium is 40, and a neutron has no charge. So rewriting the equation, we get

$$^{235}_{92}U + ^{1}_{0}n \rightarrow ^{96}_{40}Zr + ^{a}_{b}? + 2\,^{1}_{0}n$$

The sums of the superscripts are

$$235 + 1 = 96 + a + 2(1)$$

$$236 = 98 + a$$

$$a = 138$$

The sums of the subscripts then are

$$92 + 0 = 40 + b + 2(0)$$

$$92 = 40 + b$$

$$b = 92 - 40 = 52$$

Since atomic number 52 corresponds to Te, the correct symbol is $^{138}_{52}Te$. This nuclide contains 52 protons and 86 neutrons (138 − 52).

b. $^{235}U + ^{1}n \rightarrow ^{99}Nb + ? + 4\,^{1}n$

Putting in subscripts

$$^{235}_{92}U + ^{1}_{0}n \rightarrow ^{99}_{41}Nb + ^{a}_{b}? + 4\,^{1}_{0}n$$

Balancing superscripts

$$235 + 1 = 99 + a + 4(1)$$

$$b = 236 - 103 = 133$$

Balancing subscripts

$$92 + 0 = 41 + b + 4(0)$$

$$b = 92 - 41 = 51$$

Atomic number 51 corresponds to Sb, so the missing product is $^{133}_{51}Sb$. This nuclide contains 51 protons and 82 neutrons (133 − 51).

c. $^{235}U + ^{1}n \rightarrow ^{90}Rb + ? + 3\,^{1}n$

Putting in subscripts

$$^{235}_{92}U + ^{1}_{0}n \rightarrow ^{90}_{37}Rb + ^{a}_{b}? + 3\,^{1}_{0}n$$

Balancing superscripts

$$235 + 1 = 90 + a + 3(1)$$

$$a = 236 - 93 = 143$$

Balancing subscripts

$$92 + 0 = 37 + b + 3(0)$$

$$b = 92 - 37 = 55$$

Atomic number 55 corresponds to Cs, so the missing product is $^{143}_{55}Cs$. This nuclide contains 55 protons and 88 neutrons (143 − 55).

43. a. $^{210}Po \rightarrow {}^{206}Pb + ?$
Putting in subscripts

$$^{210}_{84}Po \rightarrow {}^{206}_{82}Pb + {}^{a}_{b}?$$

Balancing superscripts

$$210 = 206 + a$$
$$a = 4$$

Balancing subscripts

$$84 = 82 + b$$
$$b = 2$$

An α particle has a mass number of 4 and a charge of 2. The symbol is $^{4}_{2}He$ or $^{4}_{2}\alpha$.

b. $^{3}H \rightarrow {}^{3}He + ?$
Putting in subscripts

$$^{3}_{1}H \rightarrow {}^{3}_{2}He + {}^{a}_{b}?$$

Balancing superscripts

$$3 = 3 + a$$
$$a = 0$$

Balancing subscripts

$$1 = 2 + b$$
$$b = -1$$

A β particle has a mass of zero and a charge of -1. The symbol is $_{-1}^{0}e$ or $_{-1}^{0}\beta$.

c. $^{11}C \rightarrow {}^{11}B + ?$
Putting in subscripts

$$^{11}_{6}C \rightarrow {}^{11}_{5}B + {}^{a}_{b}?$$

Balancing superscripts

$$11 = 11 + a$$
$$a = 0$$

Balancing subscripts

$$6 = 5 + b$$
$$b = 1$$

A positron has a mass of 0 and a charge of 1. The symbol is $^{0}_{1}e$.

d. $^{111}In \rightarrow {}^{111}Cd + ?$
Putting in subscripts

$$^{111}_{49}In \rightarrow {}^{111}_{48}Cd + {}^{a}_{b}?$$

Balancing superscripts

$$111 = 111 + a$$
$$a = 0$$

Balancing subscripts

$$49 = 48 + b$$
$$b = 1$$

A positron has a mass of 0 and a charge of $+1$. The symbol is $_{+1}^{0}e$.

45. Particle accelerators (linear accelerator or cyclotron) are devices that speed of the velocity of positive particles (protons, alpha particles, and other ions) to very high speeds. The high velocities are needed to overcome the electrostatic repulsion between a positively charged particle and a positively charged nucleus.
These high-speed particles can collide with other nuclides forming heavier particles.

46. To make "superheavy" elements, scientists bombard heavy isotopes with the nuclei of medium-weight elements. The idea is to fuse the medium-weight nuclei with the heavier nuclei to make a "superheavy" one. To do this, the velocity of the medium-weight nucleus must be in a very narrow range of velocities. The medium-weight nuclei needs to be moving fast enough to overcome the electrostatic repulsion with the target nuclei; however, if the medium-weight nuclei is moving too fast, the "superheavy" nucleus that is formed will be unstable and undergo fission.

47. a. $^{32}S + {}^{1}n \longrightarrow ? + {}^{1}H$
Putting in subscripts

$$^{32}_{16}S + {}^{1}_{0}n \rightarrow {}^{a}_{b}? + {}^{1}_{1}H$$

Balancing superscripts

$$32 + 1 = a + 1$$
$$a = 32$$

Balancing subscripts

$$16 + 0 = b + 1$$
$$b = 15$$

Phosphorous (P) has an atomic number of 15. The symbol is $^{32}_{15}P$.

b. $^{55}Mn + {}^{1}H \longrightarrow {}^{52}Fe + ?$
Putting in subscripts

$$^{55}_{25}Mn + {}^{1}_{1}H \rightarrow {}^{52}_{26}Fe + {}^{a}_{b}?$$

Balancing superscripts

$$55 + 1 = 52 + a$$
$$a = 4$$

Balancing subscripts

$$25 + 1 = 26 + b$$
$$b = 0$$

There is no single species that has a mass of 4 and charge of 0. However a neutron has a charge of 0. If four neutrons were produced then the total mass would be 4 and the charge would be 0. The equation is balanced by adding to 4 neutrons ($4\,^{1}_{0}n$) to the right-hand side.

c. $^{75}As + ? \longrightarrow {}^{77}Br$
Putting in subscripts

$$^{75}_{33}As + {}^{a}_{b}? \longrightarrow {}^{77}_{35}Br$$

Balancing superscripts

$$75 + a = 77$$
$$a = 2$$

Balancing subscripts

$$33 + b = 35$$
$$b = 2$$

There is no particle which has a mass of 2 and a charge of 2, but a proton has a mass of 1 and a charge of 1, so the equation can be balanced by adding 2 protons ($2\,^{1}_{1}H$) to the right-hand side.

d. $^{124}Xe + {}^{1}n \longrightarrow ? \longrightarrow {}^{125}I + ?$
Reaction 1
Putting in subscripts

$$^{124}_{54}Xe + {}^{1}_{0}n \rightarrow {}^{a}_{b}?$$

Balancing superscripts

$$124 + 1 = a$$
$$a = 125$$

Balancing subscripts

$$54 + 0 = b$$
$$b = 54$$

The missing symbol is $^{125}_{54}Xe$.
Reaction 2
Putting in subscripts

$$^{125}_{54}Xe \rightarrow {}^{125}_{53}I + {}^{a}_{b}?$$

Balancing superscripts

$$125 = 125 + a$$
$$a = 0$$

Balancing subscripts

$$54 = 53 + b$$
$$b = 1$$

The particle that has a mass of 0 and a charge of +1 is a positron. The symbol is $^{0}_{1}e$.

49. a. $? \rightarrow {}^{122}Xe + \beta^-$
Putting in subscripts

$$^{a}_{b}? \rightarrow {}^{122}_{54}Xe + {}^{0}_{-1}e$$

Balancing superscripts

$$a = 122 + 0$$
$$a = 122$$

Balancing subscripts

$$b = 54 + -1$$
$$b = 53$$

The element with an atomic number of 53 is iodine. The symbol is $^{122}_{53}I$.

b. $? + {}^{4}He \longrightarrow {}^{13}N + {}^{1}n$
Putting in subscripts

$$^{a}_{b}? + {}^{4}_{2}\alpha \rightarrow {}^{13}_{7}N + {}^{1}_{0}n$$

Balancing superscripts

$$a + 4 = 13 + 1$$
$$a = 10$$

Balancing subscripts

$$b + 2 = 7 + 0$$
$$b = 5$$

The element with an atomic number of 5 is boron. The symbol is $^{10}_{5}B$.

c. $? + {}^{1}n \longrightarrow {}^{59}Fe$
Putting in subscripts

$$^{a}_{b}? + {}^{1}_{0}n \rightarrow {}^{59}_{26}Fe$$

Balancing superscripts

$$a + 1 = 59 + 0$$
$$a = 58$$

Balancing subscripts

$$? + 0 = 26 + 0$$
$$? = 26$$

The element with an atomic number of 26 is iron. The symbol is $^{58}_{26}Fe$.

d. $? + {}^{1}H \longrightarrow {}^{67}Ga + 2\,^{1}n$
Putting in subscripts

$$^{a}_{b}? + {}^{1}_{1}H \rightarrow {}^{67}_{31}Ga + 2\,^{1}_{0}n$$

Balancing superscripts

$$a + 1 = 67 + 2(1)$$
$$a = 68$$

Balancing subscripts

$$? + 1 = 31 + 2(0)$$

$$a = 30$$

The element with an atomic number of 30 is zinc. The symbol is $^{68}_{30}Zn$.

51. Let's start by writing out the unbalanced equation for the first nuclear reaction.

$$^{208}_{82}Pb + ^{86}_{36}Kr \rightarrow ^{293}_{118}X + ^{a}_{b}?$$

Balancing superscripts

$$208 + 86 = 293 + a$$

$$a = 1$$

Balancing subscripts

$$82 + 36 = 118 + b$$

$$b = 0$$

So the question mark corresponds to a neutron. The balanced reaction is

$$^{208}_{82}Pb + ^{86}_{36}Kr \rightarrow ^{293}_{118}X + ^{1}_{0}n$$

The unbalanced equation for the second reaction is

$$^{244}_{94}Pu + ^{48}_{20}Ca \rightarrow ^{289}_{114}X + ^{a}_{b}?$$

Balancing superscripts

$$244 + 48 = 289 + a$$

$$a = 3$$

Balancing subscripts

$$94 + 20 = 114 + b$$

$$b = 0$$

So the question mark corresponds to 3 neutrons. The balanced reaction is

$$^{244}_{94}Pu + ^{48}_{20}Ca \rightarrow ^{289}_{114}X + 3\,^{1}_{0}n$$

53. Photographic film (as in radiation dosimeters) can be used to detect radiation. The darker the film exposure the greater the amount of radiation exposure.

A Geiger counter also can be used to measure the level of radiation exposure. It detects the common products emitted by radioactivity.

A radiation dosimeter measures total radiation exposure over a given time period. A Geiger counter only measures the radiation level at the time of operation.

54. The level of radioactivity is a measure of the radioactivity present at a particular time. It is usually measured in units of radioactive decay events per unit time.

Dose of radioactivity is the quantity of ionizing radiation absorbed by a unit mass of matter (a cumulative effect).

55. The energies associated with radioactive particles can cause atoms or molecules to form free electrons and positive ions. When this occurs in the human body, it is referred to as ionizing radiation. This can lead to tissue damage, such as burns, and molecular changes that can lead to radiation sickness, altered cell growth, cancer, or birth defects.

5–25 rem	No acute effect, possible carcinogenic or mutagenic damage to DNA
25–100 rem	Temporary reduction in white blood cell count
100–200 rem	Radiation sickness, fatigue, vomiting, diarrhea, impaired immune system
200–400 rem	Severe radiation sickness: intestinal bleeding, bone marrow destruction
400–1000 rem	Death, usually through infection, within weeks
>1000 rem	Death within hours

56. Radon-222 is a gas. If it is inhaled and then exhaled before undergoing decay, there is no harm done. If it undergoes decay in your lungs, it gives off an α particle. The resulting radioactive decay product ^{218}Po is a solid that lodges in body tissue and gives off an additional α particle producing ^{214}Pb. The half-life of ^{222}Rn is 3.82 days, and that of ^{218}Po is even shorter, 3.11 minutes. These short half-lives mean that most of the ^{218}Po will decay in the lungs.

Alpha particles are very dangerous inside the body because they are capable of causing ions to form. This results in very reactive species that are capable of causing cell damage or altered cell growth.

57. a. A short half-life is desired so that the radioactivity does not remain in the body for a long period of time.
b. Beta emitters or γ-ray emitters are more desirable than α emitters because the greater ionizing ability of the α particles can cause more damage to the human body.
c. The decay products (daughter nuclei), should be non-radioactive so they cannot cause more biological damage.

58. Radioisotopes can be followed quite easily in the body. One method that does this is positron-emission topography (PET).

PET is an imaging method that is used to monitor brain function. A solution that contains a small amount of a radioactive isotope that emits positrons (man-made species that lie below the belt of stability) is injected into the bloodstream (possibly a sugar solution). The positrons emitted by the radioactive isotope are then annihilated by nearby electrons, forming two gamma rays. The production and location of the gamma rays in the brain can be monitored. Computers can then be used to produce a three-dimensional image of the gamma-ray emission in the brain.

59. After about 10 half-lives (57,000 years) almost all of the radioactive carbon-14 has decomposed to nitrogen-14. This makes it very difficult to get an accurate reading of the current level of radioactivity left in the object. If one cannot determine with any reliability the current carbon-14 activity, then one cannot determine the age of the object.

60. Only statement c is correct.

Statement a is incorrect because larger objects would contain a greater amount of carbon-14.

Statement b is false. Carbon-14 is unstable so it decays to a stable nucleus, but it is not readily lost from the atmosphere.

Statement c is true.

Statement d is false. Living tissue absorbs carbon dioxide. Any isotope of carbon present in carbon dioxide will be absorbed by living tissue.

61. The half-life for ^{40}K is 1.28×10^9 years. This is a very long half-life. It will take at least 300,000 years to get any appreciable difference in the activity levels of the potassium-40.

One of the equations that can be used to relate the activity level of a radioisotope is:

$$t = \frac{t_{1/2}}{0.693} \ln \frac{A_t}{A_0}$$

In this equation A_t and A_0 measure the activity level of a radioactive nuclide at two different times. If the two activity levels are the same then the ratio A_t/A_0 will equal 1. This happens when the two times used for measuring the activity levels is much less than the half-life of the radioactive element.

If $A_t/A_0 = 1$, then the $\ln A_t/A_0$ term equals zero and the time elapsed will equal zero. It requires more than 300,000 years for the levels of radioactivity in ^{40}K to drop to a level that would result in a lower measurable activity.

62. Treating food products with radioactive materials that emit gamma rays destroys or slows down the growth of bacteria, insects, and other microorganisms that cause food to spoil. This in turn allows food to stay fresh for a longer period of time. The mechanism by which this happens is that the energy associated with the gamma rays can tear atoms apart. This results in cell damage that can eventually lead to the death of bacteria.

63. The formula needed to determine this quantity is

$$\text{number of rems} = \text{number of rads} \times \text{RBE (relative biological effectiveness)}$$

$$1 \text{ rem} = 1000 \text{ millirem}$$

$$\text{number of rads} = \frac{\text{number of rems}}{\text{RBE}}$$

In this problem the dosage is given as 0.5 mrem so we need to convert this to rems. The RBE value for a γ ray is one. Substituting into the equation and using the conversion factor will give us the number of rads.

$$\text{number of rads} = \frac{0.5 \text{ mrem} \times \frac{1 \text{ rem}}{1000 \text{ mrem}}}{1 \text{ RBE}}$$

$$= 5 \times 10^{-4} \text{ rads}$$

To determine the amount of radiation in joules that this represents, we need to use the definition of a rad as a conversion factor.

$$1\text{rad} = 1.0 \times 10^{-2} \text{ J/kg}$$

From this we can create a formula which allows us to convert rads into joules as long as we also know the mass of interest in kilograms.

$$\text{energy in J} = (\text{number of rads })(\text{mass in kg}) \times \left(\frac{1.0 \times 10^{-2} \text{ J/kg}}{1 \text{ rad}}\right)$$

$$\text{number of rads} = 5.0 \times 10^{-4}$$

$$\text{mass in kg} = 50$$

$$\text{energy in J} = (5.0 \times 10^{-4} \text{ rads})(50 \text{ kg}) \times \left(\frac{1.0 \times 10^{-2} \text{ J/kg}}{1 \text{ rad}}\right)$$

$$= 2.5 \times 10^{-4} \text{ J}$$

65. a. To determine the number of disintegrations per second, we need to use the definition of a curie (Ci) and the definition of a becquerel (Bq).

$$1 \text{ becquerel} = 1 \text{ disintegration (decay event) per second or 1 atom decaying per second}$$

$$1 \text{ Ci} = 3.7 \times 10^{10} \text{ bq}$$

We need to convert 300 pCi/L into Bq/L

$$1 \text{ pCi (picocurie)} = 1 \times 10^{-12} \text{ Ci}$$

$$300 \text{ pCi} \times \frac{1 \times 10^{-12} \text{ Ci}}{1 \text{ pCi}} \times \frac{3.7 \times 10^{10} \text{ bq}}{1 \text{ Ci}}$$

$= 11.1$ bq (11.1 disintegrations per second per liter)

b. To determine the number of radon atoms in the sample, we need the formulas

$$\text{rate} = kN \text{ and } k = 0.693/t_{1/2}$$

where

rate $= 11.1$ disintegrations per second

$k = 0.693/3.8$ days

$N =$ the number of atoms initially present in the sample.

Combining the above two formulas gives us

$$\text{rate} = \left(\frac{0.693}{t_{1/2}}\right)(N)$$

In order for units to cancel out, we need to make sure both time units are the same. Let's change the half-life from units of days to units of seconds.

$$1 \text{ day} = 24 \text{ hr} \qquad 1 \text{ hr} = 60 \text{ min}$$

$$1 \text{ min} = 60 \text{ s}$$

$$3.8 \text{ days} \times \frac{24 \text{ hr}}{1 \text{ day}} \times \frac{60 \text{ min}}{1 \text{ hr}} \times \frac{60 \text{ s}}{1 \text{ min}}$$

$$= 3.28 \times 10^5 \text{ s}$$

Substituting into the equation

$$\frac{11.1 \text{ disintegrations}}{\text{s} \bullet \text{L}} = \left(\frac{0.693}{3.28 \times 10^5 \text{ s}}\right)(N)$$

Solving for N

$$N = \frac{(11.1 \text{ disintegrations/s} \bullet \text{L})(3.28 \times 10^5 \text{ s})}{0.693}$$

$$= 5.3 \times 10^6 \text{ atoms/L}$$

5.3×10^6 radon atoms are present in each liter of the sample and about 11 of them decay each second.

67.

Yttrium-90	$^{90}_{39}\text{Y}$	39 protons and 51 neutrons
Rhenium-188	$^{188}_{75}\text{Re}$	75 protons and 113 neutrons
Dysprosium-165	$^{165}_{66}\text{Dy}$	66 protons and 99 neutrons
Bismuth-213	$^{213}_{83}\text{Bi}$	83 protons and 130 neutrons

By using Figure 2.9 we can determine the combinations of neutrons and protons that result in stable nuclides and which combinations are radioactive (not stable).

Yttrium-90: The isotope of yttrium with 50 neutrons is stable. From Figure 2.9, we see that yttrium-90, 51 neutrons, is above the belt of stability, so we would expect it to be a β emitter. (Its neutron-to-proton ratio is too high.)

Rhenium-188: The isotopes of rhenium with 106, 108, 109, 110, and 112 neutrons are stable. From Figure 2.9, we see that rhenium-188, 113 neutrons, is above the belt of stability, so we would expect it to be a β emitter. (Its neutron-to-proton ratio is too high.)

Dysprosium-165: The isotopes of dysprosium with 90, 92, 94, 95, 96, and 97 neutrons are stable. From Figure 2.9, we see that dysprosium-165, 99 neutrons, is above the belt of stability, so we would expect it to be a β emitter. (Its neutron-to-proton ratio is too high.)

Bismuth-213: The isotope of bismuth with 126 neutrons is stable. From Figure 2.9, we see that bismuth-213, 130 neutrons, is above the belt of stability, so we would expect it to be a β emitter. (Its neutron-to-proton ratio is too high.) Being a very heavy nuclide, one might also expect this nucleus to undergo α decay.

69. a. The information given in the problem enables us to write the reaction

$$^{10}_{5}\text{B} + ^{1}_{0}\text{n} \rightarrow ^{11}_{5}\text{B} \rightarrow ^{7}_{3}\text{Li} + ^{4}_{2}\text{He}$$

This reaction is balanced as written.

b.
$$\Delta m = (10.0129 + 1.008665)$$
$$- (7.01600 + 4.00260)$$
$$= 11.0216 - 11.0186$$
$$= 0.0030 \text{ amu}$$

Note that since both sides of this equation contain 5 electrons we do not need to subtract the mass of the electrons from the atomic mass.

Now we are ready to determine the energy lost upon fission using

$$E = \Delta mc^2$$

However, before we substitute values into this equation we need to get Δm in units of grams rather than amu. To do this we need to use the conversion factor

$$1 \text{ amu} = 1.66056 \times 10^{-27} \text{ kg}$$

$$\Delta m = 0.0030 \text{ amu} \times \frac{1.66056 \times 10^{-27} \text{ kg}}{1 \text{ amu}}$$

$$= 4.982 \times 10^{-30} \text{ kg}$$

$$E = \Delta mc^2 = (4.982 \times 10^{-30}\ \text{kg}) \times (3.00 \times 10^8\ \text{m/s})^2 = 4.484 \times 10^{-13}\ \text{J}$$

c. Alpha emitters have a greater ionizing ability than do β particles or γ rays. Cancerous cells are more susceptible to ionizing radiation than normal cells. Additionally, because α emitters have low penetrating power, they can be used internally because they can be focused on a small area to selectively kill those cells.

71. The half-life is the time required for half of a radioactive sample to decay. It is important to know that this quantity is independent of the total amount of radioactivity originally present. That is, it takes the same amount of time for 100 g of a radioactive isotope to decay to 50 g as it takes for 5 g of this same isotope to decay to 2.5 g.

In a given sample, this means it takes the same amount of time for 100 g to decay to 50 g as it takes for 50 g to decay down to half of this amount (25 g).

After 1 half-life, 50% of the radioactive sample has decayed and 50% of the original radioactivity remains. After 2 half-lives, 75% of the radioactive sample has decayed and 25% of the original radioactivity remains.

73. To solve this problem, we can use

$$t = \frac{t_{1/2}}{0.693} \ln \frac{A_0}{A_t} \qquad \text{(Equation 2.28)}$$

$$t_{1/2} = 74 \text{ days}$$

$$\frac{A_0}{A_t} = \frac{1}{0.01} = 100$$

Substituting in Equation 2.28

$$t = \frac{74 \text{ days}}{0.693} \ln 100 = (107 \text{ days})(\ln 100) = (107 \text{ days})(4.61) = 493 \text{ days}$$

It would take about 493 days for 99% of the radioactivity in this sample to decay.

75. a. Iridium-92 could not possibly exist because it would have only 15 neutrons. The only known isotopes which have less neutrons than protons are all very light nuclides, such as helium-3 and beryllium-7. In these isotopes the number of neutrons is quite close to the number of protons. The known nuclides of iridium contain anywhere from 105 to 121 neutrons.

b. ${}^{192}_{77}\text{Ir} \longrightarrow {}^{0}_{-1}\text{e} + {}^{192}_{78}\text{Pt}$

c. The reader is correct about the penetrating effects of β and γ radiation. Gamma radiation is much more penetrating than β particles.

77. Use the belt of stability (Figure 2.9) to determine whether the isotope listed is found above or below the belt.

a. ${}^{197}_{80}\text{Hg}$ 117 neutrons and 80 protons

The stable isotopes of mercury contain 116, 118, 119, and 120 neutrons. ${}^{197}\text{Hg}$ (117 neutrons) is located between two stable isotopes. There is only one stable isotope with less neutrons (116), while there are three isotopes with more neutrons (118, 119, 120). We should expect this nuclide to decay using the same pathway as one that is found below (too few neutrons) the belt of stability. We should expect ${}^{197}\text{Hg}$ to undergo electron capture.

b. ${}^{75}_{34}\text{Se}$ 34 protons and 41 neutrons

From Figure 2.9, we see stable isotopes of selenium contain 40, 42, 43, 44, 46, and 48 neutrons. This nuclide lies between two stable ones. It is closer to the lower end, so we should expect it to undergo a decay pathway for a nuclide with too few neutrons. It could undergo either electron capture or positron emission. (The *Handbook of Chemistry and Physics* confirms that ${}^{75}\text{Se}$ undergoes electron capture.)

c. ${}^{113}_{49}\text{In}$ 49 protons and 64 neutrons

According to Figure 2.9, this nuclide of indium should be non-radioactive.

d. ${}^{18}_{9}\text{F}$ 9 protons and 9 neutrons

The only stable nuclide of fluorine contains 10 neutrons. This nuclide does not have enough neutrons to be stable. We should expect positron emission or electron capture. (According to the *Handbook of Chemistry and Physics* 97% of fluorine-18 nuclides do in fact decay by positron emission. The remaining three percent of the nuclides undergo electron capture.)

79. To determine the half-life, use the Equation 2.28

$$t = \left(\frac{t_{1/2}}{0.693}\right) \ln \frac{A_0}{A_t}$$

The equation contains four variable quantities. If three of them are known, we can solve for the missing one. In this problem we are asked to determine the half-life of ${}^{192}\text{Ir}$.

In this problem

$$t = 30 \text{ days} \qquad A_0 = 1.00\ \mu\text{g} \qquad A_t = 0.756\ \mu g$$

Solving Equation 2.28 for $t_{1/2}$ yields

$$t_{1/2} = \frac{(0.693)(t)}{\ln \frac{A_0}{A_t}}$$

Substituting in the equation for the half-life gives us

$$t_{1/2} = \frac{(0.693)(30 \text{ days})}{\ln \dfrac{1.00\ \mu g}{0.756\ \mu g}}$$

$$t_{1/2} = \frac{20.79 \text{ days}}{\ln 1.323}$$

$$t_{1/2} = \frac{20.79 \text{ days}}{0.280} = 74.3 \text{ days}$$

This number makes sense because after 30 days less than half of the sample had decayed (about 25% decayed). This tells us the half-life is longer than 30 days. It is important to remember that the relationship between amount of sample decay and time is not linear, so we cannot say the half-life should be about 60 days because it took 30 days for 25% to decay.

81. To determine the activity of iodine-131 after 30 days we need to use Equation 2.26

$$\frac{A_t}{A_0} = 0.5^{t/t_{1/2}}$$

We can then compare this value with the activity of the solution measured after 30 days. If the calculated activity is the same as the measured activity, then no iodine-131 was absorbed by the brain. If the calculated activity is greater, then the brain tissue must have incorporated some iodine-131.

We need to solve Equation 2.26 for A_t, which corresponds to the activity of iodine-131 after 30 days.

$A_t = ?$ $\quad A_0 = 108$ counts per minute (cpm)

$t = 30$ days $\quad t_{1/2} = 8.1$ days

Substituting these values into Equation 2.26

$$\frac{A_t}{A_0} = 0.5^{\frac{t}{t_{1/2}}}$$

$$\frac{A_t}{108 \text{ cpm}} = 0.5^{\frac{30 \text{ days}}{8.1 \text{ days}}}$$

$$= 0.5^{3.70}$$

$$= 0.0769$$

$$A_t = (108 \text{ cpm})(0.0769)$$

$$= 8.31 \text{ cpm}$$

$$A_t = 8.31 \text{ counts per minute}$$

This calculated amount is greater than the measured solution activity value, so some of the iodine-131 must have been absorbed by brain cells.

Note: We did not need to convert the two times (30 days and 8.1 days) into minutes because they canceled out in the division involving these units.

83. To solve this question, one can use Equations 2.26, 2.27, or 2.28. In this case, let's use

$$\frac{A_t}{A_0} = 0.5^{t/t_{1/2}} \qquad \text{(Equation 2.26)}$$

In this problem, we need to solve for A_t. We know

$A_0 = 10.0$ ng $\quad t = 6$ days $\quad t_{1/2} = 65$ hr

Before we can use these values in Equation 2.26, we need to make sure that both the half-life time and the length of time for the measurement are in the same units. Let's convert 6 days into hours.

$$6 \text{ days} \times \frac{24 \text{ hr}}{1 \text{ day}} = 144 \text{ hr}$$

Now we are ready to substitute values into Equation 2.26

$$\frac{A_t}{10.0 \text{ ng}} = 0.5^{\frac{144 \text{ hr}}{65 \text{ hr}}}$$

$$= 0.5^{2.22}$$

$$= 0.215$$

Solving for A_t

$$A_t = (0.215)(10.0 \text{ ng})$$

$$= 2.15 \text{ ng}$$

There are 2.15 ng of mercury-197 remaining after 6 days. *Note:* 6 days (144 hr) is between 2 and 3 half-lives. If we start with 10.0 ng, after 1 half-life, 5.00 ng remain. After 2 half-lives 2.50 ng remain and after 3 half-lives 1.25 ng remain. Our answer falls in the range between 2 and 3 half-lives and so it seems reasonable.

85. In this problem we can find the time t using Equation 2.28. Remember, quantities can be given in masses or in units of radioactivity. In this question A_t and A_0 are given in units that measure the level of radioactivity in the sample.

The quantities given are

$A_t = 3.75$ mCi $\quad A_0 = 30$ mCi

$t_{1/2} = 2.7$ days

Substituting into the equation

$$t = \left(\frac{t_{1/2}}{0.693}\right) \ln \frac{A_0}{A_t}$$

$$= \left(\frac{2.7 \text{ days}}{0.693}\right) \ln \frac{30 \text{ mCi}}{3.75 \text{ mCi}}$$

$$= (3.90 \text{ days})(\ln 8.00)$$

$$= (3.90 \text{ days})(2.08)$$

$$= 8.1 \text{ days (rounded to 2 significant figures)}$$

Alternatively one could realize that when the radioactive content drops from 30 mCi to 3.75 mCi, that 3 half-lives have passed.
After 1 half-life, the radioactivity level should 15 mCi
After 2 half-lives, the radioactivity level should be 7.5 mCi
After 3 half-lives, the radioactivity level should be 3.75 mCi
3.75 mCi is the desired level, so it should take 3 half-lives for the sample to decompose to this level. Three half-lives corresponds to 8.1 days (3 × 2.7 days).

87. In this problem, we are asked to determine how long it takes for 99% of the ^{11}C injected into a patient to decay. To calculate this we can use Equation 2.28.

$$t = \left(\frac{t_{1/2}}{0.693}\right) \ln \frac{A_0}{A_t}$$

Only 1% of the radioactivity in the original sample remains (99% of the radioactive sample has decayed).

The known information is

$$A_0 = 100\% \text{ or } 1 \quad A_t = 1\% \text{ or } 0.01 \quad t_{1/2} = 20.3 \text{ min}$$

We are now ready to substitute into the equation and solve for the time required for this process to occur.

$$t = \left(\frac{t_{1/2}}{0.693}\right) \ln \frac{A_0}{A_t}$$

$$t = \left(\frac{20.3 \text{ min}}{0.693}\right) \ln \frac{1}{0.01}$$

$$\ln \frac{1.0}{0.01} = \ln 100 = 4.61$$

$$t = \left(\frac{20.3 \text{ min}}{0.693}\right)(4.61)$$

$$= 135 \text{ min}$$

89. A ⟶ B ⟶ C

$$t_{1/2} = 4.5 \text{ s} \qquad t_{1/2} = 15.0 \text{ d}$$

After 10 half-lives, almost all the atoms in a particular decay have decomposed (more than 99.9% of the atoms have decayed). For the first decay, A to B, this occurs after 45 seconds; after 45 seconds all the A has decomposed.

We can see this by using Equation 2.25 ($n = \frac{t}{t_{1/2}}$). Once the number of half-lives is known, we can then use the equation $\frac{A_t}{A_0} = 0.5^n$. The ratio A_t/A_0 corresponds to the fraction of radioactivity remaining in the sample. Ten half-lives corresponds to $n = 10$ in the above equation.

Solving for A_t/A_0 we get

$$\frac{A_t}{A_0} = 0.5^{10}$$

$$\frac{A_t}{A_0} = 9.77 \times 10^{-4}$$

Multiplying by 100 to turn this fraction into a percentage, we find that 0.0977% of the radioactivity remains or 99.9023% of the radioactive sample has decayed.

After 30 days (which is 576,000 half-lives) there is no A remaining. Additionally since it took less than a minute for all the A to form B, we can assume that the initial amount of B is also equal to 1×10^6 atoms.

Since the half-life of B is 15 days, 30 days corresponds to 2 half-lives.

After 45 seconds (0 days) 1.0×10^6 atoms of B are present

After 15 days: 0.5×10^6 atoms of B are present (5×10^5)

After 30 days: 0.25×10^6 atoms of B remain (2.5×10^5)

(After 2 half-lives one fourth of the atoms initially present have not decayed, so three-quarters of the atoms have decayed. If we started with 1×10^6 atoms and only 2.5×10^5 atoms remain then $(1 \times 10^6) - (2.5 \times 10^5) = 7.5 \times 10^5$ have decayed to C.)

No atoms of A remain.

2.5×10^5 atoms of B remain.

7.5×10^5 atoms of C are present.

91. Since we need to find A_t/A_0, we can use

$$\frac{A_t}{A_0} = 0.5^{t/t_{1/2}} \qquad \text{(Equation 2.26)}$$

We are given

$$t_{1/2} = 15 \text{ h} \qquad t = 48 \text{ h}$$

Substituting in Equation 2.26

$$\frac{A_t}{A_0} = 0.5^{\frac{48 \text{ hr}}{15 \text{ hr}}}$$

$$= 0.5^{3.2}$$

$$= 0.1088 \text{ or } 0.11 \text{ (to 2 significant figures)}$$

The ratio $\frac{A_t}{A_0}$ is the fraction of sodium remaining after 48 hr. To convert this to a percentage, multiply by 100.

11% of the sodium atoms remain (that is, they have not decomposed).

93. a. Let's write a nuclear reaction for this decay including what we know, and using a, b, and a question mark for what we do not know.

$$^{188}_{74}W \rightarrow {}^{188}_{75}Re + {}^{a}_{b}?$$

Solve for a, b, and the question mark.

$$188 + a = 188$$
$$a = 0$$
$$75 + b = 74$$
$$b = -1$$

The particle with a mass number of 0 and an atomic number of -1 is a β particle. The balanced nuclear equation is

$$^{188}_{74}W \rightarrow {}^{188}_{75}Re + {}^{0}_{-1}e$$

b. ^{188}W (tungsten) is the isotope of tungsten that contains the greatest number of neutrons (114). All of the stable isotopes of tungsten contain less than 114 neutrons. This isotope of tungsten must be above the belt of stability. This isotope is most likely to undergo β emission as a decay mode. Electron capture and positron emission make a typical decay scheme for isotopes that lie below the belt of stability.

95. In this problem we need to solve for A_t, where A_t corresponds to the current ^{14}C:^{12}C ratio and A_0 corresponds to the ^{14}C:^{12}C ratio 8700 years ago.
We are given the following information:
The half-life of ^{14}C (from the text), $t_{1/2} = 5730$ yr.
$t_{1/2} = 5730$ yr $\quad t = 8700$ yr
$A_0 = 13.8$ counts per minute per gram of carbon-14
We can use Equation 2.26 $\quad \frac{A_t}{A_0} = 0.5^{t/t_{1/2}}$
where $\frac{t}{t_{1/2}}$ = number of half-lives that have elapsed.

$$\frac{t}{t_{1/2}} = \frac{8700}{5730} = 1.52 \text{ half-lives}$$

Substituting this value into the equation:

$$\frac{A_t}{A_0} = 0.5^{1.52}$$

Performing the operation $0.5^{1.52}$ (use the y^x calculator function) one gets:

$$0.5^{1.52} = 0.349$$
$$\frac{A_t}{A_0} = 0.349$$

Substituting in the known value of A_0 and solving

$$\frac{A_t}{13.8} = 0.349$$
$$A_t = (0.349)(13.8)$$
$$= 4.82 \text{ counts per second per gram carbon-14}$$

Or one could say that the ^{14}C:^{12}C ratio was found to be only 34.9% of the ^{14}C:^{12}C ratio that would be in a charcoal-based object made today.

97. The most recent ring (this being the year 2002) is 111 years old. The oldest ring is 1453 years old (1342 + 111). The most recent ring started to decay when the tree was cut down. The oldest ring started to decay 1453 years ago.
We can determine the number of half-lives that have passed for each ring since decay started by using the formula

$$n = \frac{t}{t_{1/2}}$$
$$= \text{number of half-lives}$$

The half-life for ^{14}C (found in Sample Exercise 2.6 in Section 2.8) is 5730 years.
For the newer ring

$$n = \frac{111}{5730}$$
$$= 0.01937$$

For the older ring

$$n = \frac{1453}{5730}$$
$$= 0.2536$$

Calculate the ratio $\frac{A_t}{A_0}$ using the equation

$$\frac{A_t}{A_0} = 0.5^n$$

For the newer ring

$$\frac{A_t}{A_0} = 0.5^{0.01937} = 0.9867$$

For the older ring

$$\frac{A_t}{A_0} = 0.5^{0.2536} = 0.8388$$

In the newer ring 0.9867 of the ^{14}C atoms remain (or 1.33% have decayed).
In the older ring 0.8388 of the ^{14}C atoms remain (or 16.12% have decayed).

ratio of ^{14}C in older ring to newer ring

$$= \frac{\frac{A_t}{A_0} \text{ for older ring}}{\frac{A_t}{A_0} \text{ for newer ring}}$$
$$= \frac{0.8388}{0.9867}$$
$$= 0.850$$

99. In this example we are given that $A_t/A_0 = 0.0119$. That is, the ^{14}C:^{12}C ratio in the tusk was only 1.19% of that in a modern living elephant.

We can find the age (time) using Equations 2.27 or 2.28. The half-life of carbon-14 is 5730 years.

$$\frac{A_t}{A_0} = 0.0119 \qquad t_{1/2} = 5730\,\text{years}$$

Substituting in Equation 2.27

$$\ln A_t/A_0 = -0.693\, t/t_{1/2}$$

$$\ln 0.0119 = -(0.693)\left(\frac{t}{5730\,\text{yr}}\right)$$

$$\ln 0.0119 = -4.43$$

Solving for t yields

$$-4.43 = -(0.693)\left(\frac{t}{5730\,\text{yr}}\right)$$

$$t = \frac{(4.43)(5730\,\text{yr})}{0.693} = 36630\,\text{yr}$$

The tusk is about 36,600 years old (to 3 significant figures).

Alternatively, using Equation 2.28

$$t = \frac{t_{1/2}}{0.693}\ln\frac{A_0}{A_t}$$

$$t = \frac{5730}{0.693}\ln\frac{1.0}{0.00119}$$

$$t = \frac{5730}{0.693}\ln 84.03$$

$$\ln 84.03 = 4.431$$

Solving for t

$$t = \frac{(5730)(4.431)}{0.693}$$

$$= 36{,}600\,\text{yr}$$

CHAPTER 3 Electrons and Electromagnetic Radiation

1. A photon is a particle of electromagnetic radiation. It corresponds to a quantum of energy.

2. An emission spectrum is produced when electrons lose energy and drop from a higher energy level to a lower one (higher n to lower n). It consists of a series of colored lines on a black background. It can be used to identify an element, as each one has its own characteristic spectrum.

   An absorption spectrum is produced when electrons gain energy and move from a lower energy level to a higher one (lower n to higher n). Light of specific wavelength is absorbed in an absorption spectrum. It is observed as a series of dark lines appearing in the continuous rainbow of colors in the spectrum.

3. Several of the Fraunhofer lines corresponded to the lines produced by the emission spectra of known elements.

4. Fraunhofer lines are the result of atomic absorption.

5. If the absorption and emission spectra of free atoms and free ions of the same element are not identical, then we can relate the differences in the spectra to the different number of electrons present in each. Therefore, the absorption and emission of electromagnetic radiation by atoms and ions must be related to increases or decreases in the energies of their electrons.

   Experiments by Ira Bowen that showed that certain Fraunhofer lines in the sun's spectrum were due to absorption by oxygen and nitrogen atoms that had lost one electron and were present as positively charged ions in the sun's atmosphere.

6. In the photoelectric effect photons strike a metal surface. The energy of the photon is transferred to a single electron. If this energy is greater than the energy needed to release the electron from its atom, a photoelectron is produced.

   The minimum energy needed to release an electron from the surface of a metal is called the work function. This can be considered a potential energy term. Any additional energy associated with the photon is transferred to the electron in the form of kinetic energy.

   The work function for each element is different. If the energy of the photons being used to dislodge electrons is the same, but the work function is different for each element, then the kinetic energy of the photoelectrons produced should be different.

7. A typical atom is about 10,000 times larger than its nucleus.

   The size of a typical atom is about 1×10^{-10} m, while the size of the nucleus is about 1×10^{-14} m.

8. Aluminum (atomic number 13) has many less protons in its nucleus than does a gold atom (79 protons). An α particle (a particle that contains 2 protons) would be deflected to a lesser extent upon striking an aluminum nucleus when compared to an α particle striking gold foil.

9. a. Quantized. One can only have a whole number of eggs. The number cannot be fractional.
   b. Continuous. The elevation on the ramp may be any value up to the maximum height of the ramp.
   c. Quantized. The elevation on the stairs can only have certain specific values.
   d. Continuous. An automobile can go any velocity up to its maximum.
   e. Quantized.

11. We need to compare the energy of a photon that has a wavelength of 500 nm to the work function. If the energy is greater than the work function, electrons will be emitted and we could make solar-powered photocells.

$$E = \frac{hc}{\lambda} = \frac{(6.626 \times 10^{-34}\ \text{J} \cdot \text{s})(3 \times 10^{8}\ \text{m/s})}{500\ \text{nm} \times \dfrac{1 \times 10^{-9}\ \text{m}}{1\ \text{nm}}} = 3.98 \times 10^{-19}\ \text{J}$$

This energy is less than the value of the work function, so sunlight cannot be used to remove electrons from a tantalum atom, and tantalum cannot be used to manufacture a solar-powered photocell.

We could also convert the work function into a wavelength. In this case, the longer the wavelength, the lower the energy. The wavelength calculated would correspond to the longest wavelength that would cause an electron to be emitted from the surface of the metal.

We can solve for the wavelength using the equation $\lambda = hc/E$

$$\lambda = \frac{hc}{E} = \frac{(6.626 \times 10^{-34}\ \text{J} \cdot \text{s})(2.998 \times 10^{8}\ \text{m/s})}{6.41 \times 10^{-19}\ \text{J}} = 3.10 \times 10^{-7}\ \text{m}$$

This wavelength (310 nm) is less than the wavelength of sunlight (500 nm). In other words, tantalum cannot be used as a possible material for a solar photocell because the energy associated with a wavelength of 500 nm is less than the energy associated with a wavelength of 310 nm, which is the minimum energy needed to remove an electron from the surface of a tantalum atom.

13. In this problem, we are given the following information:
Wavelength of light used = 162 nm
Kinetic energy of the emitted electron = 5.34×10^{-19} J

$$h\nu = \text{work function} + \tfrac{1}{2} mv^2$$

We are asked to find the work function of the metal. From the wavelength of light used, we can determine the total energy. If we know the total energy and the kinetic energy, we can find the work function.

So let's determine the total energy of the light source used in this experiment.

$$\text{total energy} = h\nu = hc/\lambda$$

$$\text{kinetic energy} = \tfrac{1}{2} mv^2$$

Before substituting values into this equation, we need to make sure all the units are consistent. In this case, we need to convert the wavelength from nanometers to meters (1 nm = 1×10^{-9} m)

$$162\ \text{nm} = 162 \times 10^{-9}\ \text{m} = 1.62 \times 10^{-7}\ \text{m}$$

$$h\nu = hc/\lambda = \frac{(6.626 \times 10^{-34}\ \text{J} \cdot \text{s})(2.998 \times 10^{8}\ \text{m/s})}{1.62 \times 10^{-7}\ \text{m}} = 1.23 \times 10^{-18}\ \text{J}$$

Now we know the total energy and the kinetic energy, so we can now determine the work function.

$$1.23 \times 10^{-18}\ \text{J} = \text{work function} + 5.34 \times 10^{-19}\ \text{J}$$

So the work function is

$$1.23 \times 10^{-18}\ \text{J} - 5.34 \times 10^{-19}\ \text{J} = 6.96 \times 10^{-19}\ \text{J}$$

15. Obviously potassium, which has a smaller work function, will emit electrons with the greater velocity (energy). In both cases the total energy imparted is the same. Since it requires less energy to eject an electron from potassium than from sodium, there is a greater amount of energy available to be absorbed by the emitted electron and hence a greater velocity.

To calculate the velocity of the emitted electron we need to use the following equation.

$$\text{total energy} = \text{work function } (\phi) + \text{kinetic energy of the electron } \left(\tfrac{1}{2} mv^2\right)$$

$$\text{total energy} = \phi + \tfrac{1}{2} mv^2$$

We are given the work function (in joules) and the wavelength of the radiation (300 nm). We need to convert the wavelength from 300 nm (3×10^{-7} m) to energy in joules using $E = hc/\lambda$.

$$E = \frac{hc}{\lambda} = \frac{(6.626 \times 10^{-34}\ \text{J} \cdot \text{s})(2.998 \times 10^{8}\ \text{m/s})}{3.00 \times 10^{-7}\ \text{m}} = 6.62 \times 10^{-19}\ \text{J}$$

Now we can calculate the energy of the emitted electron.

$$6.62 \times 10^{-19}\ \text{J} = 3.68 \times 10^{-19}\ \text{J} + \tfrac{1}{2} mv^2$$

Note: It should now be easy to see why electrons emitted from the surface of potassium should have a greater kinetic energy than those emitted from the surface of sodium atoms.

The kinetic energy of the emitted electron from potassium is

$$KE = 6.62 \times 10^{-19}\text{ J} - 3.68 \times 10^{-19}\text{ J}$$
$$= 2.94 \times 10^{-19}\text{ J}$$

Lastly determine the velocity of the electron from the kinetic energy formula ($\frac{1}{2}mv^2$).

$$KE = \frac{1}{2}mv^2$$

where v is the velocity of the emitted electron, and m is the mass of the electron (9.11×10^{-31} kg).

Solving for v gives

$$v^2 = 2\,KE/m$$
$$v = \sqrt{2\,KE/m}$$
$$= \sqrt{2(2.94 \times 10^{-19}\text{ J})/(9.11 \times 10^{-31}\text{ kg})}$$
$$= \sqrt{6.45 \times 10^{-11}\text{ m}^2/\text{s}^2}$$
$$= 8.03 \times 10^5\text{ m/s}$$

The velocity of the emitted electron from potassium is 8.03×10^5 m/s.

Next determine the velocity of the emitted electron from the sodium atom using similar calculations.

The work function for sodium is 4.41×10^{-19} J. The kinetic energy of the emitted electron from sodium is

$$KE = 6.62 \times 10^{-19}\text{ J} - 4.41 \times 10^{-19}\text{ J}$$
$$= 2.21 \times 10^{-19}\text{ J}$$

Lastly, determine the velocity of the electron from the kinetic energy formula.

$$KE = \frac{1}{2}mv^2$$

where v is the velocity of the emitted electron, and m is the mass of the electron (9.11×10^{-31} kg).

Solving for v gives

$$v^2 = 2KE/m$$
$$v = \sqrt{2KE/m}$$
$$= \sqrt{2(2.21 \times 10^{-19}\text{ J})/(9.11 \times 10^{-31}\text{ kg})}$$
$$= \sqrt{4.85 \times 10^{-11}\text{ m}^2/\text{s}^2}$$
$$= 6.97 \times 10^5\text{ m/s}$$

The velocity of the emitted electron from sodium is 6.97×10^5 m/s.

As expected, the electron emitted from the potassium atom has the greater kinetic energy.

17. Hydrogen should have the simplest atomic spectrum because it is the simplest atom. It contains only one electron, and the energy levels in the hydrogen atom depend only on the principal quantum number, n. In all other atoms, the energy levels depend on both the principal quantum number, n, and the azimuthal quantum number, l. This results in many more possible energy levels in atoms other than hydrogen.

18. The principal quantum number, n, in the Bohr hydrogen atom can be used to calculate the energy of the individual orbits.

19. The energy of light emitted by a hydrogen atom depends only on the difference between two allowed energy states.

    From an individual n value, we can determine the energy of that particular orbit ($E_n = (-2.18 \times 10^{-18}\text{ J})(\frac{1}{n^2})$). However, it is the difference in energy between two orbits that results in a spectral line observed in the spectrum associated with hydrogen ($\Delta E = (-2.18 \times 10^{-18}\text{ J})(\frac{1}{n_f^2} - \frac{1}{n_i^2})$) because energy is gained or lost only when an electron moves from one orbit to another orbit.

20. A ground state refers to an electron in its lowest energy state. An excited state refers to any higher energy state.

    In the ground state of hydrogen, its single electron is found in the energy level with $n = 1$. In an excited hydrogen atom, the single electron can be found in any energy level where n is greater than 1 (that is, 2 to infinity).

21. There are two reasons why the Bohr model works for only single electron atoms or ions.

    1. In multielectron elements and ions, electrons interact with each other. Bohr's model offers no way of dealing with the repulsive interactions between the electrons within an atom.
    2. A second reason is that electrons spin. This spinning creates tiny magnetic fields that interact with the fields produced by other electrons.

22. In the Lyman series, electrons move between the first energy level, $n = 1$, and any other energy level. For example, $n = 3$ to $n = 1$, or $n = 5$ to $n = 1$.

    On the basis of Equation 3.4, the energies associated with these transitions should be greater than in the Balmer series (visible light, Equation 3.3). Greater energy corresponds to shorter wavelengths, so the Lyman series should have wavelengths shorter than visible light. This would put these transitions in the ultraviolet region of the electromagnetic spectrum.

23. The shortest wavelength would correspond to the transition with the largest energy change. The transition from $n = 2$ to $n = 1$ has the largest ΔE and so the shortest wavelength.

24. The greater the frequency, the greater the energy difference between orbits. In general as n gets larger the energy separation between the orbits gets smaller. On this bases we would predict that the transition from $n = 12$ to $n = 15$ has the smallest energy difference and hence would be associated with radiation of the lowest frequency, and $n = 5$ to $n = 6$ has the highest frequency.

25. The emission spectra of H and He^+ are both made up of a set of characteristic lines that can be used to identify these species. The number of lines for the two species is the same but the observed wavelengths are completely different.

26. No, sodium ions do not emit the same yellow orange light that sodium atoms emit. The sodium ion has one less electron than the sodium atom. This results in a different set of electron energy levels and different emission spectrum.

27. Yes, there are, but they are not in the visible region so they cannot be observed with the naked eye.

28. No. The only visible lines associated with the spectrum of hydrogen are for electron transitions that end in the state with $n = 2$.

29. a. As Z increases, ΔE increases for the transition $n = 2$ to $n = 1$. Since wavelength is inversely proportional to energy, the higher the energy the shorter the wavelength. As Z increases, the wavelength decreases.
    b. This transition in the hydrogen atom is in the ultraviolet region (higher energy than visible light). For other atoms Z increases, thus increasing the energy, so the wavelength would be even shorter. This transition could be never be in the visible region of the hydrogen atom.

31. We need to use the equation

$$\Delta E = -(2.18 \times 10^{-18}\text{ J})\left(\frac{1}{{n_\text{f}}^2} - \frac{1}{{n_\text{i}}^2}\right)$$

where $n_\text{f} = 2$ and $n_\text{i} = 3$

Substituting these values, we get

$$\Delta E = (2.18 \times 10^{-18}\text{ J})\left(\frac{1}{2^2} - \frac{1}{3^2}\right)$$

$$= (2.18 \times 10^{-18}\text{ J})\left(\frac{1}{4} - \frac{1}{9}\right)$$

$$= 3.03 \times 10^{-19}\text{ J}$$

33. If we assume the age of the universe to be 1.1×10^{10} years old (see Chapter 1), we can solve this problem.

We can start by using the formula

$$\text{distance} = \text{velocity} \times \text{time (or } d = v \times t)$$

This allows us to determine the velocity at which this constellation is moving away from Earth.

The constellation Bootes is currently 5.00×10^6 light-years from Earth. This is a distance. We need to convert this to units of meters.

$$d\text{ (in m)} = 5.00 \times 10^6\text{ light-yr} \times \frac{9.46 \times 10^{15}\text{ m}}{1\text{ light-yr}}$$

$$= 4.73 \times 10^{22}\text{ m}$$

$$t = 1.1 \times 10^{10}\text{ yr}$$

Now we can determine the velocity at which Bootes is receding.

$$v = \frac{d}{t} = \frac{4.73 \times 10^{22}\text{ m}}{1.1 \times 10^{10}\text{ yr}} = 4.3 \times 10^{12}\text{ m/yr}$$

Next, we need to determine at what fraction of the speed of light the constellation is receding. To do this, we need to convert the velocity from m/yr to m/s.

$$4.3 \times 10^{12}\,\frac{\text{m}}{\text{yr}} \times \frac{1\text{ yr}}{365\text{ day}} \times \frac{1\text{ day}}{24\text{ hr}} \times \frac{1\text{ hr}}{60\text{ min}} \times \frac{1\text{ min}}{60\text{ s}}$$

$$= 1.36 \times 10^5\text{ m/s}$$

The constellation Bootes is receding from Earth at a velocity of 1.36×10^5 m/s.

Compare this velocity to that of the speed of light. The magnitude of the redshift can be calculated using Equation 1.4.

$$\text{magnitude of redshift} = \frac{v\text{ (velocity of constellation)}}{c\text{ (speed of light)}}$$

$$\text{magnitude of redshift} = \frac{1.36 \times 10^5\text{ m/s}}{3 \times 10^8\text{ m/s}} = 0.000455$$

This constellation is receding from Earth at only 0.000455 times the speed of light. The redshift will only differ by 0.000455 from the known spectral line for this transition of 486 nm (see Figure 3.10). This quantity is much too small to see any difference in the wavelength for this spectral line.

The wavelength will be unchanged at 486 nm.

35. The DeBroglie equation is $\lambda = h/mc$

where

λ is the wavelength of light
h is Planck's constant (6.626×10^{-34} J · s)
m is the mass of the particle
c is the speed of light

This equation relates wave and particle properties to each other, because λ (wavelength) is a wave characteristic, while the mass is a particle characteristic.

36. The diffraction patterns produced by the electron prove they behave as waves. In general, we think of light as

traveling in waves and these waves exhibit defraction. The DeBroglie equation ($\lambda = hc/m$) implies that any object should have a characteristic wavelength. In other words, matter must also travel in waves.

Electrons are particles, and if they exhibit wavelike motion then they should be limited to fixed orbits of specified radii. This is similar to waves that result when a guitar string is plucked.

If particles have wavelike motion then they should exhibit diffraction and interference patterns similar to diffraction patterns of X-rays (for example). This is the exact observation made by scientists in the late 1920's.

37. The DeBroglie wavelength can be calculated from the equation $\lambda = h/mv$.

In this equation h is Planck's constant, 6.626×10^{-34} J · s, m is the mass of the object in kilograms (in this case, the tennis ball), and v is the velocity of the object in meters per second.

Before we can substitute values into the DeBroglie equation, we need to convert the velocity of the tennis ball from miles per hour to meters per second and also convert the mass of the tennis ball from grams to kilograms.

Velocity conversion:

$$\frac{120 \text{ miles}}{1 \text{ hour}} \times \frac{1 \text{ hour}}{3600 \text{ s}} \times \frac{5280 \text{ ft}}{1 \text{ mile}} \times \frac{12 \text{ in}}{1 \text{ ft}} \times \frac{2.54 \text{ cm}}{1 \text{ in}} \times \frac{1 \text{ m}}{100 \text{ cm}} = 5.364 \times 10^{1} \text{ m/s}$$

Mass conversion:

$$56 \text{ g} \times \frac{1 \text{ kg}}{1000 \text{ g}} = 0.056 \text{ kg}$$

Now we are ready to calculate the wavelength of a tennis ball travelling at 120 mph.

$$\lambda = \frac{h}{mv}$$

$$= \frac{6.626 \times 10^{-34} \text{ J} \cdot \text{s}}{(0.056 \text{ kg})(5.36 \times 10^{1} \text{ m/s})}$$

$$= 2.21 \times 10^{-34} \text{ m}$$

39. From the DeBroglie equation ($\lambda = h/mv$), we see that that wavelength is inversely proportional to the mass of the object.
   a. False. The heavier the mass of the object, the shorter the wavelength.
   b. True
   c. True. Since velocity is also inversely proportional to the wavelength, doubling the speed should make the wavelength half as long.

41. In this problem (all parts) we need to convert the wavelength from nanometers to meters and also convert all the masses into kilograms. After this has been accomplished, we can substitute into the DeBroglie equation and solve for the velocity.
Note: All velocities will be rounded to 3 significant figures.

Converting units:

$$750 \text{ nm} \times \frac{1 \times 10^{-9} \text{ m}}{1 \text{ nm}} = 7.50 \times 10^{-7} \text{ m}$$

From the DeBroglie equation

$$v = \frac{h}{m\lambda}$$

a.
$$m = 9.10939 \times 10^{-28} \text{ g}$$
$$= 9.10939 \times 10^{-31} \text{ kg}$$
$$= 9.11 \times 10^{-31} \text{ kg}$$
$$v = \frac{6.626 \times 10^{-34} \text{ J} \cdot \text{s}}{(9.11 \times 10^{-31} \text{ kg})(7.50 \times 10^{-7} \text{ m})}$$
$$= 9.70 \times 10^{2} \text{ m/s}$$

b.
$$m = 1.67262 \times 10^{-24} \text{ g}$$
$$= 1.673 \times 10^{-27} \text{ kg}$$
$$v = \frac{6.626 \times 10^{-34} \text{ J} \cdot \text{s}}{(1.673 \times 10^{-27} \text{ kg})(7.50 \times 10^{-7} \text{ m})}$$
$$= 5.28 \times 10^{-1} \text{ m/s}$$

c.
$$m = 1.67493 \times 10^{-24} \text{ g}$$
$$= 1.675 \times 10^{-27} \text{ kg}$$
$$v = \frac{6.626 \times 10^{-34} \text{ J} \cdot \text{s}}{(1.675 \times 10^{-27} \text{ kg})(7.50 \times 10^{-7} \text{ m})}$$
$$= 5.27 \times 10^{-1} \text{ m/s}$$

As expected, the velocity of a proton and neutron are almost the same because the masses are very similar.

d.
$$m = 6.64 \times 10^{-24} \text{ g}$$
$$= 6.64 \times 10^{-27} \text{ kg}$$
$$v = \frac{6.626 \times 10^{-34} \text{ J} \cdot \text{s}}{(6.64 \times 10^{-27} \text{ kg})(7.50 \times 10^{-7} \text{ m})}$$
$$= 1.33 \times 10^{-1} \text{ m/s}$$

The velocity of an α particle is slightly slower than the velocity of the proton or neutron because it is slightly heavier.

43. An orbital represents a solution to the wave equation. It corresponds to a region in space in which there is a high

probability of finding an electron in an atom. The Bohr orbit is a series of concentric circular orbitals each with a specific energy. Each orbit is a fixed distance from the nucleus.

44. The principal quantum number n roughly determines the energy and distance from the nucleus of an orbital.

    The angular momentum quantum number l determines the three-dimensional shape of an orbital. It also affects the energy of an orbital, but to a lesser extent than n.

    The magnetic quantum number m_l determines the orientation in three-dimensional space of each orbital. When there is an external magnetic field, m_l is a factor in determining the orbital's energy level.

45. To identify a specific orbital requires three quantum numbers. We need to know n, the principal quantum number, l, the angular momentum quantum number, and m_l, the magnetic quantum number. For example, let's consider the 3d subshell. The d subshell consists of 5 orbitals. The principal quantum number, n, is 3. The angular momentum quantum, l, for a d subshell is 2. Each of the five orbitals is then identified by the magnetic quantum number, m_l. For a d subshell the five possible m_l values are 2, 1, 0, −1, and −2.

    Oftentimes, we do not need to know a specific orbital, but rather only the orbital type. In this case, we only need to know the two quantum numbers n and l. For example, to identify a 3d orbital we need to know that $n = 3$ and $l = 2$.

46. Each orbital can hold two electrons. If one wants to identify a specific electron, we need to add a fourth value, the quantum number spin, m_s, to the other three, n, l, and m_l to distinguish electrons in the same orbital from one another.

47. We can use the information that each n value contains n^2 orbitals.

    For $n = 1$, the number of orbitals is $1^2 = 1$ orbital.
    For $n = 2$, the number of orbitals is $2^2 = 4$ orbitals.
    For $n = 3$, the number of orbitals is $3^2 = 9$ orbitals.
    For $n = 4$, the number of orbitals is $4^2 = 16$ orbitals.
    For $n = 5$, the number of orbitals is $5^2 = 25$ orbitals.

    As an alternative approach, determine the subshells (l values) associated with each n value and then determine the number of orbitals present in each subshell. Lastly add up all the orbitals present.

    First we need to know the number of orbitals associated with each subshell.

    An s subshell consists of a single orbital.
    A p subshell consists of 3 orbitals.
    A d subshell consists of 5 orbitals.
    An f subshell consists of 7 orbitals.
    A g subshell consists of 9 orbitals.

    For $n = 1$, only the s subshell is present. This subshell consists of a single orbital. There is only one orbital present in $n = 1$.

    For $n = 2$, there are 2 subshells present (s and p). The s subshell consists of a single orbital and the p subshell consists of three orbitals. There are 4 orbitals present in $n = 2$.

    For $n = 3$, there are 3 subshells present (s, p, and d). There are 9 orbitals present in $n = 3$ $(1 + 3 + 5)$.

    For $n = 4$, there are 4 subshells present (s, p, d, and f). There are 16 orbitals present in $n = 4$ $(1 + 3 + 5 + 7)$.

    For $n = 5$, there are 5 subshells present (s, p, d, f, and g). There are 25 orbitals present in $n = 5$ $(1 + 3 + 5 + 7 + 9)$.

49. The angular quantum number (l) can be any integer from 0 to $n - 1$. For $n = 4$, the allowed values of l are 3, 2, 1, 0.

    Note: The number of allowed values of l is the same as the n value. So for $n = 4$, there are 4 allowed values of l (3, 2, 1, and 0).

51. The angular momentum quantum number (l) determines the orbital designation, or subshell as follows:

    $l = 0$ s orbital
    $l = 1$ p orbital
    $l = 2$ d orbital
    $l = 3$ f orbital
    $l = 4$ g orbital

    For $l = 5$ and higher, the orbital designations follow in alphabetical order, starting with h for $l = 5$.

    The n value corresponds to the principal quantum number.

    a. $n = 2, l = 0$
    This corresponds to a 2s electron because the n value tells us the principal quantum number and an l value of 0 corresponds to an s orbital.

    b. $n = 3, l = 1$
    This corresponds to a 3p orbital (any time $l = 1$, the orbital is designated as a p orbital).

    c. $n = 4, l = 2$
    This corresponds to 4d orbital.

    d. $n = 1, l = 0$
    This corresponds to a 1s orbital.

53. a. $n = 2, l = 0$
    The n and l quantum numbers refer only to a specific subshell. In this case the subshell is 2s. For an s subshell, the maximum number of orbitals is 1. Each orbital can accommodate just 2 electrons. The maximum number of electrons that can have this set of quantum numbers is 2.

b. $n = 3, l = 1, m_l = 0$
Anytime we are given a set of quantum numbers that includes an m_l value, it refers to a single orbital. This set of quantum numbers refers to one of the $3p$ orbitals. A single orbital can have a maximum of 2 electrons, so the maximum number of electrons with the same n, l, and m_l values numbers is 2.

c. $n = 4, l = 2$
This corresponds to the $4d$ subshell. The d subshell consists of 5 orbitals (the allowed m_l values are $2, 1, 0, -1, -2$) and contains a maximum of 10 electrons, so the maximum number of electrons with this set of quantum numbers is 10.

d. $n = 1, l = 0, m_l = 0$
This corresponds to a specific orbital because an m_l value is included. The subshell designation is $1s$, and the maximum number of electrons with this set of quantum numbers is 2.

55. a. not allowed, because n must be greater than l (or l must be less than n).
b. allowed
c. not allowed, because for $l = 0$ the only allowed value of m_l is 0.
d. not allowed, because for $l = 1$ the allowed values of m_l are 1, 0, and -1.

57. The orbital designations for the quantum numbers given are as follows:

a. $n = 3, l = 2$	$3d$
b. $n = 5, l = 4$	$5g$
c. $n = 3, l = 0$	$3s$
d. $n = 4, l = 1$	$4p$

The order of increasing energy is $3s < 3d < 4p < 5g$ (c < a < d < b).

59. Yes, orbitals that contain no electrons do exist. For example, when a ground-state atom absorbs energy, one of its electrons moves to an orbital that was unoccupied in the ground state.

60. All s orbitals are spherical.
All p orbitals have two balloon-shaped lobes and resemble a figure eight

Four of the five d orbitals have a cloverleaf array of four lobes that resemble two figure eights at right angles to one another. The remaining d orbital has a doughnut shape (taurus), with two large lobes resembling a p orbital above and below the lobe in the doughnut.

61. The effective nuclear charge (Z_{eff}) is the attractive force toward the nucleus felt by a particular electron in an atom.

The attractive force decreases with increasing distance from the nucleus to the electron. It also decreases when there are other electrons between the nucleus and the electron of interest.

62. The term degenerate refers to orbitals that are equal in energy. The orbitals in any subshell are degenerate; that is, they all are equivalent in energy.

63. Ne and F^- are isoelectronic because they both contain 10 electrons. The term isoelectronic refers to species that contain the same number of electrons (or have the same electronic configuration).

64. The electron configuration of an element can be predicted by the location of the element in the periodic table. Each column in the periodic table has a similar outer-shell electron configuration.

For group 1A the single valence electron occupies an s orbital. We say that its valence (or outermost electron configuration) is ns^1. This means that each element in Group 1A has its highest-energy electron in an s subshell.

The n in ns^1 corresponds to the row number of a particular element. For example, sodium is found in the third row under column 1A. Its outermost electron will be found in the $3s$ orbital, and its configuration is $3s^1$.

In general, the outermost electron configuration for the Group A elements are as follows:

1A	ns^1	5A	$ns^2\,np^3$
2A	ns^2	6A	$ns^2\,np^4$
3A	$ns^2\,np^1$	7A	$ns^2\,np^5$
4A	$ns^2\,np^2$	8A	$ns^2\,np^6$

Since the $4s$ subshell fills before the $3d$ subshell, the d subshells are shown in the fourth period. To indicate that these orbitals are $3d$ not $4d$ ones, we use the notation $n - 1$. That is, the d orbitals shown in the fourth period actually represent $3d$ orbitals (not $4d$ orbitals).

65. The first two elements in the fourth row are potassium (K) and calcium (Ca). These elements are in Groups 1A and 2A, so they must have valence electron configurations of ns^1 and ns^2, respectively. The configuration for potassium is $4s^1$ and for calcium is $4s^2$. After these elements are the d-block elements, for example, vanadium (atomic number 21) with the electronic configuration $4s^2\,3d^1$. If the $3d$ orbital were filled before the $4s$, the next element in Group 1A, after Ne (atomic number 11), would have atomic number 29 instead of atomic number 19.

66. The $4s$ orbitals are lower in energy than the $3d$ orbitals.

Electrons fill orbitals starting with the lowest energy and working up to higher energy orbitals. The fact that the $4s$ orbitals are always filled before the $3d$ ones is clear indication that the $4s$ orbitals are lower in energy than the $3d$ orbitals.

67. The abbreviation [He] $2s^1$ refers to any species that contains 3 electrons (2 electrons in the $1s$ orbital that is symbolized by [He] and 1 electron in the $2s$ subshell).

Li atom has 3 electrons	$1s^2\ 2s^1$
Li^+ ion has 2 electrons	$1s^2$
He atom has 2 electrons	$1s^2$
F^- ion has 10 electrons	$1s^2\ 2s^2\ 2p^6$
Ne atom has 10 electrons	$1s^2\ 2s^2\ 2p^6$
Na^+ ion has 10 electrons	$1s^2\ 2s^2\ 2p^6$
Mg^{2+} ion has 10 electrons	$1s^2\ 2s^2\ 2p^6$
Al^{3+} ion has 10 electrons	$1s^2\ 2s^2\ 2p^6$

The only species with the configuration [He] $2s^1$ is the Li atom.

69. Neon (Ne) has 10 electrons, while sodium (Na) has 11 electrons, so they differ by 1 electron.
    The electron configuration of neon is $1s^2\ 2s^2\ 2p^6$ or [Ne].
    The electron configuration of sodium is $1s^2\ 2s^2\ 2p^6\ 3s^1$ or [Ne] $3s^1$.

71. Any filled subshell has only paired electrons.
    The s^1 configuration has only a single electron so any atom or ion with that configuration has 1 unpaired electron. All the elements in Group 1A have 1 unpaired electron.
    The s^2 configuration has a completely filled subshell, so there are no unpaired electrons. All the elements in Group 2A have a s^2 configuration and have no unpaired electrons.
    When electrons fill degenerate orbitals (orbitals of equal energy), one electron is added to each orbital until each orbital has a single electron.
    So a p^1 configuration has 1 unpaired electron (Group 3A)
    A p^2 configuration has 2 unpaired electrons (Group 4A)
    A p^3 configuration has 3 unpaired electrons (Group 5A)
    Once each orbital has a single electron, the next electron to be added must be paired.
    So a p^4 configuration has 2 unpaired electrons (Group 6A)
    A p^5 configuration has 1 unpaired electron (Group 7A)
    A p^6 configuration has no unpaired electrons, as the p subshell is now filled (Group 8A).
    a. N — $1s^2\ 2s^2\ 2p^3$ (Group 5A), 3 unpaired electrons
    b. O — $1s^2\ 2s^2\ 2p^4$ (Group 6A), 2 unpaired electrons
    c. P^{3-} — This ion has 18 electrons. This is identical to the noble gas argon. This species has no unpaired electrons.
    d. Na^+ — This ion has 10 electrons. It is isoelectronic with the Ne atom. It has no unpaired electrons.

73. The ions K^+, S^{2-}, and I^- all contain the same number of electrons as a noble gas. (The K^+ and S^{2-} ions both contain 18 electrons; I^- has 54 electrons.) This configuration is very stable one. When atoms can either gain or lose a few electrons to reach a noble gas configuration, they usually do so.

75. a. Na (11 electrons) — $1s^2\ 2s^2\ 2p^6\ 3s^1$, or [Ne] $3s^1$
    b. Cl (17 electrons) — $1s^2\ 2s^2\ 2p^6\ 3s^2\ 3p^5$, or [Ne] $3s^2\ 3p^5$
    c. Mn (25 electrons) — $1s^2\ 2s^2\ 2p^6\ 3s^2\ 3p^6\ 4s^2\ 3d^5$, or [Ar] $4s^2\ 3d^5$
    d. Mn^{2+} (23 electrons) — $1s^2\ 2s^2\ 2p^6\ 3s^2\ 3p^6\ 3d^5$, or [Ar] $3d^5$

    When electrons are ionized from transition metals (the Group B elements), the $4s$ electrons are ionized before the $3d$ electrons.

77. a. [He] $2s^1\ 2p^5$
    This is an excited state because the $2p$ subshell starts to fill before the $2s$ subshell is completely filled. The ground-state electron configuration is [He] $2s^2\ 2p^4$.
    b. [Kr] $5s^2\ 4d^{10}\ 5p^1$
    This is a ground-state electron configuration. It has the expected electron configuration for an element with 49 electrons.
    c. [Ar] $4s^2\ 3d^{10}\ 4p^5$
    This is a ground-state electron configuration. It has the expected electron configuration for an element with 35 electrons.
    d. [Ne] $3s^2\ 3p^2\ 4s^1$
    This is an excited-state electron configuration. The $4s$ subshell contains an electron but the $3p$ subshell has not been completely filled. The ground-state electron configuration is [Ne] $3s^2\ 3p^3$.

79. To determine whether an electron configuration corresponds to the same or different atoms, we merely need to determine the number of electrons shown in that configuration. If the numbers are the same then the atoms are the same. It does not matter whether the configuration corresponds to a ground state or an excited state of the atom.

$1s^2\ 2s^2\ 2p^6\ 3s^2\ 3p^6\ 4s^2\ 3d^1$	21 electrons
[Ar] $4s^2\ 3d^{10}\ 4p^1$	31 electrons (remember [Ar] corresponds to 18 electrons)
[Ar] $4s^2\ 4p^1$	21 electrons (excited state, $4p$ fills before $3d$)
[Ne] $3s^2\ 3p^2\ 4s^2\ 4p^1$	17 electrons (excited state, [Ne] corresponds to 10 electrons)

Two of the configurations represent element 21 (Sc; the ground state and one excited state). The other configurations correspond to elements 31 (Ga) and 17 (Cl).

81. Phosphorous (P, atomic number 15) has 15 electrons. Any configuration that contains 15 electrons corresponds to that of phosphorous.

$1s^2\ 2s^2\ 2p^6\ 3s^2\ 3p^4$	16 electrons (This corresponds to sulfur.)
$1s^2\ 2s^2\ 2p^6\ 3s^2\ 3p^3$	15 electrons (This is the ground-state configuration of phosphorous.)
$[Ar]\ 3s^2\ 3p^3$	23 electrons (This configuration is impossible because it double counts electrons in the $3s$ and $3p$ subshells. The symbol [Ar] represents the configuration $1s^2\ 2s^2\ 2p^6\ 3s^2\ 3p^6$. We cannot use any of these orbitals again in the configuration.)

Only the second configuration corresponds to an allowed electron configuration for the phosphorous atom.

83. The are several different ways to determine the number of unpaired electrons contained in an atom. They are all based on knowing the electron configuration associated with a specific atom. Two methods are outlined below.

The first method starts by writing out the complete electron configuration. From this, locate the highest energy subshell and determine the electron arrangement in these orbitals.

A second method uses the periodic table. From the location of the element in the periodic table, one can determine the highest energy subshell and from this the number of unpaired electrons.

As, atomic number 33 (33 electrons). This element is found in Group 5A.
The electron configuration of As is
$1s^2\ 2s^2\ 2p^6\ 3s^2\ 3p^6\ 4s^2\ 3d^{10}\ 4p^3$
The highest energy subshell in this configuration is $4p$. It contains 3 electrons. The p subshell consists of three equivalent orbitals. Using Hund's rule each one of these electrons must be in a separate orbital.
Arsenic has 3 unpaired electrons.
Using the second method, the reasoning is as follows. The element As is in Group 5A. The valence shell (outermost) configuration for elements in Group 5A is $ns^2\ np^3$.
Completely filled subshells (s^2) have only paired electrons, so we only need to consider partially filled subshells, in this case p^3. The p subshell consists of three equivalent orbitals. Using Hund's rule each of the three electrons must be in a separate orbital, so the ground-state of As has 3 unpaired electrons.

Te, atomic number 52
(52 electrons). This element is found in Group 6A.
The electron configuration of Te is
$1s^2\ 2s^2\ 2p^6\ 3s^2\ 3p^6\ 4s^2\ 3d^{10}\ 4p^6\ 5s^2\ 4d^{10}\ 5p^4$.
The only partially filled subshell is $5p$. One p orbital has 2 electrons and the other two each have 1 electron.
Tellurium has 2 unpaired electrons.

Sn, atomic number 50, Group 4A
The electron configuration is
$1s^2\ 2s^2\ 2p^6\ 3s^2\ 3p^6\ 4s^2\ 3d^{10}\ 4p^6\ 5s^2\ 4d^{10}\ 5p^2$.
The only partially filled subshell is $5p$. One of the p orbitals is empty, while the other two each have 1 electron.
Tin has 2 unpaired electrons.

Ge, atomic number 32, Group 4A
The electron configuration is $1s^2\ 2s^2\ 2p^6\ 3s^2\ 3p^6\ 4s^2\ 3d^{10}\ 4p^2$
The only partially filled subshell is $4p$. One of the $4p$ orbitals is empty; the other two each have 1 electron.
Germanuim has 2 unpaired electrons.

85. Ti (atomic number 22, Group 4B or 4), $1s^2\ 2s^2\ 2p^6\ 3s^2\ 3p^6\ 4s^2\ 3d^2$

Titanium has 2 unpaired electrons. The only partially filled subshell in this atom is the $3d$. It has five equivalent orbitals. Since electrons fill orbitals singly, until being forced to pair, this atom should have two orbitals in the $3d$ subshell which contain 1 electron and three empty $3d$ orbitals. This atom should have 2 unpaired electrons.

Cr (atomic number 24, Group 6B), $1s^2\ 2s^2\ 2p^6\ 3s^2\ 3p^6\ 4s^1\ 3d^5$

Note: The electron configuration of chromium is an exception to the general rules for filling orbitals. This is because of the extra stability associated with half-filled subshells ($3d$ and $4s$).

Chromium contains 6 unpaired electrons. There is 1 unpaired electron in the singly occupied $4s$ subshell and 5 unpaired electrons in the $3d$ subshell. Each of the five $3d$ orbitals contains 1 unpaired electron.

Cu (atomic number 29, Group 1B), $1s^2\ 2s^2\ 2p^6\ 3s^2\ 3p^6\ 4s^1\ 3d^{10}$

Note: Copper is also an exception to the normal electron configuration rules. This is due to extra stability associated with the filled $3d$ subshell and half-filled $4s$ subshell.

Copper has 1 unpaired electron because in this atom, the only partially filled subshell is the $4s$. It consists of a single orbital which contains 1 electron.

Zn (atomic number 30, Group 2B), $1s^2\ 2s^2\ 2p^6\ 3s^2\ 3p^6\ 4s^2\ 3d^{10}$

Zinc has no unpaired electrons because all subshells are completely filled.

87. The electron configuration of ^{131}I is $1s^2\ 2s^2\ 2p^6\ 3s^2\ 3p^6\ 4s^2\ 3d^{10}\ 4p^6\ 5s^2\ 4d^{10}\ 5p^5$, or [Kr] $5s^2\ 4d^{10}\ 5p^5$

The highest energy electrons in iodine are the five $5p$ electrons.

For these $5p$ electrons:

$n = 5$ (from the 5 in $5p$)

$l = 1$ (l always equals 1 for p electrons).

$m_l = 1, 0,$ or -1

$m_s = +\frac{1}{2}$ or $-\frac{1}{2}$

A complete set of quantum numbers for these five $5p$ electrons could be

$n = 5$	$l = 1$	$m_l = 1$	$m_s = +\frac{1}{2}$
$n = 5$	$l = 1$	$m_l = 1$	$m_s = -\frac{1}{2}$
$n = 5$	$l = 0$	$m_l = 0$	$m_s = +\frac{1}{2}$
$n = 5$	$l = 0$	$m_l = 0$	$m_s = -\frac{1}{2}$
$n = 5$	$l = -1$	$m_l = -1$	$m_s = +\frac{1}{2}$

The numbers 131 and 127 in the chemical symbols ^{131}I and ^{127}I refer to the mass number, which is the sum of the number of protons and neutrons in the nucleus of an atom. In this problem we are not interested in either protons or neutrons, but in the number of electrons, which is determined by the atomic number. For iodine, the atomic number is 53, so any iodine atom contains 53 electrons. Since both ^{131}I and ^{127}I contain 53 electrons, they have the same electronic configuration and the electrons will have the same set of quantum numbers.

89. Ionization energy is the minimum energy required to remove the most loosely held electron of a gaseous atom.

90. a. As we go down a group of elements in the periodic table the ionization energies get smaller. This is mainly due to the increase in distance from the nucleus of the most loosely held electron. The increase in distance decreases the attraction between the nucleus and the electron and makes it easier to lose this electron.

b. As one moves from left to right across the periodic table the ionization energies generally increase. This is mainly a function of the larger effective nuclear charge. This results in electrons being more strongly attracted to the nucleus and makes it more difficult to lose an electron.

91. a. He (helium) and Li (lithium)

Helium (electron configuration $1s^2$)
Lithium (electron configuration $1s^2\ 2s^1$)

He has a completed shell. It should have a higher first ionization than the lithium atom. Additionally, the outermost electron in the lithium is farther from the nucleus making it easier to remove.

b. Li (lithium) and Be (beryllium)

Lithium (electron configuration $1s^2\ 2s^1$)
Beryllium (electron configuration $1s^2\ 2s^2$)

Beryllium should have the greater first ionization because of a greater effective nuclear charge felt by the outermost electron in beryllium.

c. Be (beryllium) and B (boron)

Beryllium (electron configuration $1s^2\ 2s^2$)
Boron (electron configuration $1s^2\ 2s^2\ 2p^1$)

Beryllium has the greater first ionization energy because, in the boron atom, the filled $2s$ orbital shields the more distant $2p$ electron from the nucleus. In other words, the $2p$ electron in boron feels less positive charge than does the $2s$ electron in beryllium. The $2p$ electron in the boron atom should be easier to remove.

In general, a completely filled subshell (such as the $2s^2$ in beryllium) or a half-filled subshell (such as $2p^3$ in nitrogen) are more stable configurations than only partially filled subshells.

d. N (nitrogen) and O (oxygen)

Nitrogen (electron configuration $1s^2\ 2s^2\ 2p^3$)
Oxygen (electron configuration $1s^2\ 2s^2\ 2p^4$)

The half-filled $2p$ subshell in the nitrogen atom is a more stable configuration than the $2p^4$ configuration in oxygen. It should be more difficult to remove the outermost electron in the nitrogen atom.

93. Einstein's statement refers to the idea that everything in the universe should have a definite explanation or value. Dealing in probabilities is not an acceptable explanation because it does not deal in certainties, but only with the chances of an event occurring. Einstein's concept of God led him to conclude that everything that occurs in nature should have a definite explanation.

Bohr's comment probably means that God decides how the universe is run, not Einstein.

CHAPTER 4 Stoichiometry and the Formation of Earth

1. The densest element should sink to the core of the Earth. On the basis of its location in the Earth's solid core, we would expect that Fe(*s*) is the densest substance of those listed.

2. The lower the melting point, the more easily the substance melts. Since nickel and iron are the elements in the core, we should expect these elements to have both lower melting and boiling points and for greater quantities to be present. Sulfur is a non-metal; it would not be expected to have particularly high melting and boiling points. Al_2O_3 should have the highest melting and boiling points.

3. The lighter the species, the more easily it is converted to vapor and the greater the quantity of this substance in solar wind.

   Comparing the species listed in this question, we see that He^+ is the second lightest ion (next to hydrogen). Once He^+ is vaporized, it can be carried away by solar wind. The second most abundant element in the universe is the He^+ ion. It is the element second to hydrogen ions in abundance in solar wind.

4. If the Earth had developed a solid core before Eros was formed, we should expect that many of the elements in the Earth's core would have been unavailable for incorporation into Eros. Elements such as iron (Fe) and nickel (Ni) would be present in greater abundance on Earth than on Eros. The asteroid Eros would probably have very little iron (Fe) and nickel (Ni) and would have a much greater percentage of elements such as magnesium, silicon, and aluminum.

5. The law of definite proportions and multiple proportions are compatible because they refer to different concepts.

   The law of definite proportions refers to a single substance such as water or carbon dioxide.

   A given compound (water, H_2O, for example) always has the same composition by mass regardless of how it was made or where it comes from. Water is always 88.88% oxygen by weight and 11.11% hydrogen by weight.

   The law of multiple proportions refers to two or more different compounds that are made up of the same elements. For example, carbon monoxide (CO) and carbon dioxide (CO_2) both contain the same two elements, carbon and oxygen. The law of multiple proportions states that, if the mass of one element is fixed for the two substances, the masses of the second element are in the ratio of small whole numbers. In our example, if the mass of carbon is fixed at 12 g for both substances (CO and CO_2) then:

   12 g of carbon combines with 16 g of oxygen in CO
   12 g of carbon combines with 32 g of oxygen in CO_2

   The ratio of the masses of oxygen that combine with 12 g of carbon is 32:16 or 2:1.

6. A law is a brief statement or mathematical equation that can be used to summarize scientific data. For example, the law of definite proportions summarizes information about compounds.

   A theory usually covers a wide body of knowledge, rather than a single piece of information. For example, Dalton's atomic theory accounts for a wide body of information and relies upon the law of definite proportions as one of the bases for forming its postulates.

7. Let's assume we have 1 mole of each substance, CoS and Co_2S_3.

   In CoS, 58.93 g of cobalt (Co) will combine with 32.07 g of sulfur (S).

In Co_2S_3, 117.86 g of Co will combine with 96.21 g of S.

Let's take cobalt to be the substance of constant mass and set the mass at 58.93 g. Next we need to find the mass of sulfur, in Co_2S_3, that will combine with 58.93 g of Co. We can do this by using simple ratios.

Let x be the mass of sulfur required to form Co_2S_3 from 58.93 g of Co.

$$\frac{58.93 \text{ g Co}}{117.86 \text{ g Co}} = \frac{x \text{ g S}}{96.21 \text{ g S}}$$

Solving for x grams of sulfur

$$x \text{ grams S} = 96.21 \text{ g S} \times \frac{58.93 \text{ g Co}}{117.86 \text{ g Co}}$$

$$= 48.1 \text{ g S}$$

So 58.93 g of Co will combine with 48.1 g of S in Co_2S_3.

Since 58.93 g Co will combine with 32.07 g S in CoS, the ratio of the sulfur masses is

$$\frac{48.10 \text{ g S}}{32.07 \text{ g S}} = \frac{1.5}{1.0}, \text{ or } 3{:}2$$

9. The oxoanion that has fewer oxygen atoms will have the *-ite* ending.

   XO_2^{2-}—ends in *-ite* XO_3^{2-}—ends in *-ate*

10. No prefix is required. The prefixes *hypo* and *per* are only needed when an element (such as a halogen) can form four oxoanions. If the oxoanion contains only 1 oxygen in the formula, the name of the oxoanion contains the prefix *hypo* and the ending of *-ite*. An example is hypochlorite ion (ClO^-). If the oxoanion contains 4 oxygen atoms in the formula, the prefix *per* is used and the name ends in *-ate*. An example is the perchlorate ion, ClO_4^-.

11. The Roman numerals in the names of certain compounds that contain transition metals correspond to the charge on the transition metal. For example, in iron(II) chloride ($FeCl_2$), the (II) refers to the charge on Fe; that is, the charge on the iron atom in this substance is +2 (Fe^{2+}). Since transition metals can form compounds with different charges, the Roman numerals are needed to differentiate one compound from another. The compound $FeCl_3$ is named iron(III) chloride. Without the Roman numeral designation we would not be able to distinguish $FeCl_2$ from $FeCl_3$ if the name iron chloride were used.

12. All alkali metals form ions with a +1 charge. All alkaline earth metals form ions with a +2 charge. Since there is no ambiguity in the charge on alkali or alkaline earth metals, a Roman numeral designation is unnecessary.

13. These compounds are all molecular (covalent). They contain only non-metallic elements. To name these compounds, use the full name of the first element and add the ending *-ide* to the name of the second element in the formula. Additionally one needs to specify the number of atoms of each element in the formula, by using a Greek prefix. These prefixes listed are as follows:

mono – one (usually not used for the first element)	hexa – six
di – two	hepta – seven
tri – three	octa – eight
tetra – four	nona – nine
penta – five	deca – ten

Note: Sometimes when two vowels occur next to each other in the name, one of them is dropped (usually the *a*).

a. NO_3, nitrogen trioxide

b. N_2O_5, dinitrogen pentoxide

c. N_2O_4, dinitrogen tetroxide

d. NO_2, nitrogen dioxide

e. N_2O_3, dinitrogen trioxide

f. NO, nitrogen monoxide

g. N_2O, dinitrogen monoxide

h. N_4O, tetranitrogen monoxide

15. The formulas of ionic substances for the Group A elements can be predicted by determining the ionic charges on each atom. The ionic charges can be related to the column in the periodic table in which the element is found by determining how many electrons the atom has to gain or lose to reach a noble gas number of electrons. The charges are as follows:

Group 1A	+1 (need to lose 1 electron)
Group 2A	+2 (need to lose 2 electrons)
Group 3A	+3 (need to lose 3 electrons)
Group 5A	−3 (need to gain 3 electrons)
Group 6A	−2 (need to gain 2 electrons)
Group 7A	−1 (need to gain 1 electron)

Note: Elements in Group 4A usually do not form ions because gaining or losing 4 electrons is energetically unfavorable.

To name ionic compounds that contain Group A metals:

1. First use the element name of the metal.
2. Add the name of the non-metal, changing the ending to *-ide*.
3. Prefixes are not needed because we know the ionic charges.

a. Sodium is from Group 1A, so it forms Na^+. Sulfur is from Group 6A, so it forms S^{2-}.

   It will require 2 Na^+ ions to neutralize 1 S^{2-} ion. Alternatively, we can use the "crossover rule" to determine the formula. In substances composed of ions with unequal charges, the charge on the positive ion becomes the subscript associated with the negative ion and the charge on the negative ion becomes the subscript associated with the positive ion. The $+1$ on Na^+ becomes the subscript of 1 for sulfur, while the -2 on S^{2-} becomes a subscript of 2 on the sodium.

The formula is Na_2S. The name of this compound is sodium sulfide.

b. Strontium is from Group 2A, so it forms Sr^{2+}.

Chlorine is from Group 7A, so it forms the ion Cl^-. It requires 2 Cl^- ions to neutralize 1 Sr^{2+} ion.

The $+2$ charge on Sr becomes a subscript of 2 for Cl.

The formula is $SrCl_2$. The name of this compound is strontium chloride.

c. Aluminum is from Group 3A, so it forms Al^{3+}.

Oxygen is from Group 6A, so it forms an O^{2-} ion.

Using the crossover rule, we get a subscript of 2 for the aluminum (the charge on the oxygen) and a subscript of 3 for the oxygen (the charge on the aluminum).

The formula is Al_2O_3. The name of this compound is aluminum oxide.

d. Lithium is from Group 1A, so it forms Li^+.

Hydrogen is also found in Group 1A, but is often listed in Group 7A. When hydrogen combines with a metal, it forms an ion with a -1 charge, H^-.

The formula for this compound is LiH and its name is lithium hydride.

17. To name a substance containing a transition metal such as cobalt (Co), we need to put the charge in Roman numerals in parenthesis next to the metal name.

The three compounds are CoO, Co_2O_3, and CoO_2. Since the charge on each oxygen atom is -2, we can determine the charge on Co in each of the compounds.

In CoO, there is 1 oxygen atom so the total negative charge must be -2. To balance this, the cobalt atom must have a charge of $+2$. CoO is called cobalt(II) oxide.

In Co_2O_3, there are 3 oxygen atoms having a total charge of -6. There are 2 cobalt atoms. The 2 cobalt atoms must have a charge of $+6$ to neutralize the -6 charge due to the 3 oxygen atoms. Each cobalt atom must have a charge of $+3$. The name of Co_2O_3 is cobalt(III) oxide.

In CoO_2, there are 2 oxygen atoms so the total negative charge is -4. The single cobalt atom must have a charge of $+4$. The name of CoO_2 is cobalt(IV) oxide.

19. Chlorine belongs to Group 7A. It is commonly found as the chloride ion with a charge of -1 (because it needs to gain one electron to reach a noble gas number of electrons).

The charge on the metal is determined using the same methodology, that is, from its location in the periodic table, which enables us to determine the charge on its most common ion.

Sodium and potassium are in Group 1A. The most common charge is $+1$ for both of these ions (Na^+ and K^+).

Magnesium, calcium, and strontium are found in Group 2A. The most common charge on these three species is $+2$ (Mg^{2+}, Ca^{2+}, and Sr^{2+}).

The potassium and sodium ions require 1 chloride ion for neutrality, while the magnesium, calcium, and strontium ions require 2 chloride ions.

The formulas for these five substances are as follows:

NaCl, sodium chloride — KCl, potassium chloride
$MgCl_2$, magnesium chloride — $CaCl_2$, calcium chloride
$SrCl_2$, strontium chloride

The sulfate ion (SO_4^{2-}) has a charge of negative 2.

It requires 2 potassium and 2 sodium ions for neutrality, while it requires 1 magnesium, calcium, and strontium ion for neutrality.

The formulas for these five substances are as follows:

Na_2SO_4, sodium sulfate
$MgSO_4$, magnesium sulfate
$SrSO_4$, strontium sulfate
K_2SO_4, potassium sulfate
$CaSO_4$, calcium sulfate

21. a. Sodium hypobromite is the bromine equivalent of sodium hypochlorite.

The formula for the hypobromite ion is BrO^-. Sodium, being a Group 1A element, forms an ion with a $+1$ charge. The formula for sodium hypobromite is NaBrO.

b. The formula for the sulfate ion is SO_4^{2-}.

Potassium is a Group 1A element, so it forms an ion with a $+1$ charge.

It requires 2 potassium ions to cancel out the charge on 1 sulfate ion, so the formula for potassium sulfate is K_2SO_4.

c. The formula for the iodate ion is IO_3^-. Lithium (Group 1A) forms an ion with a $+1$ charge, so the formula for lithium iodate is $LiIO_3$.

d. The formula for the nitrite ion is NO_2^-. Magnesium (Group 2A) forms an ion with a $+2$ charge. It requires 2 nitrite ions to balance the $+2$ charge on the magnesium ion, so the formula for magnesium nitrite is $Mg(NO_2)_2$.

23. a. $NiCO_3$

This is an ionic compound made up of the ions nickel(II) and carbonate. Nickel is a transition metal so we must include the charge on the nickel atom in parenthesis, using Roman numerals. We can determine the charge on the nickel ion from the charge on the carbonate ion, −2. The name of this compound is nickel(II) carbonate.

b. NaCN
This ionic compound is made up of a sodium ion (Na^+) and cyanide ion (CN^-). The name of this compound is sodium cyanide.

c. $LiHCO_3$
This ionic compound is made up of the lithium ion (Li^+) and the hydrogen carbonate or bicarbonate ion (HCO_3^-). The name of this compound is lithium hydrogen carbonate or lithium bicarbonate.

d. $Ca(ClO)_2$
This ionic compound is made up of a calcium ion (Ca^{2+}, calcium being in Group 2A) and hypochlorite ion, ClO^-. The name of this compound is calcium hypochlorite.

25. Binary acids contain hydrogen and one other element. To name a binary acid, use the following steps:

1. Name the substance as a covalent compound.
2. Change the word hydrogen to *hydro-*.
3. Change the nonmetal ending from *-ide* to *-ic*.
4. Merge the results of Steps 2 and 3 into a single word.
5. Add the word acid.

Ternary acids contain three or more elements. Usually a ternary acid contains the element hydrogen combined with a polyatomic ion. To name a ternary acid, use the following procedure.

1. Name the compound as an ionic substance.
2. Drop the word hydrogen completely from the name.
3. Change the polyatomic ion ending to from *-ate* to *-ic* or to from *-ite* to *-ous*.
4. Add the word acid.

a. HF is a binary acid.
As a covalent substance, this compound would be named hydrogen fluoride. To name this substance as an acid:
Change *hydrogen* to *hydro*.
Change *fluoride* to *fluoric*.
Merge the two to get hydrofluoric.
Add the word acid.
The name of HF is hydrofluoric acid.

b. $HBrO_3$ is a ternary acid.
Naming it as an ionic substance, we get hydrogen bromate (BrO_3^- is the bromate ion). To name this substance as an acid:
Drop the word hydrogen.
Change *bromate* to *bromic*.
Add the word acid.
The name of $HBrO_3$ is bromic acid.

c. Phosphoric acid is a ternary acid (no *hydro-* associated with the name).
The *-ic* ending means that the polyatomic ion has the *-ate* ending. The polyatomic ion must be the phosphate ion. The formula for the phosphate ion is PO_4^{3-}.
The hydrogen ion has a charge of +1.
It requires 3 hydrogen ions to balance out the −3 charge on the phosphate ion, so the formula for phosphoric acid is H_3PO_4.

d. Nitrous acid is a ternary acid (no *hydro-* associated with the name).
The *-ous* ending means that the polyatomic ion has the *-ite* ending. The polyatomic ion here must be the nitrite ion. The formula for the nitrite ion is NO_2^-.
It requires 1 hydrogen ion to balance 1 nitrite ion, so the formula for nitrous acid is HNO_2.

27. The disadvantage of using the dozen in place of the mole is that the resulting numbers would become very large. For example, a mole of atoms is 6.02×10^{23} atoms. If we used dozen as the measuring unit we would find that a mole of atoms corresponds to about 5×10^{22} dozen atoms. This number is much too big to use as a base unit. We want base units to be much more user-friendly numbers such as 1, 2, etc., not a number with a large exponential value.
Note: To convert 1 mole into dozens use the conversion factor

$$1 \text{ dozen} = 12 \text{ atoms}$$

Divide 6.02×10^{23} by 12 to get 5×10^{22} dozen

$$6.02 \times 10^{23} \text{ atoms} \times \frac{1 \text{ dozen}}{12 \text{ atoms}} = 5 \times 10^{22} \text{ dozen}$$

28. The molar mass of a molecular compound is the mass of 6.02×10^{23} molecules. The mass of one molecule is $1/6.02 \times 10^{23}$ times the mass of a mole of that substance.

29. No. It is not only the number of atoms in a molecule that determines the molar mass. The molar mass also depends on the atomic masses of each element in the molecule. For example, water (H_2O) has a molar mass of 18 g/mol. Each water molecule contains 3 atoms (2 hydrogen atoms and 1 oxygen atom). Carbon monoxide has the formula CO. It contains only 2 atoms per molecule (1 carbon atom and 1 oxygen atom). It has a molar mass of 30 g/mol. Even though carbon monoxide has fewer atoms per molecule, it has a greater molar mass than water.

30. NO_2 has 1 nitrogen atom and 2 oxygen atoms per formula unit.
N_2O has 2 nitrogen atoms and 1 oxygen atom per formula unit.
Since the only difference between the two substances is an extra oxygen atom (in NO_2) or an extra nitrogen atom (in N_2O), we should expect that the NO_2 species should have the greater molar mass because oxygen atoms are heavier than nitrogen atoms.

As a check, we can calculate the molar masses, which are 46 g/mol (NO_2) and 44 g/mol (N_2O).

31. To convert from molecules to number of moles, one divides by Avogadro's number (6.02×10^{23}). In other words, use the conversion factors

$$1 \text{ mole of molecules} = 6.022 \times 10^{23} \text{ molecules}$$

$$1 \text{ mole of atoms} = 6.022 \times 10^{23} \text{ atoms (for monatomic species)}$$

a. $$4.4 \times 10^{14} \text{ molecules Ne} \times \frac{1 \text{ mol Ne}}{6.022 \times 10^{23} \text{ molecules Ne}} = 7.3 \times 10^{-10} \text{ mol Ne}$$

b. $$4.2 \times 10^{13} \text{ molecules } CH_4 \times \frac{1 \text{ mol } CH_4}{6.022 \times 10^{23} \text{ molecules } CH_4} = 7.0 \times 10^{-11} \text{ mol } CH_4$$

c. $$2.5 \times 10^{12} \text{ molecules } O_3 \times \frac{1 \text{ mol } O_3}{6.022 \times 10^{23} \text{ molecules } O_3} = 4.2 \times 10^{-12} \text{ mol } O_3$$

d. $$4.9 \times 10^{9} \text{ molecules } NO_2 \times \frac{1 \text{ mol } NO_2}{6.022 \times 10^{23} \text{ molecules } NO_2} = 8.1 \times 10^{-15} \text{ mol } NO_2$$

33. To find the number of moles in a number of units, one divides by Avogadro's number, 6.022×10^{23}. In terms of a conversion factor

$$1 \text{ mole of bytes} = 6.02 \times 10^{23} \text{ bytes}$$

Then we need to convert moles to micromoles using the conversion factor

$$1 \text{ micromole } (\mu\text{mol}) = 1 \times 10^{-6} \text{ mole (mol)}$$

The information given in this question has a prefix in front of the bytes. We need to relate that prefix to an actual numerical value.

a. *Giga-* is the prefix which corresponds to a numerical value of 1×10^9, so a gigabyte means 1×10^9 bytes of information.

We can now determine the number of μmoles of bytes in a gigabyte.

$$10 \text{ gigabytes} \times \frac{1 \times 10^9 \text{ bytes}}{1 \text{ gigabytes}} \times \frac{1 \text{ mol of bytes}}{6.022 \times 10^{23} \text{ bytes}} \times \frac{1 \mu\text{mol}}{1 \times 10^{-6} \text{ mol}} = 1.7 \times 10^{-8} \mu\text{mol}$$

b. *Mega-* is the prefix meaning 1 million or 1×10^6, so

$$100 \text{ megabytes} \times \frac{1 \times 10^6 \text{ bytes}}{1 \text{ megabyte}} \times \frac{1 \text{ mol of bytes}}{6.022 \times 10^{23} \text{ bytes}} \times \frac{1 \mu\text{mol}}{1 \times 10^{-6} \text{ mol}} = 1.7 \times 10^{-10} \mu\text{mol}$$

c. *Kilo-* is the prefix meaning 1000, so

$$1.4 \text{ kilobytes} \times \frac{1 \times 10^3 \text{ bytes}}{1 \text{ kilobyte}} \times \frac{1 \text{ mol of bytes}}{6.022 \times 10^{23} \text{ bytes}} \times \frac{1 \mu\text{mol}}{1 \times 10^{-6} \text{ mol}} = 2.3 \times 10^{-15} \text{ mol}$$

35. A chemical formula tells us how many atoms there are in one molecule. It also tells us how many moles of atoms are in 1 mole of molecules. For example, in one molecule of glucose, $C_6H_{12}O_6$, there are

6 carbon atoms, 12 hydrogen atoms, and 6 oxygen atoms

In 1 mole of glucose molecules, there are 6 moles of carbon atoms, 12 moles of hydrogen atoms, and 6 moles of oxygen atoms

We can create a conversion factor that relates moles of atoms for each element to a mole of molecules. For glucose, these factors are as follows:

$$1 \text{ mol of glucose molecules} = 6 \text{ mol of carbon atoms}$$

$$1 \text{ mol of glucose molecules} = 12 \text{ mol of hydrogen atoms}$$

$$1 \text{ mol of glucose molecules} = 6 \text{ mol of oxygen atoms}$$

Using these conversion factors, we can convert from moles of molecules (which is what we really mean when we talk about moles) to moles of a particular element. Lastly, multiply by Avogadro's number to convert to atoms.

a. From the formula $FeTiO_3$, we see that the needed conversion factor is

$$1 \text{ mol of } FeTiO_3 \text{ formula units} = 1 \text{ mol of Ti atoms}$$

$$0.125 \text{ mol } FeTiO_3 \times \frac{1 \text{ mol Ti}}{1 \text{ mol } FeTiO_3} \times \frac{6.02 \times 10^{23} \text{ atoms Ti}}{1 \text{ mol Ti}} = 7.53 \times 10^{22} \text{ atoms of Ti}$$

b. The formula for titanium(IV) chloride is $TiCl_4$.
From the formula, the needed conversion factor is

$$1 \text{ mol } TiCl_4 \text{ formula units} = 1 \text{ mol of Ti atoms}$$

$$0.125 \text{ mol } TiCl_4 \times \frac{1 \text{ mol Ti}}{1 \text{ mol } TiCl_4} \times \frac{6.02 \times 10^{23} \text{ atoms Ti}}{1 \text{ mol Ti}} = 7.53 \times 10^{22} \text{ atoms of Ti}$$

c. The formula Ti_2O_3 tells us that there are 2 moles of Ti atoms for every mole of Ti_2O_3 formula units.
The needed conversion factor is then

$$2 \text{ mol Ti} = 1 \text{ mol } Ti_2O_3$$

$$0.125 \text{ mol } Ti_2O_3 \times \frac{2 \text{ mol Ti}}{1 \text{ mol } Ti_2O_3} \times \frac{6.02 \times 10^{23} \text{ atoms Ti}}{1 \text{ mol Ti}} = 1.51 \times 10^{23} \text{ atoms Ti}$$

d. The formula Ti_3O_5 tells us that there are 3 moles of Ti atoms for every mole of Ti_3O_5 formula units.
The needed conversion factor is

$$3 \text{ mol Ti} = 1 \text{ mol } Ti_3O_5$$

$$0.125 \text{ mol } Ti_3O_5 \times \frac{3 \text{ mol Ti}}{1 \text{ mol } Ti_3O_5} \times \frac{6.02 \times 10^{23} \text{ atoms Ti}}{1 \text{ mol Ti}} = 2.26 \times 10^{23} \text{ atoms Ti}$$

37. a. Both samples contain the same number of moles of oxygen.
From the formulas for these substances, one can see that 1 mole of formula units contain 3 moles of oxygen atoms. Since we started with 1 mole of formula units of each substance, they must both have 3 moles of oxygen atoms.

b. The sample of N_2O_4 contains more moles of oxygen atoms.
From the formula, we can create a conversion factor that shows that there are 4 moles of oxygen atoms for every mole of N_2O_4 molecules. There are only 2 moles of oxygen atoms for every mole of SiO_2 molecules.
One mole of N_2O_4 molecules contains 4 moles of oxygen atoms.
One mole of SiO_2 molecules contains only 2 moles of oxygen atoms.

c. There are more moles of oxygen atoms in the sample of CO_2 than in the sample of CO.
In the CO sample, 1 mole of CO molecules contains 1 mole of oxygen atoms, so 3 moles of CO contains 3 moles of oxygen atoms.
In the CO_2 sample 1 mole of CO_2 molecules contain 2 moles of oxygen atoms, so 2 moles of CO_2 must contain 4 moles of oxygen atoms.

39. We can use the chemical formula to relate moles of aluminum to 1 mole of the substance.

a. From the formula $Al_2Si_4O_{10}(OH)_2$, the needed conversion factor is

$$1 \text{ mol } Al_2Si_4O_{10}(OH)_2 = 2 \text{ mol Al}$$

$$1.5 \text{ mol } Al_2Si_4O_{10}(OH)_2 \times \frac{2 \text{ mol Al}}{1 \text{ mol } Al_2Si_4O_{10}(OH)_2} = 3.0 \text{ mol Al}$$

b. From the formula $KAl_3Si_3O_{10}(OH)_2$, the needed conversion factor is

$$1 \text{ mol } KAl_3Si_3O_{10}(OH)_2 = 3 \text{ mol Al}$$

$$1.5 \text{ mol } KAl_3Si_3O_{10}(OH)_2 \times \frac{3 \text{ mol Al}}{1 \text{ mol } KAl_3Si_3O_{10}(OH)_2} = 4.5 \text{ mol Al}$$

c. From the formula $NaAlSi_3O_8$, the needed conversion factor is

$$1 \text{ mol } NaAlSi_3O_8 = 1 \text{ mol Al}$$

$$1.5 \text{ mol } NaAlSi_3O_8 \times \frac{1 \text{ mol Al}}{1 \text{ mol } NaAlSi_3O_8} = 1.5 \text{ mol Al}$$

41. To convert from grams to moles (of atoms) divide by the atomic mass.
The needed conversion factor is

$$1 \text{ mol C} = 12.0 \text{ g C}$$

$$500 \text{ g C} \times \frac{1 \text{ mol C}}{12.0 \text{ g C}} = 41.7 \text{ mol C}$$

43. Calcium titanate, $CaTiO_3$, contains 1 Ca^{2+} ion and 1 $TiO_3{}^{2-}$ ion per formula unit. On a molar basis, we

can say that 1 mole of $CaTiO_3$ formula units contains 1 mole of Ca^{2+} ions and 1 mole of TiO_3^{2-} ions.

We can use the following as a conversion factor

$$1 \text{ mol } CaTiO_3 = 1 \text{ mol } Ca^{2+}$$

We need to first convert 0.25 mole of $CaTiO_3$ into moles of Ca^{2+} ions.

$$0.25 \text{ mol } CaTiO_3 \times \frac{1 \text{ mol } Ca^{2+}}{1 \text{ mol } CaTiO_3}$$

$$= 0.25 \text{ mol } Ca^{2+} \text{ ions}$$

The second part of this problem asks us to find the mass of the calcium ions in the sample. We can assume that the mass of an ion and the mass of an atom are the same. We need to convert the 0.25 mole of Ca^{2+} to grams. (To convert from moles to grams, multiply by the molar mass or atomic mass.)

The atomic mass of calcium is 40.1 g/mol, or 1 mol Ca = 40.1 g Ca.

$$0.25 \text{ mol } Ca^{2+} \times \frac{40.1 \text{ g } Ca^{2+}}{1 \text{ mol } Ca^{2+}} = 10 \text{ g } Ca^{2+} \text{ ions}$$

45. To determine the molar mass of a substance

1. Determine the number of atoms of each element in a molecule of the substance
2. Sum the masses of the individual atoms
3. The sum of the masses of the atoms equals the mass of the substance in g/mole of formula units, or amu/formula unit

The units of molar mass are g/mol.

a. SO_2 (sulfur dioxide); molar mass = 64.07 g/mol

This substance contains 1 mole of S atoms and 2 moles of O atoms per mole of SO_2 molecules.

$$1\,S \times 32.07 \text{ g} = 32.07 \text{ g}$$
$$2\,O \times 16.00 \text{ g} = 32.00 \text{ g}$$
$$64.07 \text{ g/mol}$$

b. O_3 (trioxygen or more commonly ozone); molar mass = 48.00 g/mol

This molecule consists of 3 moles of oxygen atoms per mole of molecules.

$3\,O \times 16.00 \text{ g} = 48.00 \text{ g/mol}$

c. CO_2 (carbon dioxide); molar mass = 44.01 g/mol

This molecule contains 1 mole of C atoms and 2 moles of O atoms per mole of CO_2 molecules.

$$1\,C \times 12.01 \text{ g} = 12.01 \text{ g}$$
$$2\,O \times 16.00 \text{ g} = 32.00 \text{ g}$$
$$44.01 \text{ g/mol}$$

d. N_2O_5 (dinitrogen pentoxide); molar mass = 108.02 g/mol

This substance contains 2 moles of N atoms and 5 moles of O atoms per mole of N_2O_5 molecules.

$$1\,N \times 14.01 \text{ g} = 28.02 \text{ g}$$
$$5\,O \times 16.00 \text{ g} = 80.00 \text{ g}$$
$$108.02 \text{ g/mol}$$

47. a. Vanillin, $C_8H_8O_3$; molar mass = 152.16 g/mol

Vanillin contains 8 moles of C atoms, 8 moles of H atoms, and 3 moles of O atoms per mole of vanillin molecules.

Note: Vanillin is a molecular compound, so the correct term to use is molecules rather than formula units.

$$8\,C \times 12.01 \text{ g} = 96.08 \text{ g}$$
$$8\,H \times 1.01 \text{ g} = 8.08 \text{ g}$$
$$3\,O \times 16.00 \text{ g} = 48.00 \text{ g}$$
$$152.16 \text{ g/mol}$$

b. Oil of cloves (eugenol), $C_{10}H_{12}O_2$; molar mass = 164.22 g/mol

Eugenol is the active component of the oil of cloves. It has a molecular formula of $C_{10}H_{12}O_2$. This substance contains 10 moles of C atoms, 12 moles of H atoms, and 2 moles of O atoms per mole of molecules.

$$10\,C \times 12.01 \text{ g} = 120.10 \text{ g}$$
$$12\,H \times 1.01 \text{ g} = 12.12 \text{ g}$$
$$2\,O \times 16.00 \text{ g} = 32.00 \text{ g}$$
$$164.22 \text{ g/mol}$$

c. Oil of aniseed (anethole), $C_{10}H_{12}O$; molar mass = 148.22 g/mol

Anethole is the active ingredient in oil of aniseed. It has a molecular formula of $C_{10}H_{12}O$. This substance contains 10 moles of C atoms, 12 moles of H atoms, and 1 mole of O atoms per mole of molecules.

$$10\,C \times 12.01 \text{ g} = 120.10 \text{ g}$$
$$12\,H \times 1.01 \text{ g} = 12.12 \text{ g}$$
$$1\,O \times 16.00 \text{ g} = 16.00 \text{ g}$$
$$148.22 \text{ g/mol}$$

d. Oil of cinnamon (cinnamaldehyde), C_9H_8O; molar mass = 132.17 g/mol

This substance contains 9 moles of C atoms, 8 moles of H atoms, and 1 mole of O atoms per mole of molecules.

$$9\,C \times 12.01 \text{ g} = 108.09 \text{ g}$$
$$8\,H \times 1.01 \text{ g} = 8.08 \text{ g}$$
$$1\,O \times 16.00 \text{ g} = 16.00 \text{ g}$$
$$132.17 \text{ g/mol}$$

49. The greater the molar mass, the greater the mass of each individual molecule. In samples of identical mass, there will be more molecules present in the sample that has the smaller molar mass.
   a. CO_2 (molar mass = 44.0 g/mol); NO_2 (molar mass = 46.0 g/mol)
      There are more molecules in 10.0 g of CO_2.
   b. CO_2 (molar mass = 44.0 g/mol); SO_2 (molar mass = 64.1 g/mol)
      There are more molecules in 10.0 g of CO_2.
   c. O_2 (molar mass = 32.0 g/mol); Ar (atomic mass = 39.9 g/mol)
      There are more molecules of O_2 in 10.0 g of O_2 than there are atoms of Ar in 10.0 g of Ar.

51. To convert from grams to moles, divide the mass in grams by the molar mass.
   SiO_2; molar mass = 60.1 g/mol

$$\text{moles of SiO}_2 = 45.2\,\text{g SiO}_2 \times \frac{1\,\text{mol SiO}_2}{60.1\,\text{g SiO}_2}$$
$$= 0.752\,\text{mol SiO}_2$$

53. To convert from moles to grams, multiply the number of moles by the molar mass.
   $MgCO_3$; molar mass = 84.3 g/mol

$$\text{mass of MgCO}_3 = 0.122\,\text{mol MgCO}_3 \times \frac{84.3\,\text{g MgCO}_3}{1\,\text{mol MgCO}_3}$$
$$= 10.3\,\text{g MgCO}_3$$

55. The easiest approach is to calculate the number of moles of uranium in a cube 1 cm on a side, and then do the same calculation for diamond (carbon).
   The volume of a cube 1 cm on a side is 1 cm^3 (1 cm × 1cm × 1cm = 1 cm^3).
   The mass of each cube can be calculated using $m = d \times V$

$$\text{mass of U} = \frac{19.05\,\text{g U}}{1\,\text{cm}^3} \times 1\,\text{cm}^3$$
$$= 19.05\,\text{g U}$$
$$\text{mass of diamond} = \frac{3.514\,\text{g}}{1\,\text{cm}^3} \times 1\,\text{cm}^3$$
$$= 3.514\,\text{g diamond}$$

Convert the mass (in grams) of both substances to moles by dividing by the atomic weight.

$$\text{moles of U} = 19.05\,\text{g U} \times \frac{1\,\text{mol U}}{238.0\,\text{g U}}$$
$$= 0.0800\,\text{mol U}$$

$$\text{moles of diamond} = 3.514\,\text{g diamond} \times \frac{1\,\text{mol diamond}}{12.0\,\text{g diamond}}$$
$$= 0.2928\,\text{mol diamond (C)}$$

The cube of diamond contains more moles, and so more atoms, than the cube of uranium.

A second method that uses only logical thinking and no real calculations goes as follows: Since both cubes are 1 cm on a side, the ratio of the masses of the two substances present in the cubes corresponds to the ratio of the densities; that is, the mass of uranium is roughly 6 times the mass of carbon. To have equal numbers of moles in the cube, the ratio of the masses needs to be the same as the ratio of the atomic weights. However, the atomic weights of uranium and carbon are 238 g/mol and 12 g/mol respectively; that is, 1 mole of uranium weighs about 20 times as much as 1 mole of carbon. Since the actual ratio of the masses (about 6) is a lot less than this quantity, the conclusion is that there are more moles of carbon present in a 1-cm cube.

57. No. The law of conservation of mass requires that the total mass of reactants must equal the total mass of products. Another way of stating this is that the total number of atoms of each element must be the same on the reactant and product sides.
   The total number of moles on each side does not need to be equal because atoms can combine in different proportions without violating either of the above requirements. For example, in forming water

$$2\,H_2 + O_2 \rightarrow 2\,H_2O$$

There are 4 moles of hydrogen atoms and 2 moles of oxygen atoms on both the reactant and product side. However, there are 3 moles of reactant species (2 moles of H_2 and 1 mole of O_2), but only 2 moles of product (2 moles of H_2O).

58. No, for the same reasons as given in the answer to Problem 57.
   The total number of atoms must balance. This does not require the sum of the coefficients to be the same. Different species (molecules, compounds, and elements) can all contain different numbers of atoms.

59. No. The sum of the masses of all reactants must equal the sum of the masses of all products. The states of matter are not important. If we use the formation of water as an example,

$$2\,H_2(g) + O_2(g) \rightarrow 2\,H_2O(l)$$

We find that there are only gaseous substances present on the reactant side and only liquids present on the product side. However, this reaction is balanced, so states of matter have no importance in balancing equations.

60. No. In general, there is no correlation between the volumes of reactants and the volumes of the products in a balanced equation. Later, in Chapter 8, we will learn that there is a relationship between the volumes of gaseous reactants and gaseous products in chemical reactions, but even here conservation of volume is not part of the relationship.

61. To balance reactions, we need to make the number of atoms (or moles of atoms) of each element the same on both the reactant side and the product side of the reaction.

In balancing reactions, we are only allowed to change the coefficient in front of an entire species.

We cannot change the formulas of the reactants or products. For example, we cannot change the subscripts or put coefficients in front of elements in the middle of a substance.

Note: It is usually best to balance the most complicated element last (any element present in more than one substance on either the reactant or product side of the reaction). In the reaction that follows, oxygen is present in both CO and H_2O on the product side, so it will probably be the most difficult element to balance.

a. $CH_4(g) + O_2(g) \rightarrow CO(g) + H_2O(g)$

The balanced reaction is

$$2\,CH_4(g) + 3\,O_2(g) \rightarrow 2\,CO(g) + 4\,H_2O(g)$$

An explanation follows.

Determine the number of atoms of each element on the reactant side. This can be accomplished by determining the number of atoms (or moles of atoms) of each element in a species.

The two reactants are CH_4 and O_2.

There is 1 C atom and 4 H atoms in one CH_4 molecule.
There are 2 O atoms in each O_2 molecule.
Do the same for the product side. The products are CO and H_2O.

There is 1 C atom and 1 O atom in each CO molecule.
There are 2 H atoms and 1 O atom in each H_2O molecule.

Next, compare the number of atoms in reactants and products. It is easiest to compare these values by seeing them next to one another. In other words, let's write the information under the unbalanced equation as follows:

$$CH_4(g) + O_2(g) \rightarrow CO(g) + H_2O(g)$$

Reactants	Products
1 C	1 C
4 H	2 H
2 O	2 O

We see that currently the carbon and oxygen are balanced, but not the hydrogen. To balance the hydrogen, we need to change the coefficient in front of the H_2O to 2 (so that there are 4 H atoms on both the reactant and product side). However, this unbalances the oxygen.

$$CH_4(g) + O_2(g) \rightarrow CO(g) + 2\,H_2O(g)$$

Reactants	Products
1 C	1 C
4 H	4 H
2 O	3 O

To balance the oxygen, we now need to get 3 O atoms on both the reactant and product side. The only way to get 3 O atoms on the reactant side is to put the coefficient $\frac{3}{2}$ in front of the O_2.

$$CH_4(g) + \tfrac{3}{2}\,O_2(g) \rightarrow CO(g) + 2\,H_2O(g)$$

However, a fractional number does not have any real meaning because one cannot have 1.5 molecules of O_2, and any properly balanced equation must contain whole number coefficients. To remove the fraction, we multiply all of the coefficients by 2.

The properly balanced equation for this reaction is

$$2\,CH_4(g) + 3\,O_2(g) \rightarrow 2\,CO(g) + 4\,H_2O(g)$$

Checking both sides:

Reactants	Products
2 C	2 C
8 H	8 H
6 O	6 O

b. $NH_3(g) \rightarrow N_2(g) + H_2(g)$

The balanced reaction is

$$2\,NH_3(g) \rightarrow N_2(g) + 3\,H_2(g)$$

An explanation follows.

Find the number of atoms of each element present in reactants and products:

$$NH_3(g) \rightarrow N_2(g) + H_2(g)$$

Reactants	Products
1 N	2 N
3 H	2 H

To balance the nitrogen atoms, put the coefficient 2 in front of NH_3.

$$2\,NH_3(g) \rightarrow N_2(g) + H_2(g)$$

Reactants	Products
2 N	2 N
6 H	2 H

To balance the hydrogen atoms, place the coefficient 3 in front of the H_2.

$$2\,NH_3(g) \rightarrow N_2(g) + 3\,H_2(g)$$

The reaction is now balanced.

c. $CO(g) + H_2O(g) \rightarrow CO_2(g) + H_2(g)$

The balanced reaction is

$$CO(g) + H_2O(g) \rightarrow CO_2(g) + H_2(g)$$

This reaction is balanced as written. We can check it:

$CO(g) + H_2O(g)$	$\rightarrow$	$CO_2(g) + H_2(g)$
1 C		1 C
2 O		2 O
2 H		2 H

On the reactant side, we get 1 O atom from CO and 1 O atom from H_2O, to give a total of 2 O atoms on the reactant side.

63. a. The balanced equation is

$$3\,FeSiO_3 + 4\,H_2O \rightarrow Fe_3Si_2O_5(OH)_4 + H_4SiO_4$$

Starting with the equation given

$$FeSiO_3 + H_2O \rightarrow Fe_3Si_2O_5(OH)_4 + H_4SiO_4$$

On the reactant side:

The substance $FeSiO_3$ contains 1 Fe atom, 1 Si atom, and 3 O atoms per molecule (or 1 mole of Fe atoms, 1 mole of Si atoms, and 3 moles of O atoms per mole of formula units).

Note: We use formula units instead of molecules because $FeSiO_3$ is an ionic compound.

Water (H_2O) contains 2 H atoms and 1 O atom per molecule (or 2 moles of H atoms and 1 mole of O atoms per mole of molecules).

The total number of atoms on the reactant side is

1 Fe
1 Si
4 O (3 from the $FeSiO_3$ and 1 from the H_2O)
2 H

On the product side:

$Fe_3Si_2O_5(OH)_4$ contains 3 Fe atoms, 2 Si atoms, 9 O atoms, and 4 H atoms per mole of formula units.

Note: The $(OH)_4$ in the above substance tells us that there are 4 OH units in the formula of the substance. Each of these units consists of 1 O atom and 1 H atom. In four of these OH units there are a total of 4 O atoms and 4 H atoms.

H_4SiO_4 contains 4 H atoms, 1 Si atom, and 4 O atoms per molecule.

The total number of atoms on the product side is

3 Fe
3 Si (2 from $Fe_3Si_2O_5(OH)_4$ and 1 from H_4SiO_4)
13 O (9 from $Fe_3Si_2O_5(OH)_4$ and 4 from H_4SiO_4)
8 H (4 from $Fe_3Si_2O_5(OH)_4$ and 4 from H_4SiO_4)

$FeSiO_3 + H_2O$	$\rightarrow$	$Fe_3Si_2O_5(OH)_4 + H_4SiO_4$
1 Fe		3 Fe
1 Si		3 Si
4 O		13 O
2 H		8 H

We see that none of the elements are currently balanced.

There is only 1 Fe on the reactant side, but there are 3 Fe atoms on the product side. To balance the Fe atoms put the coefficient 3 in front of $FeSiO_3$. Let's do this and recalculate the number of atoms of all the elements.

In three formula units of $FeSiO_3$, there are 3 Fe atoms, 3 Si atoms, and 9 O atoms.

$3\,FeSiO_3 + H_2O$	$\rightarrow$	$Fe_3Si_2O_5(OH)_4 + H_4SiO_4$
3 Fe		3 Fe
3 Si		3 Si
10 O		13 O
2 H		8 H

Notice that this balanced the Si atoms as well as the Fe atoms. It just worked out this way. There was no intent here to make this happen. Often when you balance one element, another element is balanced along with it.

The next element to balance is the oxygen. It will be very difficult to balance the oxygen because this element is present in all four substances in this reaction. One would not know which one of these substances needs to have its coefficient changed. When this occurs, skip over that element and try a different one. The hope is that when all other elements have been balanced, the tough one will also become balanced.

That leaves only the element hydrogen to balance. This element is present in only one substance on each of side of the equation. It should not be a problem to balance.

Since there are 8 H atoms on the product side, but only 2 H atoms on the reactant side, we need more H atoms on the reactant side. To get 8 H atoms on the reactant side, put the coefficient 4 in front of the H_2O. In $4\,H_2O$ molecules, there are 8 H atoms and 4 O atoms. Let's change the coefficient for H_2O to 4 and recalculate the number of atoms.

$3\,FeSiO_3 + 4\,H_2O$	$\rightarrow$	$Fe_3Si_2O_5(OH)_4 + H_4SiO_4$
3 Fe		3 Fe
3 Si		3 Si
13 O		13 O
8 H		8 H

Note: From $3\,FeSiO_3$ there are 9 O atoms, and from $4\,H_2O$ there are 4 O atoms, a total of 13 O atoms on the reactant side.

This reaction is now balanced.

b. The balanced equation is

$$Fe_3Si_2O_5(OH)_4 + 3\,CO_2 + 2\,H_2O \rightarrow 3\,FeCO_3 + 2\,H_4SiO_4$$

Starting with the equation given

$$Fe_3Si_2O_5(OH)_4 + CO_2 + H_2O \rightarrow FeCO_3 + H_4SiO_4$$

3 Fe	1 Fe
2 Si	1 Si
12 O	7 O
6 H	4 H
1 C	1 C

To balance the Fe and Si, place the coefficient 3 in front of $FeCO_3$ and the coefficient 2 in front of H_4SiO_4. Since we need to put a 3 in front of the $FeCO_3$ (there must be 3 carbon atoms on the product side), we can balance the carbon atoms by placing a 3 in front the CO_2.

Let's check and see which elements are balanced.

$$Fe_3Si_2O_5(OH)_4 + 3\,CO_2 + H_2O \rightarrow 3\,FeCO_3 + 2\,H_4SiO_4$$

3 Fe	3 Fe
2 Si	2 Si
16 O	17 O
6 H	8 H
3 C	3 C

All atoms are balanced except the hydrogen and oxygen. We need an additional 2 H atoms and 1 O atom on the reactant side. We can accomplish this by adding 1 H_2O molecule. Change the coefficient in front of the H_2O to a 2.

$$Fe_3Si_2O_5(OH)_4 + 3\,CO_2 + 2\,H_2O \rightarrow 3\,FeCO_3 + 2\,H_4SiO_4$$

This reaction is now balanced.

c. The balanced equation is

$$Fe_2SiO_4 + 2\,CO_2 + 2\,H_2O \rightarrow 2\,FeCO_3 + H_4SiO_4$$

Starting with the reaction given

$$Fe_2SiO_4 + CO_2 + H_2O \rightarrow FeCO_3 + H_4SiO_4$$

2 Fe	1 Fe
1 Si	1 Si
7 O	7 O
1 C	1 C
2 H	4 H

To balance the Fe, put a 2 in front of $FeCO_3$. This results in 2 carbon atoms on the product side, so we need to place a 2 in front of the CO_2 to balance the carbon atoms. Let's check to see which elements are still not balanced.

$$Fe_2SiO_4 + 2\,CO_2 + H_2O \rightarrow 2\,FeCO_3 + H_4SiO_4$$

2 Fe	2 Fe
1 Si	1 Si
9 O	10 O
2 C	2 C
2 H	4 H

The only elements left unbalanced are hydrogen and oxygen. We need to add 2 H atoms and 1 O atom on the reactant side. Add 1 H_2O molecule to the reactant side. Do this by changing the coefficient to 2 in front of the H_2O.

$$Fe_2SiO_4 + 2\,CO_2 + 2\,H_2O \rightarrow 2\,FeCO_3 + H_4SiO_4$$

This reaction is now balanced.

65. a. The balanced equation is

$$N_2(g) + O_2(g) \rightarrow 2\,NO(g)$$

Starting with the given reaction:

$$N_2 + O_2 \rightarrow NO$$

2 N	1 N
2 O	1 O

To balance the nitrogen and oxygen, put the coefficient 2 in front of the NO species.

$$N_2 + O_2 \rightarrow 2\,NO$$

b. The balanced equation is

$$2\,NO(g) + O_2(g) \rightarrow 2\,NO_2(g)$$

Starting with the given reaction:

$$NO + O_2 \rightarrow NO_2$$

1 N	1 N
3 O	2 O

The nitrogen atoms are balanced, but the oxygen atoms are not. As written, there are 3 oxygen atoms on the reactant side and 2 on the product side. There is no whole number coefficient that we can place in front of the O_2 that will result in the equation being balanced. As long as the coefficients in front of NO and NO_2 are the same, the nitrogen atoms will always balance. Let's put 2's in front of both NO and NO_2.

$$2\,NO(g) + O_2(g) \rightarrow 2\,NO_2(g)$$

2 N	2 N
2 O + 2 O = 4 O	4 O

This equation is now balanced.

c. The balanced reaction is

$$NO(g) + NO_3(g) \rightarrow 2\,NO_2(g)$$

Starting with the given reaction:

$$NO + NO_3 \rightarrow NO_2$$

2 N	1 N
4 O	2 O

To balance this equation, put the coefficient 2 in front of the NO_2.

$$NO + NO_3 \rightarrow 2\,NO_2$$

d. The balanced reaction is

$$2\,N_2(g) + O_2(g) \rightarrow 2\,N_2O(g)$$

$$N_2 + O_2 \rightarrow N_2O$$

2 N	2 N
2 O	1 O

To balance the oxygen atoms, put the coefficient 2 in front of the N_2O.

$$N_2 + O_2 \rightarrow 2\,N_2O$$

2 N	4 N
2 O	2 O

To balance the nitrogen atoms, put the coefficient 2 in front of the N_2.

$$2\,N_2 + O_2 \rightarrow 2\,N_2O$$

67. The starting point in each part in this problem is to convert a name into a formula. Next, write the unbalanced equation, and then balance it.
   a. Dinitrogen pentoxide, $N_2O_5(g)$
      Sodium metal, $Na(s)$
      Sodium nitrate, $NaNO_3$
      Nitrogen dioxide, NO_2
      The unbalanced reaction is

$$N_2O_5(g) + Na(s) \rightarrow NaNO_3(s) + NO_2(g)$$

2 N	2 N
5 O	5 O
1 Na	1 Na

   This reaction is balanced as written.
   b. Nitric acid, HNO_3
      Nitrous acid, HNO_2
      Water, H_2O
      Dinitrogen tetroxide, N_2O_4

$$N_2O_4(g) + H_2O(l) \rightarrow HNO_3(aq) + HNO_2(aq)$$

2 N	2 N
5 O	5 O
2 H	2 H

   This reaction is balanced as written.
   c. Nitrogen monoxide, NO
      Dinitrogen monoxide, N_2O
      Nitrogen dioxide, NO_2

   The unbalanced reaction is

$$NO(g) \rightarrow N_2O(g) + NO_2(g)$$

1 N	3 N
1 O	3 O

   To balance both the N and O atoms put the coefficient 3 in front of the NO.
   The balanced equation is

$$3\,NO(g) \rightarrow N_2O(g) + NO_2(g)$$

69. A combustion process is a reaction with oxygen (O_2). If the material being combusted contains only the elements carbon and hydrogen, the products are carbon dioxide (CO_2) and water (H_2O).
   The unbalanced reaction for the combustion of acetylene (C_2H_2) is

$$C_2H_2(g) + O_2(g) \rightarrow CO_2(g) + H_2O(g)$$

2 C	1 C
2 H	2 H
2 O	3 O

   To balance the carbon atoms put the coefficient 2 in front of the CO_2.

$$C_2H_2(g) + O_2(g) \rightarrow 2\,CO_2(g) + H_2O(g)$$

   There are 2 O atoms on the reactant side and 5 O atoms on the product side (4 O atoms from the 2 CO_2 molecules and 1 O atom from H_2O).
   To balance the oxygen atoms put the coefficient $\frac{5}{2}$ in front of the O_2.

$$C_2H_2(g) + \tfrac{5}{2}\,O_2(g) \rightarrow 2\,CO_2(g) + H_2O(g)$$

   To remove the fractional coefficient, we need to double each coefficient. The properly balanced equation is

$$2\,C_2H_2(g) + 5\,O_2(g) \rightarrow 4\,CO_2(g) + 2\,H_2O(g)$$

71. An empirical formula is the smallest whole number ratio of the different atoms in a substance. The molecular formula gives us the actual number of atoms of each element in a molecule of a substance. For example, glucose has the molecular formula $C_6H_{12}O_6$. The empirical formula of glucose is CH_2O. The ratio of the carbon to hydrogen to oxygen atoms in glucose is 1:2:1.

72. Yes. Since the ratio of the atoms does not change, the elemental percentages must be the same.

73. No. Two factors contribute to the elemental percentage.

   1. The mass of the atom
   2. The number of atoms of the element

   The second factor can and often does make this factor the more important of the two factors. For example, in $NaNO_3$, the sodium atoms are the heaviest ones, but the 3 oxygen atoms make it the element in the greatest percentage in this substance.

74. The quantity mole percent and atom percent should be identical for a given compound.

For example, in 1 molecule of glucose, $C_6H_{12}O_6$, there are 24 atoms: 6 C atoms, 12 H atoms, and 6 O atoms.

The atom percentages are 25% C (6/24 × 100), 50% H (12/24 × 100), and 25% O (6/24 × 100).

In 1 mole of glucose molecules there are 6 moles of C atoms, 12 moles of H atoms, and 6 moles of O atoms. These numbers are identical to the number of atoms, so mole percentage will be the same as the atom percentage.

75. The percent composition is given by the formula

$$\% \text{ element} = \frac{(\text{number of atoms of element})(\text{atomic weight of element})}{\text{molar mass of substance}} \times 100$$

By determining the molar mass of the substance, we can determine the masses of each of the elements present in 1 mole of the compound.

a. Na_2O

$$2\,\text{Na} \times 23.0 = 46.0\,\text{g}$$
$$1\,\text{O} \times 16.0 = 16.0\,\text{g}$$
$$\text{Molar mass} = 62.0\,\text{g/mol}$$

$$\%\,\text{Na} = \frac{46.0\,\text{g}}{62.0\,\text{g}} \times 100 = 74.2\%\,\text{Na}$$

$$\%\,\text{O} = \frac{16.0\,\text{g}}{62.0\,\text{g}} \times 100 = 25.8\%\,\text{O}$$

b. NaOH

$$1\,\text{Na} \times 23.0 = 23.0\,\text{g}$$
$$1\,\text{O} \times 16.0 = 16.0\,\text{g}$$
$$1\,\text{H} \times 1.0 = 1.0\,\text{g}$$
$$\text{Molar mass} = 40.0\,\text{g/mol}$$

$$\%\,\text{Na} = \frac{23.0\,\text{g}}{40.0\,\text{g}} \times 100 = 57.5\%\,\text{Na}$$

$$\%\,\text{O} = \frac{16.0\,\text{g}}{40.0\,\text{g}} \times 100 = 40.0\%\,\text{O}$$

$$\%\,\text{H} = \frac{1.0\,\text{g}}{40.0\,\text{g}} \times 100 = 2.5\%\,\text{H}$$

c. $NaHCO_3$

$$1\,\text{Na} \times 23.0 = 23.0\,\text{g}$$
$$1\,\text{H} \times 1.0 = 1.0\,\text{g}$$
$$1\,\text{C} \times 12.0 = 12.0\,\text{g}$$
$$3\,\text{O} \times 16.0 = 48.0\,\text{g}$$
$$\text{Molar mass} = 84.0\,\text{g}$$

$$\%\,\text{Na} = \frac{23.0\,\text{g}}{84.0\,\text{g}} \times 100 = 27.4\%\,\text{Na}$$

$$\%\,\text{O} = \frac{48.0\,\text{g}}{84.0\,\text{g}} \times 100 = 57.1\%\,\text{O}$$

$$\%\,\text{H} = \frac{1.0\,\text{g}}{84.0\,\text{g}} \times 100 = 1.2\%\,\text{H}$$

$$\%\,\text{C} = \frac{12.0\,\text{g}}{84.0\,\text{g}} \times 100 = 14.3\%\,\text{C}$$

d. Sodium carbonate, Na_2CO_3

$$2\,\text{Na} \times 23.0 = 46.0\,\text{g}$$
$$1\,\text{C} \times 12.0 = 12.0\,\text{g}$$
$$3\,\text{O} \times 16.0 = 48.0\,\text{g}$$
$$\text{Molar mass} = 106\,\text{g}$$

$$\%\,\text{Na} = \frac{46.0\,\text{g}}{106.0\,\text{g}} \times 100 = 43.4\%\,\text{Na}$$

$$\%\,\text{O} = \frac{48.0\,\text{g}}{106.0\,\text{g}} \times 100 = 45.3\%\,\text{O}$$

$$\%\,\text{C} = \frac{12.0\,\text{g}}{106.0\,\text{g}} \times 100 = 11.3\%\,\text{C}$$

77. The substance with the greatest carbon-to-hydrogen atom ratio should have the greatest percentage of carbon, ie, pentacene, $C_{24}H_{12}$.

Let's check this, by actually calculating the percent carbon in each substance.

a. naphthalene, $C_{10}H_8$ (molar mass 128 g/mol; 120 g C/mol $C_{10}H_8$).

$$\% C = \frac{120.0\,g}{128.0\,g} \times 100$$
$$= 93.8\% C$$

b. chrysene, $C_{18}H_{12}$ (molar mass 228 g/mol; 216 g C/mol $C_{18}H_{12}$)

$$\% C = \frac{216.0\,g}{228.0\,g} \times 100$$
$$= 94.7\% C$$

c. pentacene, $C_{24}H_{12}$ (molar mass 300 g/mol; 288 g C/mol $C_{24}H_{12}$)

$$\% C = \frac{288.0\,g}{300.0\,g} \times 100$$
$$= 96.0\% C$$

d. pyrene, $C_{16}H_{10}$ (molar mass 202 g/mol; 192 g C/mol $C_{16}H_{10}$)

$$\% C = \frac{192.0\,g}{202.0\,g} \times 100$$
$$= 95.0\% C$$

79. Mole basis: N_2O_3 and NO_2 are over 50% oxygen; NO is exactly 50% oxygen.

Weight percent: NO, N_2O_3, and NO_2 are all more than 50% oxygen by weight.

N_2O

N_2O contains 2 moles of nitrogen atoms and 1 mole of oxygen atoms for every mole of molecules. Since this substance contains a total of 3 moles of atoms and only 1 mole of oxygen atoms, N_2O contains less than 50% oxygen on a molar basis (only 33.3% oxygen).

The weight percent of oxygen in N_2O can be determined by determining the molar mass of N_2O and using this to determine the mass percentages.

$$2\,N \times 14.0\,g = 28.0\,g$$
$$1\,O \times 16.0\,g = 16.0\,g$$
$$\text{Molar mass} = 44.0\,g/mol$$

$$\% O = \frac{16.0\,g}{44.0\,g} \times 100$$
$$= 36.4\% O$$

This substance contains less than 50% oxygen by mass.

NO

This substance contains 1 mole of nitrogen atoms and 1 mole of oxygen atoms per mole of molecules. The mole percent of oxygen is exactly 50%.

The molar mass of NO is 30.0 g. The weight percent of oxygen is

$$\% O = \frac{16.0\,g}{30.0\,g}$$
$$= 53.3\% O$$

This substance contains more than 50% oxygen by mass.

N_2O_3

This substance contains 2 moles of nitrogen atoms and 3 moles of oxygen atoms for every mole of N_2O_3. This species contains more than 50% oxygen on a molar basis, since there are 3 moles of oxygen atoms and a total of 5 moles of atoms; that is, 60% (3/5 × 100) of the moles of atoms are oxygen.

The molar mass of N_2O_3 is 76.0 g/mol

$$\% O = \frac{48.0\,g}{76.0\,g} \times 100$$
$$= 63.2\% O$$

This substance contains more than 50% oxygen by mass.

NO_2

This substance contains 2 moles of oxygen atoms and 1 mole of nitrogen atoms per mole of molecules. NO_2 contains 66.6% oxygen on a mole basis because there is a total of 3 moles of atoms and 2 moles of oxygen atoms (2/3 × 100 = 66.6%).

The molar mass of NO_2 is 46.0 g/mol. The percentage oxygen by weight is

$$\% O = \frac{32.0\,g}{46.0\,g} \times 100$$
$$= 69.6\% O$$

81. No, all of these substances have different empirical formulas. (The empirical formula is the smallest whole number ratio of the atoms in a substance.)

Compound	Molecular Formula	Empirical Formula
Naphthalene	$C_{10}H_8$	C_5H_4
Chrysene	$C_{18}H_{12}$	C_3H_2
Anthracene	$C_{14}H_{10}$	C_7H_5
Pyrene	$C_{16}H_{10}$	C_8H_5
Benzoperylene	$C_{22}H_{12}$	$C_{11}H_6$
Coronene	$C_{24}H_{12}$	C_2H

83. This question asks us to determine the empirical formula of a substance from the percent composition.

Step 1: Determine the number of grams of each element in the sample.

In this problem, the only information given is the percent by weight of each element.

One can assume any sample weight and then determine how much of each element is the sample from the weight percentages. The easiest value to choose is 100.00 g. In this case, the weight percentages correspond to the mass of each element in grams.

Assuming a 100.00-g sample of zircon, we find that it contains 49.76 g of zirconium (Zr), 15.32 g of silicon (Si), and 34.92 g of oxygen (O). The mass of oxygen is determined by difference between 100.00 g and the mass of the other two elements (100.00 g − 49.76 g − 15.32 g = 34.92 g).

Step 2: Determine the number of moles of each element. To do this, divide the mass of each element by its atomic weight in units of g/mol.

$$\text{Moles of Zr} = 49.76\,\text{g Zr} \times \frac{1\,\text{mol Zr}}{91.22\,\text{g Zr}}$$

$$= 0.5455\,\text{mol Zr}$$

$$\text{Moles of Si} = 15.32\,\text{g Si} \times \frac{1\,\text{mol Si}}{28.09\,\text{g Si}}$$

$$= 0.5454\,\text{mol Si}$$

$$\text{Moles of O} = 34.92\,\text{g O} \times \frac{1\,\text{mol O}}{16.00\,\text{g O}}$$

$$= 2.183\,\text{mol O}$$

Step 3: Divide each of the moles by the smallest number of moles of the three. In this problem, 0.5454 mole is the smallest quantity.

$$\text{Zr}_{\frac{0.5455}{0.5454}}\text{Si}_{\frac{0.5454}{0.5455}}\text{O}_{\frac{2.183}{0.5454}} = \text{Zr}_1\text{Si}_1\text{O}_4,\ \text{or ZrSiO}_4$$

Step 4: If all subscripts are not whole numbers, then multiply each subscript by 2. If this does not transform all subscripts into whole numbers, then multiply by 3, then 4, etc., until all subscripts are whole numbers.

Note: It is usually acceptable to round off if the values are within 0.05 units of a whole number. You may round 4.95 to 5, but it is unacceptable to round 4.75 to 5.

In this problem we did not need to perform Step 5 because all the subscripts were indeed whole numbers after Step 4.

The empirical formula of zircon is $ZrSiO_4$.

85. In this problem we are given the actual masses of each element in the compound to be analyzed. We have 2.43 g of Mg and 1.60 g of O (forming 4.03 g of magnesium oxide). As expected the law of conservation of mass is observed.

a. Step 1: 2.43 g Mg and 1.60 g O

Step 2:

$$\text{Moles of Mg} = 2.43\,\text{g Mg} \times \frac{1\,\text{mol Mg}}{24.3\,\text{g Mg}}$$

$$= 0.100\,\text{mol Mg}$$

$$\text{Moles of O} = 1.60\,\text{g O} \times \frac{1\,\text{mol O}}{16.0\,\text{g O}}$$

$$= 0.100\,\text{mol O}$$

Step 3: Since the subscripts for both elements are the same, the empirical formula of magnesium oxide is MgO.

b. The balanced equation is:

$$2\,\text{Mg}(s) + \text{O}_2(g) \rightarrow 2\,\text{MgO}(s)$$

87. To determine the empirical formula from an elemental analysis, use the three steps (maybe four in some examples) outlined in the answer to Problem 83.

Step 1: Determine the mass of each element in the substance. Assuming a 100.00-g sample these values are

28.03 g Mg   21.60 g Si   1.16 g H

The percentage of oxygen is not given. However, it can be determined because the sum of the percentages of the four elements (Mg, Si, H, and O) must add up to 100%. The percentage of oxygen must be 49.21% (100.00 g − 28.03 g −21.60 g − 1.16 g).

The mass of oxygen is 49.21 g O.

Step 2: Determine the number of moles of each element.

$$\text{Moles of Mg} = 28.03\,\text{g Mg} \times \frac{1\,\text{mol Mg}}{24.31\,\text{g Mg}}$$

$$= 1.153\,\text{mol Mg}$$

$$\text{Moles of Si} = 21.60\,\text{g Si} \times \frac{1\,\text{mol Si}}{28.09\,\text{g Si}}$$

$$= 0.7690\,\text{mol Si}$$

$$\text{Moles of H} = 1.16\,\text{g H} \times \frac{1\,\text{mol H}}{1.008\,\text{g H}}$$

$$= 1.151\,\text{mol H}$$

$$\text{Moles of O} = 49.21\,\text{g O} \times \frac{1\,\text{mol O}}{16.00\,\text{g O}}$$

$$= 3.076\,\text{mol O}$$

Step 3: Divide each subscript in Step 3 by the smallest subscript. In this case, divide each by 0.7690.

$$\text{Mg}_{\frac{1.153}{0.7690}}\text{Si}_{\frac{0.7690}{0.7690}}\text{H}_{\frac{1.151}{0.7690}}\text{O}_{\frac{3.076}{0.7690}} = \text{Mg}_{1.50}\text{SiH}_{1.50}\text{O}_4$$

Step 4: Since all subscripts are not whole numbers we need to multiply each subscript through by a multiplier until all whole numbers are attained. First, start with 2. If this does not produce all whole numbers, go on to 3, and then 4, etc., until only whole numbers are attained.

$$Mg_{1.5\times2}Si_{1\times2}H_{1.5\times2}O_{4\times2} = Mg_3Si_2H_3O_8$$

The empirical formula of chrysolite (a form of asbestos) is $Mg_3Si_2H_3O_8$.

Molecular Formula:

Step 1: Determine the mass corresponding to the of the empirical formula $Mg_3Si_2H_3O_8$

$$\begin{aligned} 3\,\text{Mg} \times 24.31\,\text{g} &= 72.93\,\text{g} \\ 2\,\text{Si} \times 28.09\,\text{g} &= 56.18\,\text{g} \\ 3\,\text{H} \times 1.01\,\text{g} &= 3.03\,\text{g} \\ 8\,\text{O} \times 16.00\,\text{g} &= \underline{128.00\,\text{g}} \\ &\ 260.14\,\text{g} \end{aligned}$$

Step 2: Divide the molar mass (520.27 g/mol) by the empirical formula mass. This value results in a whole number value.

$$\frac{\text{molar mass}}{\text{empirical formula mass}} = \text{whole number}$$

$$\frac{520.27\,\text{g}}{260.14\,\text{g}} = 2$$

The molecular formula for chrysolite is twice its empirical formula, that is, $Mg_6Si_4H_6O_{16}$.

89. No. The coefficients of the balanced equation allow us to convert between two substances in a chemical reaction. An unbalanced equation would imply that all substances react in a 1-to-1 mole ratio. This is most often not the case.

90. No. The ratios of all the coefficients are still the same in these two equations. For example, the ratio of moles of water to moles of carbon dioxide in the top reaction is 3 moles of water for every 2 moles of carbon dioxide. In the bottom equation, the ratio is 6 moles of water for every 4 moles of carbon dioxide. Reducing the ratio 6:4, we get the same value 3:2.

91. a. We need to convert kilograms of carbon into moles of carbon.

1. Convert kilograms of C to grams of C (1 kg = 1000)
2. Convert grams of C to moles of C (divide by the atomic weight of C; 1 mole C = 12 g C)

Performing these steps yields

$$5.4 \times 10^9\,\text{kg C} \times \frac{1000\,\text{g C}}{1\,\text{kg C}} \times \frac{1\,\text{mol C}}{12\,\text{g C}}$$

$$= 4.5 \times 10^{11}\,\text{mol C}$$

b. Convert 4.5×10^{11} moles of carbon to kilograms of carbon dioxide, CO_2.

From the molecular formula of carbon dioxide, CO_2, there is 1 mole C for every mole of carbon dioxide (since there is 1 C atom in each CO_2 molecule). We can use the molecular formula to create a conversion factor between the number of moles of each element and number of moles of the compound. For carbon dioxide, we get

$$1\,\text{mol CO}_2 = 1\,\text{mol C atoms} = 2\,\text{mol O atoms}$$

We can use this conversion factor to convert from moles of C to moles of CO_2. Next convert moles of CO_2 to grams of CO_2 by multiplying by the molar mass of CO_2 (44.0 g/mol). Lastly, convert grams of CO_2 to kilograms of CO_2.

$$4.5 \times 10^{11}\,\text{mol C} \times \frac{1\,\text{mol CO}_2}{1\,\text{mol C}} \times \frac{44.0\,\text{g CO}_2}{1\,\text{mol CO}_2}$$

$$\times \frac{1\,\text{kg CO}_2}{1000\,\text{g CO}_2} = 1.98 \times 10^{10}\,\text{kg CO}_2$$

93. a. The unbalanced equation is

$$NaHCO_3(s) \rightarrow Na_2CO_3(s) + CO_2(g) + H_2O(g)$$

To balance the sodium atoms, put the coefficient 2 in front of $NaHCO_3$.

$$2\,NaHCO_3(s) \rightarrow Na_2CO_3(s) + CO_2(g) + H_2O(g)$$

This reaction is now balanced. There are 2 Na atoms, 2 H atoms, 2 C atoms, and 6 O atoms on both sides of the arrow.

b. Convert 25.0 g $NaHCO_3$ to grams of CO_2.
To do this:

1. Convert grams of $NaHCO_3$ to moles. (Divide the mass of $NaHCO_3$ in grams by the molar mass of $NaHCO_3$, 84.0 g/mol.)
2. Convert moles of $NaHCO_3$ to moles of CO_2. (Use the coefficients of the balanced equation as a conversion factor. In this case, 2 mol $NaHCO_3$ = 1 mol CO_2.)
3. Convert moles of CO_2 to grams of CO_2. (Multiply the moles of CO_2 by the molar mass of CO_2, 44.0 g/mol.)

$$25.0\,\text{g NaHCO}_3 \times \frac{1\,\text{mol NaHCO}_3}{84.0\,\text{g NaHCO}_3}$$

$$\times \frac{1\,\text{mol CO}_2}{2\,\text{mol NaHCO}_3} \times \frac{44.0\,\text{g CO}_2}{1\,\text{mol CO}_2} = 6.55\,\text{g CO}_2$$

95. Convert 1.00 kg of Na_3AlF_6 to grams of $NaAlO_2$.

1. Convert 1.00 kg to grams (1 kg = 1000 g)
2. Convert grams of Na_3AlF_6 to moles of Na_3AlF_6. (Divide by the molar mass of Na_3AlF_6, 210 g/mol.)
3. Convert moles of Na_3AlF_6 to moles of $NaAlO_2$. (Use the coefficients of the balanced equation, 1 mol Na_3AlF_6 = 3 mol $NaAlO_2$)

4. Convert moles of $NaAlO_2$ to grams of $NaAlO_2$. (Multiply by the molar mass of $NaAlO_2$, 82.0 g/mol.)

$$1.00\,\text{kg } Na_3AlF_6 \times \frac{1000\,\text{g } Na_3AlF_6}{1.00\,\text{kg } Na_3AlF_6}$$

$$\times \frac{1\,\text{mol } Na_3AlF_6}{210\,\text{g } Na_3AlF_6} \times \frac{3\,\text{mol } NaAlO_2}{1\,\text{mol } Na_3AlF_6}$$

$$\times \frac{82.0\,\text{g } NaAlO_2}{1\,\text{mol } NaAlO_2} = 1.17 \times 10^3\,\text{g } NaAlO_2$$

or 1.17 kg (to 3 significant figures)

97. We start this problem with 25 tons of coal. However, we are not interested in the coal, but the sulfur in the coal. The problem informs us that the coal contains 3.0% sulfur. Using this percentage, we can calculate the amount of sulfur in 25 tons of coal.

Step 1: Determine the amount of sulfur (in tons) in 25 tons of coal. To do this use the percentage formula

$$\%\,S = \frac{\text{tons S}}{\text{ton coal}} \times 100$$

Step 2: Convert tons of S to grams of S (1 ton = 2000 lbs and 1 lb = 454 g)

Step 3: Convert grams of S to moles of S. (Divide by the atomic weight of S, 32.1 g/mol.)

Step 4: Convert moles of S to moles of SO_2. (Use the molecular formula of SO_2 to create a conversion factor between moles of SO_2 and moles of S; 1 mol SO_2 = 1 mol S.)

Step 5: Convert moles of SO_2 to tons of SO_2. (Multiply by the molar mass of SO_2 (64.1 g/mol) to convert to grams and then convert to tons using the conversion factors listed in Step 2.)

The amount of sulfur in 25 tons of coal:

$$3.0\% = \frac{\text{tons S}}{25\,\text{tons}} \times 100$$

$$\text{tons S} = \frac{(3.0)(25\,\text{tons})}{100} = 0.75\,\text{ton S}$$

Next, convert 0.75 ton S to grams of S:

$$0.75\,\text{ton S} \times \frac{2000\,\text{lb S}}{1\,\text{ton S}} \times \frac{454\,\text{g S}}{1\,\text{lb S}} = 6.81 \times 10^5\,\text{g S}$$

Convert grams of S to grams of SO_2:

$$6.81 \times 10^5\,\text{g S} \times \frac{1\,\text{mol S}}{32.1\,\text{g S}} \times \frac{1\,\text{mol } SO_2}{1\,\text{mol S}}$$

$$\times \frac{64.1\,\text{g } SO_2}{1\,\text{mol } SO_2} = 1.36 \times 10^6\,\text{g } SO_2$$

Convert grams of SO_2 to tons of SO_2:

$$1.36 \times 10^6\,\text{g } SO_2 \times \frac{1\,\text{lb } SO_2}{454\,\text{g } SO_2} \times \frac{1\,\text{ton } SO_2}{2000\,\text{lbs } SO_2}$$

$$= 1.5\,\text{ton } SO_2$$

Note: The final answer should have only 2 significant figures because the percentage of sulfur and the amount of coal initially given had 2 significant figures.

99. a. Convert 5.00 kg UO_2 to kilograms of HF (5.00 kg = 5000 g)

Step 1: Convert 5.00 kg UO_2 to moles of UO_2. (Divide by the molar mass UO_2, 270.0 g/mol.)

Step 2: Convert moles of UO_2 to moles of HF. (From the balanced equation, 1 mol UO_2 = 4 mol HF.)

Step 3: Convert moles of HF to kilograms of HF. (Multiply by the molar mass of HF, 20.0 g/mol.)

$$5.0 \times 10^3\,\text{g } UO_2 \times \frac{1\,\text{mol } UO_2}{270.0\,\text{g } UO_2} \times \frac{4\,\text{mol HF}}{1\,\text{mol } UO_2}$$

$$\times \frac{20.0\,\text{g HF}}{1\,\text{mol HF}} \times \frac{1\,\text{kg}}{1000\,\text{g}} = 1.48\,\text{kg HF}$$

b. This question requires us to first convert 850 g UO_2 to grams of UF_4 and then use the calculated amount of UF_4 to determine how much UF_6 is formed.

Convert 850 g UO_2 to grams of UF_4 using the balanced equation

$$UO_2 + 4\,HF \rightarrow UF_4 + 2\,H_2O$$

Step 1: Convert 850 g UO_2 to moles of UO_2. (Divide by the molar mass of UO_2, 270.0 g/mol.)

Step 2: Convert moles of UO_2 to moles of UF_4. (From the balanced equation, 1 mol UO_2 = 1 mol UF_4.)

$$850\,\text{g } UO_2 \times \frac{1\,\text{mol } UO_2}{270.0\,\text{g } UO_2} \times \frac{1\,\text{mol } UF_4}{1\,\text{mol } UO_2}$$

$$= 3.15\,\text{mol } UF_4$$

Now, convert moles of UF_4 to grams UF_6 using the balanced equation

$$UF_4 + F_2 \rightarrow UF_6$$

Step 1: Convert moles of UF_4 to moles of UF_6. (From the balanced equation, 1 mol UF_4 = 1 mol UF_6.)

Step 2: Convert moles of UF_6 to grams of UF_6. (Multiply by the molar mass of UF_6, 352 g/mol.)

$$3.15\,\text{mol } UF_4 \times \frac{1\,\text{mol } UF_6}{1\,\text{mol } UF_4} \times \frac{352\,\text{g } UF_6}{1\,\text{mol } UF_6}$$

$$= 1.11 \times 10^3\,\text{g } UF_6$$

1.11×10^3 g UF_6 can be produced from 850. g of UO_2.

101. a. There are two methods we can use to answer this question.

Method 1: Use the chemical formula $CuFeS_2$ to determine the moles of copper in 1 mole of chalcopyrite ($CuFeS_2$). The conversion factor is

$$1 \text{ mol } CuFeS_2 = 1 \text{ mol Cu}$$

Method 2: Use the weight percent formula to determine the percentage of Cu in $CuFeS_2$.

Solving Using Method 1:

i. Convert 1.00 kg $CuFeS_2$ to moles of $CuFeS_2$ (1 kg = 1000 g; molar mass $CuFeS_2$ = 183.5 g/mol)

ii. Convert moles of $CuFeS_2$ to moles of Cu. (From the formula, 1 mol $CuFeS_2$ = 1 mol Cu.)

iii. Convert moles of Cu to grams of Cu. (Multiply by the atomic weight of Cu, 63.5 g/mol.)

$$1.00 \text{ kg } CuFeS_2 \times \frac{1000 \text{ g } CuFeS_2}{1.00 \text{ kg } CuFeS_2} \times \frac{1 \text{ mol } CuFeS_2}{183.5 \text{ g } CuFeS_2} \times \frac{1 \text{ mol Cu}}{1 \text{ mol } CuFeS_2} \times \frac{63.5 \text{ g Cu}}{1 \text{ mol Cu}} = 346 \text{ g Cu}$$

1.00 kg of $CuFeS_2$ can produce 346 g of Cu.

Solving Using Method 2:

The percent copper in chalcopyrite is given by the equation:

$$\% \text{ Cu in chalcopyrite} = \frac{\text{g Cu (in } CuFeS_2)}{\text{molar mass } CuFeS_2} \times 100$$

We have 1.00 kg (1000 g) $CuFeS_2$.

The molar mass of $CuFeS_2$ is 183.5 g/mol and the atomic weight of Cu is 63.5 g/mol.

$$\% \text{ Cu} = \frac{63.55}{183.5} \times 100 = 34.63\% \text{ Cu}$$

The percent copper in the mineral $CuFeS_2$ is 34.63. We can now use this percentage and the 1000 g $CuFeS_2$ to find out how much copper is in 1.00 kg (1000 g) of $CuFeS_2$

$$\text{g Cu} = \frac{(\text{g } CuFeS_2)(\% \text{ Cu})}{100} = \frac{(1000)(34.63)}{(100)} = 346 \text{ g Cu}$$

As expected, both methods give the same value for the mass of copper in 1.00 kg of $CuFeS_2$.

b. If the process is only 82% efficient, only 82% of the original sample of $CuFeS_2$ is converted into copper metal; that is, 18% of the original sample is not converted to copper metal. We can again use the percentage formula to determine how much Cu could be produced if the process is 100% efficient.

Then we can work backward from grams of Cu to grams of $CuFeS_2$ required, as we did in part *a* of this problem.

1. Convert grams of Cu to moles of Cu. (Divide by the atomic weight of Cu.)
2. Convert moles of Cu to moles of $CuFeS_2$ (1 mol Cu = 1 mol $CuFeS_2$).
3. Convert grams of $CuFeS_2$ to grams of $CuFeS_2$. (Multiply by the molar mass of $CuFeS_2$, 183.5 g/mol.)

$$\% \text{ yield Cu} = \frac{\text{actual yield Cu}}{\text{theoretical yield}} \times 100$$

The theoretical yield is the amount of copper metal recovered if the process is 100% efficient.

The 100. g of copper is the actual yield of copper metal.

We know the actual yield and the percentage yield (82%), so we can solve for the theoretical yield (this is what the question actually asked).

$$\text{theoretical yield} = \frac{\text{actual yield Cu}}{\% \text{ yield Cu}} \times 100 = \frac{100}{82} \times 100 = 122 \text{ g Cu}$$

If this process is 100% efficient, we could have recovered 122 g Cu instead of only 100 g Cu. Next convert the 122 g Cu into g $CuFeS_2$ (using the steps outlined). This will correspond to the amount of $CuFeS_2$ needed to get 100. g Cu.

$$122 \text{ g Cu} \times \frac{1 \text{ mol Cu}}{63.5 \text{ g Cu}} \times \frac{1 \text{ mol } CuFeS_2}{1 \text{ mol Cu}} \times \frac{183.5 \text{ g } CuFeS_2}{1 \text{ mol } CuFeS_2} = 353 \text{ g } CuFeS_2$$

If the recovery process is 82% efficient, we would recover 100. g of Cu metal if we started with 353 g of $CuFeS_2$.

103. Answer *c* is correct. The balanced equation for this process is

$$Fe + S \rightarrow FeS$$

This balanced equation tells us that 1 mole of Fe is required for every mole of S that reacts. Since the atomic weights of Fe (55.85 g/mol) and S (32.07 g/mole) are not equal, the masses of the two substances that react cannot be equal. The balanced equation implies that 55.85 g of Fe reacts with 32.07 g of S.

If we start with equal masses of Fe and S before the reaction, some of the S must be left over unreacted. If 55.85 g Fe and 55.85 g S are reacted, 55.85 g Fe reacts with 32.07 g S, leaving 23.78 g S (55.85 g − 32.07 g) unreacted.

The mass of FeS that could theoretically be formed has to be less than the sum of the masses of Fe and S.

104. The balanced equation for this process is

$$2\,\mathrm{Mg}(s) + \mathrm{O_2}(g) \rightarrow 2\,\mathrm{MgO}(s)$$

This equation requires that 2 moles of magnesium metal react with every mole of O_2 gas to form 2 moles of MgO (for every mole of O_2 gas).

The atomic weights of magnesium and oxygen are 24.3 g/mol and 16.00 g/mol, respectively.

2 moles of magnesium = 2 mol × 24.31 g/mol = 48.62 g Mg

1 mole of O_2 = 2 mol × 16.0 g/mol = 32.0 g O_2

The molar mass of MgO equals 40.31 g/mol (24.31 + 16.00).

2 moles of MgO = 2 mol × 40.31 g/mol = 80.62 g MgO so 48.62 g Mg + 32.0 g O_2 must react to form 80.62 g MgO.

If we had equal masses of both Mg metal and O_2, then we could have started with 48.62 g of each. Since we only needed 32.0 g of the O_2 gas, there must be some O_2 gas left (48.62 − 32.00 = 16.62 g left unreacted).

This does not violate the law of conservation of mass.

Mass of Mg reacted + mass of O_2 reacted = mass of MgO formed + mass of O_2 unreacted.

105. The theoretical yield is the maximum amount of a substance that can be produced from the given amounts of reactants. The percent yield is the percentage of the theoretical yield that is actually formed in a reaction.

106. No. Since the theoretical yield is the maximum amount, the percent yield can never be greater than 100%.

107. Some reasons why the actual yield is less than the theoretical yield (that is, the process is not 100% efficient) are as follows:

    1. The reactants may undergo more than one reaction, yielding different sets of products.
    2. Sometimes the rate of reaction is so slow that reactants remain unreacted even after an extended period of time.
    3. Some reactions do not go to completion no matter how long they are allowed to run, yielding a mixture of reactants and products whose composition does not change with time.

108. No. For any of the reasons listed in the previous question, reactions may not be 100% efficient and less product than expected is formed.

109. We can use each of the quantities given to create a set of conversion factors, similar to those we have used with balanced equations.

For hollandaise sauce we get:

1 cup sauce = $\frac{1}{2}$ cup butter = $\frac{1}{4}$ hot H_2O = 4 egg yolks = 1 lemon

Create an individual conversion factor between each item and the hollandaise sauce.

1 cup hollandaise sauce = $\frac{1}{2}$ cup butter
1 cup hollandaise sauce = $\frac{1}{4}$ cup hot water
1 cup hollandaise sauce = 4 egg yolks
1 cup hollandaise sauce = 1 medium lemon

We are given the conversion factor, 1 pound of butter = 2 cups of butter.

Our objective here is to determine which one of the ingredients runs out first. Once an ingredient is gone, we cannot make any more sauce.

Let's determine how many cups of hollandaise sauce can be made from the given quantities of each recipe component.

From 1 pound of butter (use the conversion factor created between the butter and the hollandaise sauce)

$$1\text{ lb butter} \times \frac{2\text{ cups butter}}{1\text{ lb butter}} \times \frac{1\text{ cup hollandaise sauce}}{1/2\text{ cup butter}}$$
$$= 4\text{ cups hollandaise sauce}$$

From the butter available we could make 4 cups of hollandaise sauce.

Do the same calculations using a dozen (12) eggs and 4 lemons. Since there is unlimited hot water we do not need to perform a calculation using the water.

$$12\text{ eggs} \times \frac{1\text{ yolk}}{1\text{ egg}} \times \frac{1\text{ cup hollandaise sauce}}{4\text{ yolks}}$$
$$= 3\text{ cups hollandaise sauce}$$

$$4\text{ lemons} \times \frac{1\text{ cup hollandaise sauce}}{1\text{ lemon}}$$
$$= 4\text{ cups hollandaise sauce}$$

From these calculations we see that we have used up all of the eggs after making 3 cups of hollandaise sauce. The number of eggs limits the amount of sauce that can be made. The eggs are called the limiting reactant.

3 cups of hollandaise sauce can be made.

There remains unused 1 lemon and a $\frac{1}{2}$ cup (or $\frac{1}{4}$ pound) of butter.

111. There are several methods one can use to determine this quantity. Probably the best is to perform two different stoichiometric calculations and determine separately the theoretical yield of the desired product based on the

amount of each starting material. Then compare the two yield values. The smaller yield is always the correct one.

The stoichiometric conversions to be performed in this problem are:

1. Convert 2.50 g of KO_2 to grams of O_2.
2. Convert 4.50 g of CO_2 to grams of O_2.

Compare the yields of O_2 for both of these processes.

The substance that produces the smaller yield of O_2 is the limiting reactant and the quantity of O_2 produced by the limiting reactant is the theoretical yield of O_2 for this process using the starting quantities given.

Convert 2.50 g KO_2 to grams of O_2:

Step 1: Convert 2.50 g KO_2 to moles of O_2. (Divide by the molar mass of KO_2, 71.1 g/mol.)

Step 2: Convert moles of KO_2 to moles of O_2. (Use the coefficients of the balanced equation, 4 mol KO_2 = 3 mol O_2.)

Step 3: Convert moles of O_2 to grams of O_2. (Multiply by the molar mass of O_2, 32.0 g/mol.)

$$2.50\,\text{g}\,KO_2 \times \frac{1\,\text{mol}\,KO_2}{71.1\,\text{g}\,KO_2} \times \frac{3\,\text{mol}\,O_2}{4\,\text{mol}\,KO_2} \times \frac{32.0\,\text{g}\,O_2}{1\,\text{mol}\,O_2}$$

$$= 0.844\,\text{g}\,O_2$$

Using up all the KO_2 produces 0.844 g of O_2.

Convert 4.5 g CO_2 to grams of O_2:

Step 1: Convert 4.50 g CO_2 to moles of O_2. (Divide by the molar mass of CO_2, 44.0 g/mol.)

Step 2: Convert moles of CO_2 to moles of O_2. (Use the coefficients of the balanced equation, 2 mol CO_2 = 3 mol O_2.)

Step 3: Convert moles of O_2 to grams of O_2. (Multiply by the molar mass of O_2, 32.0 g/mol.)

$$4.50\,\text{g}\,CO_2 \times \frac{1\,\text{mol}\,CO_2}{44.0\,\text{g}\,CO_2} \times \frac{3\,\text{mol}\,O_2}{2\,\text{mol}\,CO_2} \times \frac{32.0\,\text{g}\,O_2}{1\,\text{mol}\,O_2}$$

$$= 4.91\,\text{g}\,O_2$$

Using up all the CO_2 produces 4.91 g of O_2.

0.844 g O_2 is less than 4.91 g O_2, so the maximum amount of O_2 that can be produced is 0.844 g and KO_2 is the limiting reactant.

113. Formulas of the compounds:
Ammonia, NH_3
Hydrogen chloride, HCl
Ammonium chloride, NH_4Cl
The balanced equation is

$$NH_3 + HCl \rightarrow NH_4Cl$$

We need to determine the limiting reactant. From this information, we can then determine how much of the reactant in excess is used up. Finally, from the starting amount of the reactant in excess and the amount of this reactant consumed, we can determine how much of this substance remains unreacted.

Let's determine the theoretical yield of NH_4Cl based on the amounts given for each starting material. From this we can determine the limiting reactant.

1. Convert 3.0 g NH_3 to grams of NH_4Cl.
2. Convert 5.0 g HCl to grams of NH_4Cl.

To convert 3.0 g NH_3 to g NH_4Cl:

Step 1: Convert 3.0 g NH_3 to moles of NH_3. (Divide by the molar mass of NH_3, 17.0 g/mol.)

Step 2: Convert moles of NH_3 to moles of NH_4Cl (1 mol NH_3 = 1 mol NH_4Cl).

Step 3: Convert moles of NH_4Cl to grams of NH_4Cl. (Multiply by the molar mass of NH_4Cl, 53.5 g/mol.)

$$3.0\,\text{g}\,NH_3 \times \frac{1\,\text{mol}\,NH_3}{17.0\,\text{g}\,NH_3} \times \frac{1\,\text{mol}\,NH_4Cl}{1\,\text{mol}\,NH_3}$$

$$\times \frac{53.5\,\text{g}\,NH_4Cl}{1\,\text{mol}\,NH_4Cl} = 9.4\,\text{g}\,NH_4Cl$$

To convert 5.0 g HCl to grams of NH_4Cl:

Step 1: Convert 5.0 g HCl to moles of HCl. (Divide by the molar mass of HCl, 36.5 g/mol.)

Step 2: Convert moles of HCl to moles of NH_4Cl (1 mol HCl = 1 mol NH_4Cl).

Step 3: Convert moles of NH_4Cl to grams of NH_4Cl. (Multiply by the molar mass of NH_4Cl, 53.5 g/mol.)

$$5.0\,\text{g}\,HCl \times \frac{1\,\text{mol}\,HCl}{36.5\,\text{g}\,HCl} \times \frac{1\,\text{mol}\,NH_4Cl}{1\,\text{mol}\,HCl}$$

$$\times \frac{53.5\,\text{g}\,NH_4Cl}{1\,\text{mol}\,NH_4Cl} = 7.3\,\text{g}\,NH_4Cl$$

7.3 g NH_4Cl is smaller than 9.4 g. The limiting reactant is HCl (this substance produced the smaller quantity of NH_4Cl) and the theoretical yield of NH_4Cl is 7.3 g.

Since all of the HCl is reacted (limiting reactant), we know that the ammonia must be present in excess (some ammonia remains unreacted).

Next, we need to determine how much ammonia is unreacted. To do this, convert the starting amount of the limiting reactant (5.0 g HCl) to grams of NH_3 (the substance present in excess). This is the amount of the NH_3 that reacts. Finally, subtract the amount of NH_3 reacted from the starting amount to determine how much NH_3 is left unreacted.

Step 1: Convert 5.0 g HCl to moles of HCl. (Divide by the molar mass of HCl, 36.5 g/mol.)

Step 2: Convert moles of HCl to moles of NH_3 (1 mol HCl = 1 mol NH_3).

Step 3: Convert moles of NH_3 to grams of NH_3. (Multiply by the molar mass of NH_3, 17.0 g/mol.)

$$5.0\,\text{g HCl} \times \frac{1\,\text{mol HCl}}{36.5\,\text{g HCl}} \times \frac{1\,\text{mol NH}_3}{1\,\text{mol HCl}}$$

$$\times \frac{17.0\,\text{g NH}_3}{1\,\text{mol NH}_3} = 2.3\,\text{g NH}_3$$

This tells us that 2.3 g of the NH_3 reacted.

Since we started with 3.0 g NH_3 and 2.3 g reacted, then 0.7 g (3.0 g − 2.3 g) of NH_3 is left unreacted.

We could have also done this problem based on the conservation of mass. We started with 5.0 g HCl and 3.00 g NH_3 or a total of 8.0 g of reactants. We know that the theoretical yield of NH_4Cl is 7.3 g and that the HCl is the limiting reactant. So some ammonia must be left unreacted.

Since 8.0 g of reactants formed 7.3 g of products, 0.7 g (8.0 g – 7.3 g) of the NH_3 must remain unreacted.

115. First, we need to write the balanced equation for the reaction that occurs when carbon and oxygen react.

$$C + O_2 \rightarrow CO_2$$

The problem tells us that there is excess oxygen (the carbon is the limiting reactant), so we need to determine the theoretical yield of CO_2 for this reaction when 3.0 g of C is completely used.

Step 1: Convert 3.0 g of C to moles of C (atomic weight of C, 12.0 g/mol).

Step 2: Convert moles of C to moles of CO_2 (1 mol C = 1 mol CO_2).

Step 3: Convert moles of CO_2 to grams of CO_2 (molar mass of CO_2, 44.0 g/mol)

$$3.0\,\text{g C} \times \frac{1\,\text{mol C}}{12.0\,\text{g C}} \times \frac{1\,\text{mol C}}{1\,\text{mol CO}_2} \times \frac{44.0\,\text{g CO}_2}{1\,\text{mol CO}_2}$$

$$= 11\,\text{g CO}_2$$

The theoretical yield for the complete reaction of 3.0 g of carbon with excess oxygen is 11 g.

The actual yield of CO_2 is 6.5 g.

To find the percent yield, use the equation

$$\%\ \text{yield} = \frac{\text{actual yield}}{\text{theoretical yield}} \times 100$$

Substituting the known values in the preceding equation gives the percent yield

$$\frac{6.5\,\text{g}}{11\,\text{g}} \times 100 = 59\%$$

117. a. $C_6H_{12}O_6(aq) \rightarrow 2\,C_2H_5OH(l) + 2\,CO_2(g)$

b. The 100.0 g of glucose ($C_6H_{12}O_6$) is the starting quantity.

The 50 mL of ethanol is the actual yield. We will need to convert this from a volume to a mass using the density of ethanol (0.789 g/mL).

To find the percent yield, we need to calculate the theoretical yield of ethanol, C_2H_5OH. In other words, we need to convert the 100.0 g of glucose to grams of ethanol.

Step 1: Convert 100.0 g $C_6H_{12}O_6$ (glucose) to moles of $C_6H_{12}O_6$ (molar mass $C_6H_{12}O_6$, 180.0 g/mol).

Step 2: Convert moles of $C_6H_{12}O_6$ to moles of C_2H_5OH (1 mol $C_6H_{12}O_6$ = 2 mol C_2H_5OH).

Step 3: Convert moles of C_2H_5OH to grams of C_2H_5OH (molar mass C_2H_5OH, 46.0 g/mol).

$$100.0\,\text{g C}_6\text{H}_{12}\text{O}_6 \times \frac{1\,\text{mol C}_6\text{H}_{12}\text{O}_6}{180.0\,\text{g C}_6\text{H}_{12}\text{O}_6}$$

$$\times \frac{2\,\text{mol C}_2\text{H}_5\text{OH}}{1\,\text{mol C}_6\text{H}_{12}\text{O}_6} \times \frac{46.0\,\text{g C}_2\text{H}_5\text{OH}}{1\,\text{mol C}_2\text{H}_5\text{OH}}$$

$$= 51.1\,\text{g C}_2\text{H}_5\text{OH}$$

The theoretical yield of C_2H_5OH from 100.0 g of glucose is 51.1 g.

We can calculate the actual yield of ethanol in grams (convert 50.0 mL to grams) using the relationship mass = density × volume

$$50.0\,\text{mL} \times \frac{0.789\,\text{g}}{1\,\text{mL}} = 39.4\,\text{g ethanol}$$

The percent yield is

$$\frac{39.4\,\text{g}}{51.1\,\text{g}} \times 100 = 77.1\%$$

CHAPTER 5 Solution Chemistry and the Hydrosphere

1. The component that is present in the greater quantity, measured in moles, is considered to be the solvent.

2. A 50:50 mixture by mass corresponds to a solution that has the same amount of both water and methanol by mass. If we assume that the total solution mass is 100 g, then the solution contains 50.0 g of water and 50.0 g of methanol (CH_3OH). To determine which component is the solvent, convert the mass of each component into moles of each substance (molar mass $H_2O = 18.0$ g/mol and $CH_3OH = 32.0$ g/mol).

   Without doing a calculation, we should realize that the substance with the greater molar mass has the smaller number of moles. There are more moles of water than methanol in this solution. The water is the solvent and the methanol is the solute.

   Crunching the numbers:

$$\text{moles of } H_2O = 50.0 \text{ g } H_2O \times \frac{1 \text{ mol } H_2O}{18.0 \text{ g } H_2O}$$
$$= 2.78 \text{ mol } H_2O$$
$$\text{moles of } CH_3OH = 50.0 \text{ g } CH_3OH \times \frac{1 \text{ mol } CH_3OH}{32.0 \text{ g } CH_3OH}$$
$$= 1.56 \text{ mol } CH_3OH$$

   There are more moles of water. It is the solvent.

3. Yes, a solid can be considered a solvent. In a solid–solid solution, such as an alloy, a solid must be the solvent. Many of the minerals that make up the Earth's crust are examples of solid solutions (mixtures of substances in the solid state that do not have the fixed the composition of a pure substance).

4. No, a saturated solution is not always a concentrated solution. Since the maximum amount of solute that can be dissolved in a given solvent varies from solute to solute, certain substances may not be very soluble in the solvent of interest. For example, silver chloride is not very soluble in water. Typically at room temperature only about 0.002 g of silver chloride can dissolve in a liter of water. A saturated solution of silver chloride contains only 0.002 g/L. This would not usually be considered to be a concentrated solution, but it is saturated.

5. Molarity is defined as the number of moles of solute per liter of solution.

6. A solution that contains 1.00 millimole per milliliter of solution has a molarity of 1.00 *M*. Both 1.00 millimole and 1.00 milliliter are one thousandth of their base unit. If we scale both units up by a factor of 1000, we get 1.00 mol/L.

   That is, we can convert 1.00 millimole to moles and 1 milliliter to liters and divide these two quantities to get molarity (*M*).
   1 millimole = 0.001 mole   1 milliliter = 0.001 liter
   molarity = 0.001 mol/0.001 L = 1.00 *M*

7. The molarity of a solution is independent of the volume of that solution. The solution in this case will always have a concentration of 1.00 *M* regardless of how much of this solution is taken (1.00 ml to 99 mL).

If we were to add more water (dilute) to the solution, this would change the concentration of the solution.

8. a. A closed can of soda would most likely be classified as a homogeneous mixture because the dissolved carbon dioxide gas is not observed as a separate phase.
   b. A freshly opened can of soda is a heterogeneous mixture because some of the carbon dioxide comes out of solution and gas bubbles are clearly seen.
   c. A soda can left open for 3 days would be a homogeneous mixture because most of the carbon dioxide would have escaped and the solution would appear uniform throughout.

9. Molarity is defined as moles of solute divided by liters of solution. In each part of this question convert the volume from milliliters to liters. In part *b* the millimoles (mmoles) need to be converted into moles. Conversion factors are

$$1\text{ L} = 1000\text{ mL; } 1\text{ mole} = 1000\text{ mmole}$$

a.
$$\text{moles} = 0.56\text{ mol BaCl}_2$$
$$\text{volume} = 0.150\text{ L}$$
$$\text{molarity} = \frac{0.56\text{ mol BaCl}_2}{0.150\text{ L}} = 3.7M$$

b.
$$20\text{ mmoles Na}_2\text{CO}_3 = 0.020\text{ mol Na}_2\text{CO}_3$$
$$\text{volume} = 0.750\text{ L}$$
$$\text{molarity} = \frac{0.020\text{ mol Na}_2\text{CO}_3}{0.750\text{ L}} = 0.027M$$

c.
$$\text{moles} = 0.325\text{ mol C}_6\text{H}_{12}\text{O}_6$$
$$\text{volume} = 0.100\text{ L}$$
$$\text{molarity} = \frac{0.325\text{ mol C}_6\text{H}_{12}\text{O}_6}{0.100\text{ L}} = 3.25M$$

d.
$$\text{moles} = 1.48\text{ mol KNO}_3$$
$$\text{volume} = 0.250\text{ L}$$
$$\text{molarity} = \frac{1.48\text{ mol KNO}_3}{0.250\text{ L}} = 5.92\ M$$

11. In this problem, we need to convert the mass of each ion into moles and then divide by the volume in liters (1000 mL = 1.00 L).

    To find the number of moles of each ion, divide the mass of each ion by its molar mass.

    a. 0.33 g sodium ions per 100. mL of water

$$\text{molariy} = \frac{0.33\text{ g Na}^+}{0.100\text{ L}} \times \frac{1\text{ mol Na}^+}{23.0\text{ g Na}^+} = 0.14\ M\text{ Na}^+$$

   b. 0.38 g of chloride ions per 100. mL of water

$$\text{molarity} = \frac{0.38\text{ g Cl}^-}{0.100\text{ L}} \times \frac{1\text{ mol Cl}^-}{35.45\text{ g Cl}^-} = 0.11M\text{Cl}^-$$

   c. 0.26 g sulfate (SO_4^{2-}) ion per 50. mL of water

$$\text{molarity} = \frac{0.46\text{ g SO}_4^{2-}}{0.050\text{ L}} \times \frac{1\text{ mol SO}_4}{96.0\text{ g SO}_4^{2-}}$$
$$= 0.096\ M\text{ SO}_4^{2-}$$

   d. 0.40 g of calcium ion per 50. mL of water

$$\text{molarity} = \frac{0.40\text{ g Ca}^{2+}}{0.050\text{ L}} \times \frac{1\text{ mol Ca}^{2+}}{40.1\text{ g Ca}^{2+}}$$
$$= 0.20\ M\text{ Ca}^{2+}$$

13. In this problem, the information given is the volume of the solution and the molarity of the solution. From this, we can calculate the moles of solute from the relationship

$$\text{moles} = \text{molarity} \times \text{liters}$$

The mass of each solute in grams can then be calculated by multiplying by the molar mass of the solute.

a. 1.00 L of 0.092 *M* NaCl (molar mass NaCl = 58.5 g/mol)

$$\text{moles of NaCl} = \frac{0.092\text{ mol NaCl}}{1.00\text{ L}} \times 1.00\text{ L}$$
$$= 0.092\text{ mol NaCl}$$
$$\text{mass of NaCl} = 0.092\text{ mol NaCl} \times \frac{58.5\text{ g NaCl}}{1\text{ mol NaCl}}$$
$$= 5.4\text{ g NaCl}$$

b. 300. mL of 0.125 *M* $CuSO_4$ (molar mass $CuSO_4$ = 159.6 g/mol)

$$\text{moles of CuSO}_4 = \frac{0.125\text{ mol CuSO}_4}{1.00\text{ L}} \times 0.300\text{ L}$$
$$= 0.0375\text{ mol CuSO}_4$$
$$\text{mass of CuSO}_4 = 0.0375\text{ mol CuSO}_4 \times \frac{159.6\text{ g CuSO}_4}{1\text{ mole CuSO}_4}$$
$$= 5.98\text{ g CuSO}_4$$

c. 250. mL of 0.400 *M* CH_3OH (molar mass CH_3OH = 32.0 g/mol)

$$\text{moles of CH}_3\text{OH} = \frac{0.400\text{ mol CH}_3\text{OH}}{1.00\text{ L}} \times .250\text{ L}$$
$$= 0.100\text{ mol CH}_3\text{OH}$$
$$\text{mass of CH}_3\text{OH} = 0.100\text{ mol CH}_3\text{OH} \times \frac{32.0\text{ g CH}_3\text{OH}}{1\text{ mol CH}_3\text{OH}}$$
$$= 3.2\text{ g CH}_3\text{OH}$$

15. In this problem we are given the concentrations of several solutes in units of mM (millimoles/liter) and the

total volume of solution present. From this information, we can find the millimoles (mmoles) of each solute using the formula

$$\text{mmoles} = \text{mM} \times \text{liters}$$

To find the mass of each solute, convert the mmoles to moles and then multiply the number of moles by the molar mass (1 mole = 1000 mmole) as follows:

$$\text{mass of solute} = \text{mmoles} \times \frac{1 \text{ mole}}{1000 \text{ mmoles}} \times \text{molar mass}$$

Let's do the calculations:

Concentration of each solute: 0.820 mM Ca^{2+}, 0.430 mM Mg^{2+}, 0.300 mM Na^{+}, 0.0200 mM K^{+}, 0.250 mM Cl^{-}, 0.380 mM SO_4^{2-}, and 1.82 mM HCO_3^{-}

Volume of solution = 2.75 liters

Mass of Ca^{2+} ions (molar mass of Ca^{2+} = 40.08 g/mol):

$$\text{grams } Ca^{2+} = 2.75 \text{ L} \times \frac{0.820 \text{ mmol}}{\text{L}} \times \frac{1 \text{ mole } Ca^{2+}}{1000 \text{ mmol } Ca^{2+}} \times \frac{40.08 \text{ g } Ca^{2+}}{1 \text{ mol } Ca^{2+}} = 0.0904 \text{ g } Ca^{2+}$$

Mass of Mg^{2+} ions (molar mass of Mg^{2+} = 24.31 g/mol):

$$\text{grams } Mg^{2+} = 2.75 \text{ L} \times \frac{0.430 \text{ mmol}}{\text{L}} \times \frac{1 \text{ mol } Mg^{2+}}{1000 \text{ mmol } Mg^{2+}} \times \frac{24.31 \text{ g } Mg^{2+}}{1 \text{ mol } Mg^{2+}} = 0.0288 \text{ g } Mg^{2+}$$

Mass of Na^{+} ions (molar mass of Na^{+} = 23.0 g/mol):

$$\text{grams } Na^{+} = 2.75 \text{ L} \times \frac{0.300 \text{ mmol}}{\text{L}} \times \frac{1 \text{ mol } Na^{+}}{1000 \text{ mmol } Na^{+}} \times \frac{23.00 \text{ g } Na^{+}}{1 \text{ mol } Na^{+}} = 0.0190 \text{ g } Na^{+}$$

Mass of K^{+} ions (molar mass of K^{+} = 39.1 g/mol):

$$\text{grams } K^{+} = 2.75 \text{ L} \times \frac{0.0200 \text{ mmol}}{\text{L}} \times \frac{1 \text{ mol } K^{+}}{1000 \text{ mmol } K^{+}} \times \frac{39.1 \text{ g } K^{+}}{1 \text{ mol } K^{+}} = 0.00215 \text{ g } K^{+}$$

Mass of Cl^{-} ions (molar mass of Cl^{-} = 35.45 g/mol):

$$\text{grams } Cl^{-} = 2.75 \text{ L} \times \frac{0.250 \text{ mmol}}{\text{L}} \times \frac{1 \text{ mol } Cl^{-}}{1000 \text{ mmol } Cl^{-}} \times \frac{35.45 \text{ g } Cl^{-}}{1 \text{ mol } Cl^{-}} = 0.0244 \text{ g } Cl^{-}$$

Mass of SO_4^{2-} ions (molar mass of sulfate, SO_4^{2-} = 96.1 g/mol):

$$2.75 \text{ L} \times \frac{0.380 \text{ mmol}}{\text{L}} \times \frac{1 \text{ mol } SO_4^{2-}}{1000 \text{ mmol } SO_4^{2-}} \times \frac{96.1 \text{ g } SO_4^{2-}}{1 \text{ mol } SO_4^{2-}} = 0.100 \text{ g } SO_4^{2-}$$

Mass of HCO_3^{-} ions (molar mass of bicarbonate, HCO_3^{-} = 61.0 g/mol):

$$\text{grams } HCO_3^{-} = 2.75 \text{ L} \times \frac{1.82 \text{ mmol}}{\text{L}} \times \frac{1 \text{ mol } HCO_3^{-}}{1000 \text{ mmol } HCO_3^{-}} \times \frac{61.00 \text{ g } HCO_3^{-}}{1 \text{ mol } HCO_3^{-}} = 0305 \text{ g } HCO_3^{-}$$

Total mass of dissolved minerals

$$= \text{g } Ca^{2+} + \text{g } Mg^{2+} + \text{g } Na^{+} + \text{g } K^{+} + \text{g } Cl^{-} + \text{g } SO_4^{2-} + \text{g } HCO_3^{-}$$

$$\text{mass of solutes} = 0.0904 + 0.0288 + 0.0190 + 0.00215 + 0.0244 + 0.100 + 0.305$$

Total mass of dissolved minerals in 2.75 L of river water is 0.570 g (to 3 decimal places).

17. Since we are given the molarity and the volume of solution, we can use the molarity formula to calculate the number of moles of solute in each solution.

$$\text{moles} = \text{molarity}(M) \times \text{volume (in liters)}$$

a. Volume in liters = 0.400 L; concentration = 0.024 *M* Lindane

$$\text{moles of Lindane} = 0.400 \text{ L} \times \frac{0.024 \text{ mol Lindane}}{\text{L}} = 0.0096 \text{ mol Lindane}$$

b. Volume in liters = 1.65 *L*; concentration = 0.473 m*M* Dieldrin

The unit mM means mmole of solute per liter of solution.

$$\text{mmole} = \text{mM} \times V$$

To convert from millimoles to moles, use the conversion factor 1 mole = 1000 mmole.

$$\text{moles of Dieldrin} = 1.65\ \text{L} \times \frac{0.473\ \text{mmol Dieldrin}}{\text{L}} \times \frac{1\ \text{mol}}{1000\ \text{mmol}}$$

$$= 7.80 \times 10^{-4}\ \text{mol Dieldrin}$$

c. Volume in liters = 25.8 L; concentration = 3.4 μM DDT

The symbol μ stands for the metric prefix micro (1×10^{-6}). The conversion factor between micromolarity and molarity is

$$1\text{micromolar} = 1 \times 10^{-6}\ \text{molar}$$

$$\text{moles of DDT} = 25.8\ \text{L} \times \frac{3.4\ \mu\text{mol DDT}}{1\ \text{L}} \times \frac{1 \times 10^{-6}\ \text{mol}}{1\ \mu\text{mol}}$$

$$= 8.8 \times 10^{-5}\ \text{mol DDT}$$

d. Volume in liters = 154 L; concentration = 27.4 μM Aldrin

$$1\ \text{micromolar} = 1 \times 10^{-6}\ \text{molar}$$

$$\text{moles of Aldrin} = 154\ \text{L} \times \frac{27.4\ \mu\text{mol Aldrin}}{1\ \text{L}} \times \frac{1 \times 10^{-6}\ \text{mol}}{1\ \mu\text{mol}}$$

$$= 4.22 \times 10^{-3}\ \text{mol Aldrin}$$

19. In this problem we are given the volume and mass of each sample and asked to find the molarity of the sample. To determine the molarity, we need to find the number of moles of the pesticide DDT in each sample. Since we are given the mass of DDT in each sample and the molecular formula, we can calculate the number of moles of DDT from the relation

$$\text{moles of DDT} = \frac{\text{mass DDT}}{\text{molar mass DDT}}$$

We will need the conversion factor

$$1\ \mu\text{g} = 1 \times 10^{-6}\text{g}$$

The molar mass of DDT ($C_{14}H_9Cl_5$) = 354.5 g/mol.

Orchard concentration:

Volume in liters = 0.250 L; mass in μg = 0.030 μg

$$\text{moles of DDT} = 0.030\ \mu\text{g DDT} \times \frac{1 \times 10^{-6}\ \text{g}}{1\ \mu\text{g}} \times \frac{1\ \text{mol DDT}}{354\ \text{g DDT}}$$

$$= 8.5 \times 10^{-11}\ \text{mol DDT}$$

$$\text{molarity of DDT} = \frac{\text{moles DDT}}{\text{L}} = \frac{8.5 \times 10^{-11}\ \text{mol DDT}}{0.250\ \text{L}}$$

$$= 3.4 \times 10^{-10} M\ \text{DDT}$$

Residential concentration:

Volume in liters = 1.75; mass in μg = 0.035 μg

$$\text{molarity of DDT} = 0.035\mu\text{g DDT} \times \frac{1 \times 10^{-6}\text{g}}{1\mu\text{g}} \times \frac{1\ \text{mol DDT}}{354.5\ \text{g DDT}} \times \frac{1}{1.75\ \text{L}}$$

$$= 5.7 \times 10^{-11} M\ \text{DDT}$$

Residential concentration after a storm:

Volume in liters = 0.050 L; mass in μg = 0.57 μg

$$\text{molarity of DDT} = 0.57\mu\text{g DDT} \times \frac{1 \times 10^{-6}\ \text{g}}{1\ \mu\text{g}} \times \frac{1\ \text{mol DDT}}{354.5\ \text{g DDT}} \times \frac{1}{0.050\ \text{L}}$$

$$= 3.2 \times 10^{-8} M\ \text{DDT}$$

21. To convert from units of μg/L to molarity (mole/L):

Convert μg to g (1 μg = 1×10^{-6} g)

Convert mass in grams to moles (divide by the molar mass of solute)

Hexachlorobenzene, C_6Cl_6

Concentration = 0.55 μg/L; molar mass C_6Cl_6 = 285 g/mol

$$\text{molarity of } C_6Cl_6 = 0.55\ \mu\text{g}\ C_6Cl_6 \times \frac{1 \times 10^{-6}\ \text{g}}{1\ \mu\text{g}} \times \frac{1\ \text{mol}\ C_6Cl_6}{285\ \text{g}\ C_6Cl_6} \times \frac{1}{1\ \text{L}}$$

$$= 1.9 \times 10^{-9} M\ C_6Cl_6$$

Dieldrin, $C_{12}H_8Cl_6O$

Concentration = 0.06 μg/L; molar mass $C_{12}H_8Cl_6O$ = 381 g/mol

$$\text{molarity of } C_{12}H_8Cl_6O = 0.06\ \mu\text{g}\ C_{12}H_8Cl_6O \times \frac{1 \times 10^{-6}\ \text{g}}{1\ \mu\text{g}} \times \frac{1\ \text{mol}\ C_{12}H_8Cl_6O}{381\ \text{g}\ C_{12}H_8Cl_6O} \times \frac{1}{1\ \text{L}}$$

$$= 1.6 \times 10^{-10} M\ C_{12}H_8Cl_6O,\ \text{or}\ 2 \times 10^{-10} M$$

Hexachlorocyclohexane, $C_6H_6Cl_6$

Concentration = 1.02 μg/L; molar mass $C_6H_6Cl_6$ = 291 g/mol

$$\text{molarity of } C_6H_6Cl_6 = 1.02\mu g\ C_6H_6Cl_6 \times \frac{1 \times 10^{-6}\ g}{1\ \mu g}$$

$$\times \frac{1\ \text{mol}\ C_6H_6Cl_6}{291\ g\ C_6H_6Cl_6} \times \frac{1}{1\ L}$$

$$= 3.5 \times 10^{-9} M\ C_6H_6Cl_6$$

23. First, we need to determine how many grams of copper sulfate are present in 20.0 g of the fertilizer sample.
To do this use the percentage formula

$$\%\ CuSO_4 = \frac{g\ CuSO_4}{g\ \text{fertilizer}} \times 100$$

% $CuSO_4$ = 0.07; mass of fertilizer = 20.0 g
Solving for g $CuSO_4$ and substituting

$$0.07 = \frac{g\ CuSO_4}{20.0\ g} \times 100$$

$$g\ CuSO_4 = \frac{(20.0)(0.07)}{100} = 0.014\ g\ CuSO_4$$

This tells us the mass of $CuSO_4$ in the sample of fertilizer. We actually need to determine the mass of copper(II) ions in the sample of the fertilizer.
Next determine the moles of $CuSO_4$ (molar mass of $CuSO_4$ = 159.6 g/mol) and then determine the concentration (M) of $CuSO_4$ in this solution.

$$\text{mol of } CuSO_4 = 0.014\ g\ CuSO_4 \times \frac{1\ \text{mol}\ CuSO_4}{159.6\ g\ CuSO_4}$$

$$= 8.77 \times 10^{-5}\ \text{mol}\ CuSO_4$$

$$M \text{ of } CuSO_4 = \frac{8.77 \times 10^{-5}\ \text{mol}\ CuSO_4}{2.0\ L}$$

$$= 4.4 \times 10^{-5}\ M\ CuSO_4$$

From the formula of $CuSO_4$, we can see that there is 1 Cu^{2+} ion per formula. We can use this information to determine the molarity of Cu^{2+} ions in this solution.

$$4.4 \times 10^{-5}\ M\ CuSO_4 \times \frac{1\ M\ Cu^{2+}}{1\ M\ CuSO_4}$$

$$= 4.4 \times 10^{-5}\ M\ Cu^{2+}$$

25. In this problem, convert the mass of gold (6×10^9) into moles by dividing by the atomic weight of gold (197 g/mol). To find the molarity, divide by the total volume of all the oceans combined (1.5×10^{21} L)

$$\text{Molarity of gold} = \frac{6 \times 10^9\ g\ Au}{1.5 \times 10^{21}\ L} \times \frac{1\ \text{mol Au}}{197\ g\ Au}$$

$$= 2 \times 10^{-14} M$$

27. In each part of this problem, we need to determine the number of moles of solute present in the given volume and then use this number to determine the new volume. We can use the dilution formula

$$M_i V_i = M_f V_f \qquad \text{(moles initially present = moles present after dilution)}$$

In each part of this problem, solve for the final volume. However, the question does not ask for the final volume; instead we are asked to determine how much water needs to be added to each solution. To determine the amount of water to be added, subtract the initial volume from the final volume.

$$\text{volume } H_2O \text{ added} = \text{final Volume} - \text{initial Volume}$$

a. $M_i = 0.024\ M$ Lindane $\quad V_i = 1.00$ mL
$M_f = 1.00\ \mu M \quad V_f = ?$

Since the molarity units are not the same, we need to convert μM to M.

$$1.00\ \mu M = 1.00 \times 10^{-6} M$$

$$M_f = 1.00 \times 10^{-6} M$$

Now we can substitute into the dilution formula and solve for V_f.

$$(0.024M)(1.00\ \text{mL}) = (1.00 \times 10^{-6}\ M)(V_f)$$

$$V_f = \frac{(0.024M)(1.00\ \text{mL})}{(1.00 \times 10^{-6} M)}$$

$$= 2.4 \times 10^4\ \text{mL, or } 24{,}000\ \text{mL}$$

The final volume is 24,000 mL.
We need to add 24,000 mL − 1.0 mL = 23,999 mL or 24,000 mL to the required number of significant figures.

b. $M_i = 7.16 \times 10^{-3} mM\ Cu^{2+} \quad V_i = 15.5$ mL
$M_f = 1.00\ \mu M \quad V_f = ?$

The molarity units are not the same. We need to convert μM and mM to M.

$$1.00\ \mu M = 1.00 \times 10^{-6}\ M$$

$$7.16 \times 10^{-3}\ mM \times \frac{1M}{1000\ mM} = 7.16 \times 10^{-6} M$$

Now we can substitute into the dilution formula and solve for V_f.

$$(7.16 \times 10^{-6} M)(15.5\ \text{mL}) = (1.00 \times 10^{-6} M)(V_f)$$

$$V_f = \frac{(7.16 \times 10^{-6} M)(15.5\ \text{mL})}{(1.00 \times 10^{-6} M)} = 111\ \text{mL}$$

The final volume is 111 mL.
We need to add 111 mL − 15.5 mL = 95.5 mL or 96 mL (to the correct number of significant figures).

c. $M_i = 75.\ \mu M\ PO_4^{3-}$ $\quad V_i = 25.0\ \mu L$
$M_f = 1.00\ \mu M$ $\quad V_f = ?$

Both concentrations have the same units, so no conversion is needed.

$$(75.\ \mu M)(25.0\ \mu L) = (1.00\ \mu M)(V_f)$$

$$V_f = \frac{(75\ \mu M)(25.0\ \mu L)}{(1.00\ \mu M)} = 1875\ \mu L$$

The final volume is 1875 μL. The starting volume was 25.0 μL.

We need to add 1875 μL – 25.0 μL = 1850 μL of water.

29. This is also a dilution problem. We need to determine how much of the standard solution is needed to make 500.0 mL of a solution that has a solute concentration of 2.25 mg/L. We can do this using the relationship

$$C_i V_i = C_f V_f$$

$C_i = 1000.$ mg/L $\quad V_i = ?$
$C_f = 2.25$ mg/L $\quad V_f = 500.0$ mL

$$(1000.\ \text{mg/L})(V_i) = (2.25\ \text{mg/L})(500.0\ \text{mL})$$

$$V_i = \frac{(2.25\ \text{mg/L})(500.0\ \text{mL})}{(1000.\ \text{mg/L})} = 1.125\ \text{mL}$$

We need to measure out 1.125 mL of the standard solution and add water until the final volume is 500.0 mL.

31. Strong electrolytes completely dissociate in solution. Strong electrolytes include all soluble ionic substances (see solubility rules, Table 5.4), strong acids (HCl, HBr, HI, HNO_3, $HClO_4$, and H_2SO_4), and strong bases (the soluble Group 1A hydroxides LiOH, NaOH, KOH, RbOH, CsOH, and the souluble Group 2A hydroxides $Ca(OH)_2$, $Sr(OH)_2$, and $Ba(OH)_2$).

Weak electrolyes include all weak acids and weak bases.

Table salt or NaCl: This is a soluble ionic substance. It is a strong electrolyte and an excellent conductor of electricity in water.

Table sugar (sucrose) or $C_{12}H_{22}O_{11}$: This substance is made up of only nonmetallic elements. Since it is not a weak acid or base, it must be a nonelectrolyte; that is, it does not conduct electricity.

Formic acid or HCOOH: This substance is a weak acid so it must be classified as a weak electrolyte. It conducts electricity only weakly.

Methanol or CH_3OH: This substance is a nonelectrolyte. It contains only nonmetallic elements and is not a weak acid or base. It does not conduct electricity.

Table salt and formic acid increase the conductivity of water, whereas table sugar and methanol do not.

32. NaCl and $MgCl_2$ are both soluble ionic compounds. As such they are both strong electrolytes and dissociate into ions when dissolved in water. The solution that has the greater total ion concentration has the greater conductivity.

The concentration of both solutions is 1.0 *M*. However, when dissolved in water, NaCl forms two ions (Na^+ and Cl^-) for every formula unit, whereas $MgCl_2$ forms three ions (Mg^{2+} and 2 Cl^-) per formula unit.

The total ion concentration in a solution of 1.0 *M* NaCl is 2.0 *M* (1.0 *M* Na^+ ions and 1.0 *M* Cl^- ions).

The total ion concentration in a solution of 1.0 *M* $MgCl_2$ is 3.0 *M* (1.0 *M* in Mg^{2+} ions and 2.0 *M* in Cl^- ions).

The $MgCl_2$ solution should be the better conductivity of electricity.

33. In this problem, determine how many Na^+ ions form for each substance when dissolved in water, and then determine the concentration of the Na^+ ion from the number of ions of Na^+ per formula unit.

a. $NaBr \rightarrow Na^+ + Br^-$
In this substance one sodium ion (Na^+) forms per formula unit of NaBr.
The concentration of $Na^+ = 0.025$ M.

b. $Na_2SO_4 \rightarrow 2\ Na^+ + SO_4^{2-}$
Since two sodium ions form per formula unit of Na_2SO_4, the sodium ion concentration should be twice the Na_2SO_4 concentration.
The concentration of Na^+ is $0.050\ M\ (2 \times 0.025M)$.

c. $Na_3PO_4 \rightarrow 3\ Na^+ + PO_4^{3-}$
Three sodium ions form per formula unit of Na_3PO_4. The sodium ion concentration should be three times that of the Na_3PO_4 solution.
The sodium ion concentration is $0.075\ M(3 \times 0.025M)$.

35. a. The greater the solute concentration, the greater the osmotic pressure. Dissolving a solute in a solvent results in an increase in osmotic pressure.
b. Dissolving a solute in a solvent results in a solution that freezes at a lower temperature than the pure solvent. This is known as freezing point depression.
c. Dissolving a solute in a solvent results in a solution that boils at a higher temperature than the pure solvent. This is known as boiling point elevation.

36. A semipermeable membrane is one that allows certain substances to pass through but not others.

37. On one side of a semipermeable membrane is pure solvent; on the other side is a solution containing a solute dissolved in the same solvent. The solvent molecules can pass through the membrane, while the larger solute molecules cannot. The solvent flows from the side with

higher solvent concentration to the side with lower solvent concentration, or in this case from the pure solvent side of the membrane to the solution side, because there is a tendency to equalize the concentration of solute on both sides of the membrane. As the solvent molecules move through the membrane to the solution side, the concentration of the solution must decrease (the moles of solute remain the same, but total volume increases).

38. Since only solvent molecules can pass through the membrane, the flow of solvent molecules is always from the solution with lower solute concentration to the solution with greater solute concentration.

    The driving force for osmosis to occur is to equalize the solute concentration in each compartment. Since only solvent molecules can pass through the semipermeable membrane, they must flow from lower solute concentration to higher solute concentration. As the solvent molecules move through the membrane, the concentration of solute (on the lower concentration side) must increase because the moles of solute remain the same, but the volume of solution must decrease. This increases the solute concentration. On the other side of the membrane pure water is increasing the total solution volume. Again, the moles of solute remain the same but the total volume increases, so the solute concentration decreases.

39. The formula for osmotic pressure is $\pi = MRT$, where π is the osmotic pressure, M is the molarity, R is a constant, and T is the temperature in Kelvins. From this equation we see that osmotic pressure is directly proportional to both the molarity of the solute and the temperature of the solution.

40. Reverse osmosis is a process that changes the direction of solvent flow. In reverse osmosis the solvent molecules flow from the region of higher solute concentration to the region of lower solute concentration (in osmosis, the opposite is true).

41. Colligative properties depend on the concentration of solute particles. The van't Hoff factor allows us to determine the actual number of solute particles present per formula unit and so the total concentration of solute particles.

    In ionic substances, each ion behaves as a separate particle. If a solute completes dissociates, the van't Hoff number is the number of ions formed by this process. However, there are many solutions in which ionic dissociation is not complete. The van't Hoff factor is a measure of the fraction of the ionic particles that dissociate. In addition, as solute concentration increases, ions may interact with each other to form ion pairs. This reduces the number of solute particles present and lowers the van't Hoff factor below its theoretical value. For nonelectrolyte solutes, the van't Hoff factor is 1 because no dissociation occurs.

42. With each of these techniques we can use experimentally measured data to find the concentration of the unknown solute. From this information and either the volume of solution or the mass of solvent used, one can find the moles of solute in the sample. If the mass of solute is measured at the beginning of the experiment, we also know this quantity. If we know both the mass of solute and moles of solute, we can determine the molar mass (molar mass = grams of solute/mole of solute).

    It is important to know whether the solute is an electrolyte or a nonelectrolyte. If the solute is a nonelectrolyte, then the measured colligative property (osmotic pressure, freezing point depression, or boiling point elevation) can be used to determine the solute concentration (molarity or molality). However, if the solute is an electrolyte, then the measured solute concentration will be greater than its molar concentration because dissociation of the solute increases the number of particles in solution.

43. This statement is false.

    NaCl dissociates into two ions in water (Na^+ and Cl^-).

    $CaCl_2$ dissociates into three ions in water (Ca^{2+} and $2\ Cl^-$).

    The formula for osmotic pressure is $\pi = iMRT$ where i is the van't Hoff factor

    If π and T are the same, then the product of $M \times i$ (molarity $\times$ van't Hoff factor) must also be the same.

    Since the van't Hoff factor is greater for $CaCl_2 (i = 3)$ than for NaCl ($i = 2$), the concentration of $CaCl_2$ must be lower than the concentration of NaCl solution.

44. The osmotic pressure can be determined using the formula $\pi = iMRT$. Since the molarity and temperature for each solution are the same, the only factor that allows for the three solutions to have a different osmotic pressure is the degree of dissociation. The van't Hoff factor, i, measures this value.

    Glucose, $C_6H_{12}O_6$, is a nonelectrolyte ($i = 1$). It does not dissociate to any degree.

    Acetic acid, $C_2H_3O_2H$, is a weak acid (i slightly greater than 1). It partially dissociates.

    Sodium chloride, NaCl, is a strong electrolyte ($i = 2$). It completely dissociates.

    In terms of total particle concentration, the strong electrolyte should have the highest value and therefore the greatest osmotic pressure. The weak electrolyte acetic acid will have a slightly greater total particle concentration than the nonelectrolyte, glucose.

    The sodium chloride solution has the greatest osmotic pressure, and the glucose solution has the lowest osmotic pressure.

45. Molality is defined as moles of solute divided by kilograms of solvent, whereas molarity is defined as moles of solute divided by liters of solution. That is, molarity is a mass/volume ratio, while molality is a mass/mass ratio. The other difference is that the denominator in molarity is a quantity associated with the total solution, while in molality the quantity is associated with the solvent only.

46. As a solution becomes more concentrated, the density of the solution increases. When the solute concentration is very small, the density of the solution is very close to 1.0 g/mL; that is, the density of the solution is the same as the density of water. The result is that the volume of solution (in liters) and the mass of the solvent (in kilograms) are just about equal numerically, so the molality and molarity are very close in value. As the solution becomes more concentrated, the density of the solution increases. The result is that a liter of solution contains more solute and less solvent. Since the density is greater than 1.0 g/mL, the total mass of the solution is greater than the total volume of the solution. This is only possible if the amount of water is less than 1 kilogram. As the density of the solution increases the amount of water in kilograms needed to make 1.0 liter of solution must decrease. The difference between molarity and molality increases as a solution becomes more concentrated.

    Let's consider a 20.0% NaCl solution that has a density of 1.15 g/mL. One liter of this solution has a mass of 1150 g (1.15 g/mL × 1000 mL). The mass of NaCl can be determined by finding 20.0% of 1150 g, which is equal to 230 g (0.20 × 1150 g = 230 g). Now we can compare the molarity and molality of this solution. We start by calculating the moles of NaCl

$$\text{moles of NaCl} = 230\text{ g NaCl} \times \frac{1\text{ mol NaCl}}{58.5\text{ g NaCl}} = 3.93\text{ mol NaCl}$$

    The molarity of a 20% NaCl solution is 3.93 *M* (3.93 mol/1 L).

    The kilograms of solvent can be calculated as follows:

$$\text{mass of solution} = \text{mass of solute} + \text{mass of solvent}$$

    We know the mass of solution (1150 g) and the mass of solute, NaCl (230 g).

$$\text{the mass of solvent} = \text{mass of solution} - \text{mass of solute} = 1150\text{ g} - 230\text{ g} = 920\text{ g}$$

    The mass of solvent in kilograms is 0.92 (920 g/1000).

    The molality of a 20% NaCl solution is 4.27 *m* (3.93 mol/0.92 kg).

    The molarity and molality of a 20.0% NaCl solution differ significantly.

47. The change in the freezing point of a solution can be calculated using the formula $\Delta T_f = iK_f m$. The greater the total particle concentration (van't Hoff factor × molality, $i \times m$), the more the freezing point is lowered.

    In this scenario, each of the three aqueous solutions listed has the same molality. The only factor that allows each of the three solutions to have different freezing points is the number of dissolved particles each mole of solute produces (the values of their van't Hoff factors). The greater the van't Hoff factor, the more the freezing point is lowered and the lower the freezing point of the solution.

    Glucose, $C_6H_{12}O_6$, is a nonelectrolyte. It has a van't Hoff factor, i, of 1.

    Sodium chloride is a soluble ionic substance that theoretically should completely dissociate to form Na^+ and Cl^- ions. This makes NaCl a strong electrolyte. It should have a van't Hoff factor, i, of close to 2.

    Calcium chloride, $CaCl_2$, is a soluble ionic substance that theoretically should completely dissociate to Ca^{2+} and two Cl^- ions per formula unit. It should have a van't Hoff factor close to 3.

    In terms of total particles, the $CaCl_2$ has the greatest total solute concentration and the glucose solution has the lowest value.

    The greater the total solute molality the higher the boiling point and the lower the freezing point. For this reason $CaCl_2$ has the lowest freezing point and the glucose the highest.

48. The change in the boiling point of a solution can be calculated using the formula $\Delta T_b = iK_b m$. The greater the total particle concentration (van't Hoff factor × molality, $i \times m$), the more the boiling point is increased.

    In this scenario, each of the three aqueous solutions listed has the same molality. The only factor that allows each of the three solutions to have different boiling points is the value of the van't Hoff factor. The greater the van't Hoff factor, the more the boiling point is raised and the higher the boiling point of the solution.

    The $CaCl_2$ solution has the highest boiling point and the glucose solution has the lowest.

    See the answer to Problem 47 for an explanation.

49. Seawater contains dissolved salts. This lowers the freezing point of the solution.

50. The total solute concentration in the regular nondiet soft drink must be greater than that of the diet drinks.

    The change in the freezing point of a solution can be calculated using the formula $\Delta T_f = iK_f m$. The greater the total particle concentration (van't Hoff factor × molality, $i \times m$), the more the freezing point is lowered.

51. Osmosis is the movement of solvent molecules through a semipermeable membrane from the region of lower solute concentration to the region of higher solute concentration. This allows the molarity in the two solutions to eventually equalize. As the solvent molecules

leave the less concentrated solution, the volume of the solution decreases and the molarity increases, because no solute molecules pass through the semipermeable membrane. In the other compartment, the opposite is true. Since solvent molecules pass in, the volume of the solution increases causing the molarity to decrease.

Remember osmosis is a colligative property. This means that the driving force is dependent on the total molarity of solute particles. For an ionic species the molarity we need to use is the total ion molarity (also called the colligative molarity, or osmolarity).

$$\text{total ion molarity} = M \times i$$

where i (van't Hoff factor) is the number of particles formed when the ionic species forms individual ions.

Assume an idealized value of the van't Hoff factor for each ionic solute.

a. When NaCl is dissolved in water, it forms Na^+ and Cl^- ions. For NaCl the van't Hoff factor, i, = 2.

When KCl is dissolved in water, it has a van't Hoff factor of 2 (K^+ and Cl^- ions).

The total ion molarity of NaCl = 1.25 $M \times 2$ = 2.50 M.

The total ion molarity of KCl = 1.50 $M \times 2$ = 3.00 M.

The NaCl compartment has the lower total ion molarity, so solvent (water) molecules flow from the NaCl solution to the KCl solution.

b. $CaCl_2$ $\quad i = 3$ $\quad$ (1 Ca^{2+} ion and 2 Cl^- ions)
NaBr $\quad i = 2$ $\quad$ (1 Na^+ ion and 1 Br^- ion)

The total ion molarity of $CaCl_2$ = 3.45 $M \times 3$ = 10.35 M.

The total ion molarity of NaBr = 3.45$M \times 2$ = 6.90 M.

The NaBr compartment has the lower total ion molarity, so solvent (water) molecules flow from the NaBr solution to the $CaCl_2$ solution.

c. Dextrose (a molecular substance) forms no ions, so $i = 1$.

For NaCl, $i = 2$ (1 Na^+ ion and 1 Cl^- ion).

The molarity of dextrose = 4.68 M.

The total ion molarity of NaCl = 3.00 × 2 = 6.00 M.

The dextrose compartment has the lower concentration, so solvent (water) molecules move from the dextrose solution to the NaCl solution.

53. Osmotic pressure is calculated using the formula $\pi = MRT$, where M is the total ion concentration, R is a constant having a value of 0.0821 L • atm/mol • K, and T is the temperature in degrees Kelvin.

The temperature for all of the solutions is 20°C, or 293 K.

Assume an idealized value of the van't Hoff factor for any ionic solute.

a. Methanol, CH_3OH, is a nonelectrolyte. It has a van't Hoff factor, i, equal to 1.

$$\pi = (2.39\ M)(0.0821\ \text{L} \bullet \text{atm/mol} \bullet K)(293\ \text{K}) = 57.5\ \text{atm}$$

b. 9.45 mM $MgCl_2$
$MgCl_2$ forms 1 Mg^{2+} and 2 Cl^- ions, so $i = 3$.

We need to convert mM to M (1 M = 1000 mM).

$$9.45\ \text{m}M = 0.00945\ M$$

$$\pi = (0.00945\ M \times 3) \times (0.0821\ \text{L} \bullet \text{atm/mol} \bullet \text{K})(293\ \text{K}) = 0.682\ \text{atm}$$

c. 40.0 mL glycerol in 250 mL of solution.

We are given the volume of glycerol and the density of pure glycerol. From the volume and density values we can calculate the mass of glycerol present in the solution.

$$\text{mass} = d \times V = \frac{1.265\ \text{g}\ C_3H_8O_3}{1\ \text{mL}} \times 40.0\ \text{mL} = 50.6\ \text{g}\ C_3H_8O_3$$

Now we can calculate the molarity of glycerol, $C_3H_8O_3$, in the solution.

$$\text{molarity of } C_3H_8O_3 = \frac{50.6\ \text{g}\ C_3H_8O_3}{250\ \text{mL}} \times \frac{1000\ \text{mL}}{1\ \text{L}} \times \frac{1\ \text{mol}\ C_3H_8O_3}{92.0\ \text{g}\ C_3H_8O_3} = 2.20\ M\ C_3H_8O_3$$

Lastly, we can calculate the osmotic pressure of this solution.

Glycerol, $C_3H_8O_3$, is not an ionic compound, so $i = 1$.

$\pi = (2.20M \times 1)(0.0821\ \text{L} \bullet \text{atm/mol} \bullet \text{K})$
(293 K) = 52.9 atm

d. 25 g $CaCl_2$ in 350 mL of solution

$$\text{molarity of } CaCl_2 = \frac{25\ \text{g}\ CaCl_2}{350\ \text{mL}} \times \frac{1000\ \text{mL}}{1\ \text{L}} \times \frac{1\ \text{mol}\ CaCl_2}{111\ \text{g}\ CaCl_2} = 0.64\ M\ CaCl_2$$

$CaCl_2$ forms 1 Ca^{2+} and 2 Cl^- ions in solution, so $i = 3$.

$\pi = (0.64\ M \times 3)(0.0821\ \text{L} \bullet \text{atm/mol} \bullet \text{K})$
(293 K) = 46. atm

55. In this problem, we are asked to calculate the molarity of a particular solution, given the osmotic pressure and the temperature (in kelvins). Remember the calculated

molarity for an electrolyte is the total ion molarity and not the actual molarity of the solute described in each part. The experimentally determined value of the van't Hoff factor (i) is given for each ionic solute.

The formula for osmotic pressure is $\pi = MRT$.

Solving for total ion molarity, we get

$$M = \frac{\pi}{RT}$$

To determine solute molarity from total ion molarity (for an electrolyte) use the formula

$$\text{total ion molarity} = (\text{solute molarity})(i)$$

$$\text{solute molarity} = \frac{\text{total ion molarity}}{i}$$

a. $\pi = 0.674$ $T = 25°\text{C} = 298\text{ K}$
$R = 0.0821\text{ L} \bullet \text{atm/mol} \bullet \text{K}$

Ethanol, C_2H_5OH, is a nonelectrolyte, so the value of $i = 1$. For a nonelectrolyte the molarity calculated is the same as the solute molarity.

$$\text{molarity} = \frac{0.674\text{ atm}}{0.0821\text{ L} \bullet \text{atm/mol} \bullet \text{K}(298\text{ K})} = 0.0275M$$

The molarity of ethanol in this solution is $0.0275M$.

b. $\pi = 0.0271$ atm $T = 298$ K
Aspirin, $C_9H_8O_4$, is a nonelectrolyte. The value of $i = 1$, so the molarity calculated is the solute molarity.

$$\text{molarity} = \frac{0.0271\text{ atm}}{(0.0821\text{ L} \bullet \text{atm/mol} \bullet \text{K})(298\text{ K})} = 0.00111\ M$$

The molarity of aspirin in this solution is $0.00111\ M$.

c. $\pi = 0.605$ atm $T = 298$ K
$i = 2.47$

$CaCl_2$ is an electrolyte. The molarity calculated below is the total ion molarity.

$$\text{total ion molarity} = \frac{0.605\text{ atm}}{(0.0821\text{ L} \bullet \text{atm/mol} \bullet \text{K})(298\text{ K})} = 0.0247M$$

We now convert from total ion molarity to molarity of calcium chloride, $CaCl_2$

$$\text{solute molarity} = \frac{\text{total ion molarity}}{i} = \frac{0.0247M}{2.47} = 0.0100\ M\ CaCl_2$$

The molarity of $CaCl_2$ in this solution is $0.0100\ M$.

57. Before we can determine the osmotic pressure of this solution, we need to determine the molarity of salt in the final solution.

The original solution of 100 mL contained 0.92% NaCl by weight. Because this solution is rather dilute we can assume that the density of the solution is very close to 1.00 g/mL. (Since the actual density is about 1.007 g/mL, this is a reasonable assumption.)

The mass of 100 mL is about 100 g. The mass of NaCl in the solution is 0.92% of 100 g, or 0.92 g.

Next we can calculate the number of moles of NaCl present in the original solution. The molar mass of NaCl is 58.5 g/mol.

$$\text{moles of NaCl} = 0.92\text{ g NaCl} \times \frac{1\text{ mol NaCl}}{58.5\text{ g NaCl}} = 0.0157\text{ mol NaCl}$$

Adding 250 mL of pure water to the original solution does not change the number of moles of NaCl present. The volume of the final solution is 350 mL (100 mL + 250 mL).

The molarity of NaCl in the final solution is

$$\frac{0.0157\text{ mol NaCl}}{0.350\text{ L}} = 0.0449\ M$$

NaCl is an ionic substance; we need to find the total ion molarity before determining the osmotic pressure of the solution. NaCl forms 2 ions (Na^+ and Cl^-). The theoretical value of the van't Hoff factor (i) for NaCl is 2.

$$\text{total ion molarity} = 0.0449 \times 2 = 0.0898\ M$$

We can now calculate the osmotic pressure. The temperature in kelvins is 310 K (37 K + 273 K).

$\pi = (0.0898\ M)(0.0821\text{ L} \bullet \text{atm/mol} \bullet \text{K})$
$(310\text{ K}) = 2.29\text{ atm}$

59. The formula for molality is $m = \frac{\text{moles of solute}}{\text{kilograms of solvent}}$
This question simply asks us to substitute numbers into this equation.

a. moles of glucose (solute) = 0.875 mol
kilograms water (solvent) = 1.5 kg

$$\text{molality} = \frac{0.875\text{ mol glucose}}{1.5\text{ kg}} = 0.58\ m$$

b. millimoles of vinegar (solute) = 11.5 mmol
grams of water (solvent) = 65 g

Before substituting this data into the molality formula we need to convert these quantities to the correct units, using the following conversion factors
1 mole = 1000 mmole   1 kg = 1000 g
11.5 mmole = 0.0115 mole   65 g = 0.065 kg

$$\text{molality} = \frac{0.0115\text{ mol vinegar}}{0.065\text{ kg}} = 0.18\ m$$

An alternate method allows us to calculate the molality using the units given. Molality is defined as moles of solute per kilogram of solvent. If we divide each of these units by 1000 we should get the same value. That is, molality can be calculated by dividing the number of millimoles (1/1000 of a mole) by g (1/1000 of a kilogram). We can restate the formula for molality as

$$\text{molality} = \frac{\text{mmol of solute}}{\text{g of solvent}}$$

$$\text{molality} = \frac{11.5\ \text{mmol vinegar}}{65\ \text{g}} = 0.18\,m$$

Note that although we started with different units, we arrive at the same numerical value and units for the answer.

c. amount of baking soda, $NaHCO_3$ (solute) = 0.325 mol
mass of water (solvent) = 290 g

In this case, we need to convert the mass of water in grams to kilograms before we can substitute into the formula.

290 g water = 0.290 kg water

$$\text{molality} = \frac{0.325\ \text{mol baking soda}}{0.290\ \text{kg}} = 1.12\,m$$

61. This question is a little tricky in that it asks for the mass of solution. From the information given (molality and moles of solute) we can calculate the kilograms of solvent. To find the mass of solution we need to add together the mass of the solute and the mass of solvent. To find the mass of solute, we need to convert the moles of solute to grams of solute (grams = moles × molar mass).

a. moles of solute (NH_4NO_3) = 0.100; molality = 0.334 mol/kg

We can find the mass of solvent in kilograms from the relation

$$\text{kg solvent} = \frac{\text{moles of solute}}{\text{molality}}$$

$$\text{mass of solvent} = \frac{0.100\ \text{mol}\ NH_4NO_3}{\dfrac{0.334\ \text{mol}\ NH_4NO_3}{\text{kg}}} = 0.299\ \text{kg}$$

This solution contains 299 g of solvent.

To find the mass of solution, we need to calculate the mass of the solute, NH_4NO_3 (mass solute = moles × molar mass).

$$\text{mass of}\ NH_4NO_3 = 0.100\ \text{mol}\ NH_4NO_3 \times \frac{80.0\ \text{g}\ NH_4NO_3}{1\ \text{mol}\ NH_4NO_3} = 8.00\ \text{g}\ NH_4NO_3$$

The mass of solution is 299 g solvent + 8 g solute = 307 g of solution.

b. moles of solute ($C_2H_6O_2$) = 0.100 mol; molality = 1.24 mol/kg

$$\text{mass of solvent} = \frac{0.100\ \text{mol}\ C_2H_6O_2}{\dfrac{1.24\ \text{mol}\ C_2H_6O_2}{\text{kg}}} = 0.0806\ \text{kg}$$

This solution contains 80.6 g of solvent.

Next, find the mass of solute:

$$\text{mass of}\ C_2H_6O_2 = 0.100\ \text{mol}\ C_2H_6O_2 \times \frac{62.0\ \text{g}\ C_2H_6O_2}{1\ \text{mol}\ C_2H_6O_2} = 6.20\ \text{g}\ C_2H_6O_2$$

The mass of solution is 80.6 g solvent + 6.2 g solute = 86.8 g solution.

c. moles of solute ($CaCl_2$) = 0.100; molality = 5.65 mol/kg

$$\text{mass of solvent} = \frac{0.100\ \text{mol}\ CaCl_2}{\dfrac{5.65\ \text{mol}\ CaCl_2}{\text{kg}}} = 0.0177\ \text{kg}$$

The solution contains 17.7 g of solvent.

$$\text{mass of}\ CaCl_2 = 0.100\ \text{mol}\ CaCl_2 \times \frac{111\ \text{g}\ CaCl_2}{1\ \text{mol}\ CaCl_2} = 11.1\ \text{g}\ CaCl_2$$

The mass of solution is 17.7 g solvent + 11.1 g solute = 28.8 g of solution.

63. In this problem, we need to convert from the units mg/L to molality (moles/kg). This can be done in three steps:

Since concentration is given in mg/L, we need to first convert the mass in milligrams to grams and then convert this mass to number of moles.

Second, we need to determine the mass of water (solvent) present (mass of water = mass of solution − mass of solute).

Finally, we can find the molality by dividing the number of moles by the mass of water in kilograms.

Ammonia, NH_3:
Concentration = 1.1 mg/L
Molar mass NH_3 = 17.0 g/mol
Conversion factor: 1 g = 1000 mg

$$\text{moles of}\ NH_3 = 1.1\ \text{mg}\ NH_3 \times \frac{1.00\ \text{g}\ NH_3}{1000\ \text{mg}\ NH_3} \times \frac{1\ \text{mol}\ NH_3}{17.0\ \text{g}\ NH_3} = 6.5 \times 10^{-5}\ \text{mol}\ NH_3$$

The mass of solution is 1000 g because the concentration is given in mg/L and the density of the solution is 1.00 g/mL (1 L = 1000 mL = 1000 g).

$$\text{mass of water} = \text{mass of solution} - \text{mass of solute} = 1000\ \text{g} - 0.0011\ \text{g} = 1000\ \text{g, or}\ 1.00\ \text{kg}$$

The molality of NH_3 is

$$\frac{6.5 \times 10^{-5} \text{ mol } NH_3}{1 \text{ kg}} = 6.5 \times 10^{-5}\, m$$

Nitrite ion, NO_2^-:
Concentration = 0.40 mg/L
Molar mass NO_2^- = 46.0 g/mol

$$\text{moles of } NO_2^- = 0.40 \text{ mg } NO_2^- \times \frac{1.00 \text{ g } NO_2^-}{1000 \text{ mg } NO_2^-}$$

$$\times \frac{1 \text{ mol } NO_2^-}{46.0 \text{ g } NO_2^-} = 8.7 \times 10^{-6} \text{ mol } NO_2^-$$

As in part *a*, the mass of solution is 1000 g.

$$\text{mass of water} = \text{mass of solution} - \text{mass of solute}$$
$$= 1000 \text{ g} - 0.0004 \text{ g} = 1000 \text{ g},$$
$$\text{or } 1.00 \text{ kg}$$

The molality of NO_2^- is

$$\frac{8.7 \times 10^{-6} \text{ mol } NO_2^-}{1 \text{ kg}} = 8.7 \times 10^{-6}\, m$$

Nitrate ion, NO_3^-:
Concentration = 1360 mg/L
Molar mass NO_3^- = 62.0 g/mol

$$\text{moles of } NO_3^- = 1360 \text{ mg } NO_3^- \times \frac{1.00 \text{ g } NO_3^-}{1000 \text{ mg } NO_3^-}$$

$$\times \frac{1 \text{ mol } NO_3^-}{62.0 \text{ g } NO_3^-} = 0.02190 \text{ mol } NO_3^-$$

as in part *a*, the mass of solution is 1000 g.

$$\text{mass of water} = \text{mass of solution} - \text{mass of solute}$$
$$= 1000 \text{ g} - 1.36 \text{ g}$$
$$= 998.64 \text{ g, or } 0.99864 \text{ kg}$$

The molality of NO_3^- is

$$\frac{0.0219 \text{ mol } NO_3^-}{0.99864 \text{ kg}} = 0.0220\, m$$

65. From the information given, we can determine the change in the boiling point of this solution, using the formula $\Delta T_b = k_b m$.

Information given:
The mass of solute (cinnamaldehyde, C_9H_8O) is 100 mg (0.100 g).
The mass of solvent (carbon tetrachloride) is 1.00 g (0.00100 kg).
The boiling point of pure carbon tetrachloride is 76.8°C.
K_b for carbon tetrachloride is 5.02 °C/m.
From the information given, we can calculate the molality of the cinnamaldehyde solute.
Molar mass of cinnamaldehyde, C_9H_8O = 132 g/mol (9 C × 12.0 + 8 H × 1.0 + 1 O × 16.0)

$$\text{molality of } C_9H_8O = \frac{0.100 \text{ g } C_9H_8O \times \dfrac{1 \text{ mol } C_9H_8O}{132 \text{ g } C_9H_8O}}{0.00100 \text{ kg}}$$

$$= 0.758\, m \; C_9H_8O$$

Cinnamaldehyde, C_9H_8O, is a nonelectrolyte because it is a covalent substance that does not produce any ions. For a nonelectrolyte the van't Hoff factor is 1.
Now we can calculate the boiling point elevation:

$$\Delta T_b = (K_b)(m) = (5.02°\text{C}/m)(0.758\, m) = 3.81°\text{C}$$

The boiling point is increased by 3.81°C above its normal boiling point of 76.8°C.

67. To solve this problem we can use the freezing point depression formula.

$$\Delta T_f = K_f m$$

$$\Delta T_f = 1.000°\text{C};\ K_f = 39.7\ °\text{C}/m$$

Since we know both ΔT_f and K_f, we can solve for the molality of the solution. The solute is a nonelectrolyte, so the van't Hoff factor is 1.
Solving for the molality

$$m = \frac{\Delta T_f}{K_f} = \frac{1.000°\text{C}}{39.7\ °\text{C}/m} = 0.252\, m$$

The molality of the unknown nonelectrolyte solution is 0.252 *m*.

69. Information given:
Mass of solute (saccharin, $C_7H_5O_3NS$) is 186 mg (0.186 g).
Volume of solvent = 1.00 mL

Element	***Milligrams***	***Grams***	***At. Wt.***	***Moles***	***Molality***
Al (aluminum)	0.050	0.000050	27.0	1.9×10^{-6}	1.9×10^{-6}
Fe (iron)	0.040	0.000040	55.85	7.2×10^{-7}	7.2×10^{-7}
Ca (calcium)	13.4	0.0134	40.1	3.34×10^{-4}	3.34×10^{-4}
Na (sodium)	5.2	0.0052	23.0	2.3×10^{-4}	2.3×10^{-4}
K (potassium)	1.3	0.0013	39.1	3.3×10^{-5}	3.3×10^{-5}
Mg (magnesium)	3.4	0.0034	24.31	1.4×10^{-4}	1.4×10^{-4}

Density of solvent = 1.00 g/mL
$K_f = 1.86\ °\text{C}/m$
From the information given, we know that the mass of solvent is

$$1.00\ \text{mL} \times \frac{1.00\ \text{g}}{\text{mL}} = 1.00\ \text{g}$$

Saccharin is a molecular substance; thus it is a nonelectrolye. The van't Hoff factor for saccharin is 1.

$$\Delta T_f = K_f m$$

We are asked to find the melting point of the solution. (The freezing point and melting point refer to the same temperature.)

From the information given, we can calculate the molality of saccharin in this solution.

The molar mass of saccharin, $C_7H_5O_3NS$, is 183.1g/mol.

(7 C × 12.0 + 5 H × 1.0 + 3 O × 16.0 + 1 N × 14.0 + 1 S × 32.1)

$$m = 0.186\ \text{g saccharin} \times \frac{1\ \text{mol saccharin}}{183.1\ \text{g saccharin}}$$

$$\times \frac{1}{0.00100\ \text{kg}} = 1.02\,m$$

Now we can determine the freezing point depression.

$$\Delta T_f = K_f m = (1.86\ °\text{C}/m)(1.02\,m) = 1.90°\text{C}$$

The melting point (freezing point) of this solution has been lowered by 1.90°C to −1.90°C.

71. When a solute is dissolved in water (or any other solvent) the net effect is an increase in the boiling point of the solvent. The change in boiling point can be calculated using the boiling point elevation formula

$$\Delta T_b = K_b m i$$

where m refers to the solute molality and i is the van't Hoff factor. The product of the molality and the van't Hoff factor ($m \times i$) is the total particle molality.

We can assume an idealized value of the van't Hoff factor (that is, 100% dissociation of ions) if it is not given in the problem.
For CH_3OH, $i = 1$ (CH_3OH is a nonelectrolyte)
For KCl, $i = 2$ (1 K^+ and 1 Cl^-)
For $Ca(NO_3)_2$, $i = 3$ (1 Ca^{2+} and 2 NO_3^-)
K_b is the same for each solution (they are all aqueous solutions and K_b is a function of the solvent not the solute). The solution with largest total particle molality ($m \times i$) will produce the greatest effect (the largest increase in boiling point temperature) and the highest boiling point.
Total particle molalities:
CH_3OH, $0.0200\ m \times 1 = 0.0200\ m$
KCl, $0.0125\ m \times 2 = 0.0250\ m$
$Ca(NO_3)_2$, $0.0100\ m \times 3 = 0.0300\ m$

The $Ca(NO_3)_2$ solution is the solution with the greatest total particle molality and should have the highest boiling point. The CH_3OH solution should have the lowest boiling point.

73. This problem is similar to problems 71 and 72, except the experimentally determined van't Hoff factor is given. Use the experimentally determined values of the van't Hoff factor to determine the total particle molality.
Total particle molalities:
$FeCl_3$, $0.06\ m \times 3.4 = 0.204\ m$
$MgCl_2$, $0.10\ m \times 2.7 = 0.27\ m$
KCl, $0.20\ m \times 1.9 = 0.38\ m$
The order of total particle molalities is $FeCl_3 < MgCl_2 <$ KCl.

From the boiling point elevation formula, $\Delta T_b = K_b m i$, we see that the change in boiling point is directly proportional to the total particle molality ($m \times i$). The solution with the lowest total particle molality will have the lowest elevation of the boiling point and the solution with the highest total particle molality will have the highest increase in boiling point.

The order of increasing boiling points is $FeCl_3 < MgCl_2 <$ KCl.

75. From the information given, we can calculate the freezing point depression for a saturated solution of $CaCl_2$. The solvent is water, which normally freezes at 0°C. If the freezing point depression is greater than 20°C, a saturated solution of $CaCl_2$ will lower the freezing point of ice to a temperature less than –20°C and $CaCl_2$ will melt the ice.
Information given:
The solubility of $CaCl_2$ is 70. g $CaCl_2$ per 100. mL of water.
The value of i (the van't Hoff factor) for $CaCl_2$ is 2.5.
K_f for water is 1.86 °C/m.
The freezing point depression can be calculated using $\Delta T_f = K_f m i$.

The molality of $CaCl_2$ can be calculated using the solubility information.

The molality of the $CaCl_2$ solution is number of moles of $CaCl_2$ per kilogram of water. We can calculate this quantity as follows: Molar mass of $CaCl_2$ is 111 g/mol; mass of H_2O is 100 g (0.1 00 kg).

$$\text{molality } CaCl_2 = \frac{\text{mol } CaCl_2}{\text{kg water}} = 70\ \text{g } CaCl_2$$

$$\times \frac{1\ \text{mol } CaCl_2}{111\ \text{g } CaCl_2} \times \frac{1}{0.100\ \text{kg}}$$

$$= 6.3\,m\ CaCl_2$$

Now we can calculate the freezing point depression of this solution.

$$\Delta T_f = K_f m i = (1.86\ °\text{C}/m)(6.3\,m)(2.5) = 29.3°\text{C}$$

The freezing point of this solution is −29.3°C. This is lower than −20°C, so $CaCl_2$ will melt ice at this temperature.

77. The easiest way to determine the van't Hoff factor is to include it in the freezing point depression formula as follows:

$$\Delta T_f = K_f m i$$

Since we are given the values of ΔT_f, K_f, and m for these solutions it becomes a simple problem to solve for the van't Hoff factor.

(*Note:* The problem does not gives us the K_f value for water, but it can be found in the text.)

K_f for water is 1.86 °C/m.

For the ammonium chloride solution:

$K_f = 1.86\,°C/m$; $\Delta T_f = 0.332°C$; $[NH_4Cl] = 0.0935\,m$

Note: The freezing point of pure water is 0°C, so the freezing point of any aqueous solution is a negative temperature when measured on the Celsius scale.

Substituting and solving for i yields

$$i = \frac{\Delta T_f}{K_f m} = \frac{0.322°C}{(1.86\,°C/m)(0.935\,m)} = 1.85$$

The value of the van't Hoff factor in this ammonium chloride solution is 1.85.

For the ammonium sulfate, $(NH_4)_2SO_4$, solution

$K_f = 1.86\,°C/m$; $\Delta T_f = 0.173°C$; $m = 0.0378\,m$

Substituting and solving for i yields

$$i = \frac{\Delta T_f}{K_f m} = \frac{0.173°C}{(1.86\,°C/m)(0.0378\,m)} = 2.46$$

The value of the van't Hoff parameter in this ammonium sulfate solution is 2.46.

79. Information given:

Mass of solute = 188 mg.

Volume of solution = 10.0 mL.

The osmotic pressure, $\pi = 4.89$ atm.

The temperature is 25°C, or 298 K.

We are asked to find the molar mass of the nonelectrolyte. A nonelectrolyte has a van't Hoff factor, i, of 1.

From the information given we can use the osmotic pressure formula, $\pi = MRT$, to find the molarity of the solution.

For a nonelectrolyte, the van't Hoff factor, i, is 1, we do not need to include it in the calculation. That is, for a nonelectrolyte, molarity of the solute is the same as the total particle molarity.

$$M = \frac{\pi}{RT} = \frac{4.89\text{ atm}}{(0.0821\text{ L} \bullet \text{atm/mol} \bullet \text{K})(298\text{ K})}$$

$$= 0.200M$$

We need to determine the molar mass of the solute. If we go back to the definition of molarity we find that

$$\text{molarity of solute} = \frac{\text{moles of solute}}{\text{liters of solution}}$$

We know both the concentration (0.200 M) and the volume of solution (10 mL = 0.0100 L), so we can calulate the number of moles of solute.

$$\text{moles of solute} = \text{molarity} \times \text{volume}$$

$$= \frac{0.200\text{ mol}}{\text{L}} \times 0.0100\text{ L}$$

$$= 2.00 \times 10^{-3}\text{ mol}$$

Using the definition of the mole

$$\text{moles of solute} = \frac{\text{mass of solute}}{\text{molar mass of solute}}$$

We can calculate the molar mass of the solute because we know the mass of solute (188 mg = 0.188 g) and the number of moles of solute (2.00×10^{-3} mol).

$$\text{molar mass} = \frac{\text{mass of solute}}{\text{moles of solute}} = \frac{0.188\text{ g}}{2.00 \times 10^{-3}\text{ mol}}$$

$$= 94.0\text{ g/mol}$$

81. Information given:

Mass of solute (eugenol) = 111 mg (0.111 g)

Mass of chloroform (solvent) = 1.00 g (0.001 kg)

K_b (for chloroform) = 3.63 °C/m

Boiling point elevation = 2.45°C

The solute, eugenol, is a nonelectrolyte, so $i = 1$

Let's start by writing down the boiling point elevation formula, $\Delta T_b = K_b m$.

From the information given, we know ΔT_b (boiling point elevation) and K_b (the boiling point elevation constant for chloroform). We can solve for the molality of eugenol in this solution.

$$m = \frac{\Delta T_b}{K_b} = \frac{2.45°C}{3.63\,°C/m} = 0.675\,m$$

Next we can use the definition of molality to find the number of moles of eugenol present in the solution.

$$m = \frac{\text{moles of eugenol}}{\text{kg chloroform}}$$

We know the molality and the mass of chloroform so we can solve for the number of moles of eugenol.

$$\text{molality} \times \text{kilogram of chloroform}$$

$$= (0.675\text{ mol/kg})(0.001\text{ kg})$$

$$= 0.000675\text{ mol eugenol}$$

Lastly, using the molar mass formula we can determine the molar mass of eugenol.

$$\text{molar mass eugenol} = \frac{\text{g eugenol}}{\text{moles eugenol}}$$

$$= \frac{0.111\text{ g}}{0.000675\text{ mol}} = 164\text{ g/mol}$$

The next part of this question asks us to determine the molecular formula of eugenol.

If we assume that we have a 100-g sample of eugenol, then it contains 73.17 g C, 7.32 g H, and 19.51 g O.

The number of moles of each element present in the sample is

$$\text{mol C} = 73.17 \text{ g C} \times \frac{1 \text{ mol C}}{12 \text{ g C}} = 6.10 \text{ mol C}$$

$$\text{mol H} = 7.32 \text{ g H} \times \frac{1 \text{ mol H}}{1.007 \text{ g H}} = 7.27 \text{ mol H}$$

$$\text{mol O} = 19.51 \text{ g O} \times \frac{1 \text{ mol O}}{16.0 \text{ g O}} = 1.22 \text{ mol O}$$

Next, we divide each number of moles by the smallest value (1.22) to get a whole number ratio of the three elements. That ratio is 5 C : 6 H : 1 O.

The empirical formula of eugenol is C_5H_6O.

The molar mass of eugenol is 164 g/mol. The empirical formula mass of C_5H_6O is 82 g/mol. The ratio of the masses is 2 (164/82), so the molecular formula of eugenol is $C_{10}H_{12}O_2$.

83. A half-reaction is an equation that describes either the oxidation or the reduction part of a redox reaction. The main benefit of a half-reaction is that it includes the electrons either gained or lost in the oxidation or reduction part of the process.

84. reduction (gain of electrons); oxidation (loss of electrons)

85. The number of electrons gained by an atom or an ion when it is reduced corresponds to the numerical decrease in the oxidation number of that atom. For example, when Cr^{6+} is reduced to Cr^{3+}, a chromium ion with a + 6 charge gains three electrons; its oxidation number decreases by 3 units.

    The number of electrons lost by an atom or ion when it is oxidized corresponds to the numerical increase in oxidation number on that species.

86. Molten NaCl is the liquid form of NaCl. The only species present are sodium ions (Na^+) and chloride ions (Cl^-). In electrolysis, energy is used to power a redox reaction that would not occur spontaneously. In this case, sodium ions are converted into sodium atoms and chloride ions are converted into chlorine gas (Cl_2).

    The two half-reactions are

$$Na^+ + e^- \rightarrow Na(s) \qquad \text{(reduction)}$$

$$2\,Cl^- \rightarrow Cl_2(g) + 2\,e^- \qquad \text{(oxidation)}$$

87. The sum of the oxidation numbers on all the atoms in a species must add up to the overall charge on the species.

    Since a molecule is a neutral species, the sum of the oxidation numbers must be zero.

88. If the species is ionic, then the sum of the oxidation numbers must equal the charge on the ion.

OH^-	-1
NH_4^+	$+1$
SO_4^{2-}	-2
PO_4^{3-}	-3

SPECIAL TIPS: GENERAL RULES FOR DETERMINING OXIDATION NUMBERS

The rules for determining oxidation number need to be used in the order given below; that is, rule 4 takes precedence over rule 5, etc.

1. All pure elements have an oxidation number of zero; that is, any neutral form of any element, whether it is atomic or molecule has an oxidation number of zero. Some examples are atomic oxygen, O, diatomic oxygen, O_2, and ozone, O_3. All have oxygen atoms with an oxidation number of zero.
2. In compounds, the Group 1A elements have an oxidation number of +1 and the Group 2A elements have an oxidation number of +2.
3. The oxidation number of any monoatomic ion is the same as the charge on the ion.
4. The oxidation of fluorine in any compound is always −1.
5. The oxidation number of oxygen is −2 and the oxidation number of hydrogen is usually +1, unless rules 2–4 require otherwise.
6. The sum of the oxidation numbers on all of the atoms in the species must add up to the overall charge on the species. This sum is zero if the substance is neutral and equals the charge on the ion if it is ionic.
7. Determine the oxidation number of all other atoms in the substance, after assigning oxidation numbers using rules 1–6.

As an example of how to apply these rules, let's look at the polyatomic sulfate ion (SO_4^{2-}), whose net charge is −2. The oxidation number associated with the sulfur atom in the sulfate ion is +6. This is determined as follows:

The oxidation number of the sulfur atom + the oxidation numbers of the four oxygen atoms = −2.

The oxidation number on each oxygen atom must be −2 (since rules 1–4 do not apply), so

$$S + 4(-2) = -2$$
$$S - 8 = -2$$
$$S = -2 + 8$$
$$S = +6$$

Note: Oxidation numbers do not necessarily need to be whole numbers; they can be fractional in some cases.

89. a. Hypochlorous acid, HClO
This is a neutral species. The oxidation numbers on the hydrogen atom, chlorine atom, and oxygen atom must add up to zero.

$$\mathrm{H} + \mathrm{Cl} + \mathrm{O} = 0$$

Hydrogen has an oxidation number of +1 and oxygen has an oxidation number of −2 (rules 2–4 do not apply).

$$+1 + \mathrm{Cl} + (-2) = 0$$
$$\mathrm{Cl} = 2 - 1 = +1$$

The oxidation number for hydrogen is +1, for chlorine is +1, and for oxygen is −2.

b. Chloric acid, $HClO_3$

$$\mathrm{H} = +1 \text{ and } \mathrm{O} = -2 \text{ (rules 2–4 do not apply)}$$
$$\mathrm{H} + \mathrm{Cl} + 3\,\mathrm{O} = 0$$
$$+1 + \mathrm{Cl} + 3(-2) = 0$$
$$+1 + \mathrm{Cl} + (-6) = 0$$
$$\mathrm{Cl} = 6 - 1 = +5$$

The oxidation number for hydrogen is +1, for chlorine is +5, and for oxygen is −2.

c. Perchlorate ion, ClO_4^-
This is an ionic species. The sum of the oxidation numbers must add up to the charge on the ion (−1).

$$\mathrm{Cl} + 4\,\mathrm{O} = -1$$

Each oxygen has an oxidation number of −2 (rules 2–4 do not apply).

$$\mathrm{Cl} + 4(-2) = -1$$
$$\mathrm{Cl} - 8 = -1$$
$$\mathrm{Cl} = 8 - 1 = +7$$

The oxidation number for chlorine is +7 and for oxygen is −2.

91. Oxidation can be defined and/or described in several different ways:
Loss of electrons
Electrons are found on the product side in a balanced half-reaction
A numerical increase in oxidation number.
Reduction can be defined and/or described in several different ways:
Gain of electrons
Electrons are found on the reactant side in a balanced half-reaction
A numerical decrease in oxidation number

There are two requirements in order to have a properly balanced half-reaction:
Atoms must be balanced.
Charges must be balanced (to balance charge add electrons to the side of the half-reaction with the more positive total charge)
In the half-reactions given in this problem, the atoms are already balanced, so all we need to do is to balance the overall charge for the given process. If we have to add electrons to the reactant side, the process is a reduction process. If the electrons are added to the product side, the process is an oxidation.

a. $Br_2(l) \rightarrow 2\,Br^-(aq)$
In this half-reaction the reactant (Br_2) is a neutral species; it has no charge. On the product side Br^- is an ionic species with a charge of −1. Since there are two bromide ions the total charge on the product side is −2, while there is a charge of zero on the reactant side.
To balance the charge, add electrons (species with a −1 charge) to the side of the half-reaction that has the more positive total charge. In this half-reaction, zero is more positive than −2, so add 2 electrons to the reactant side. The balanced half-reaction is

$$\mathrm{Br_2}(l) + 2\,\mathrm{e^-} \rightarrow 2\,\mathrm{Br^-}(aq)$$

This is a reduction.

b. $Pb(s) + 2\,Cl^-(aq) \rightarrow PbCl_2(s)$
The total charge on the reactant side is −2 (Pb is neutral; 2 chloride ions each with a charge of −1).
The total charge on the product side is 0 ($PbCl_2$ is a neutral compound).
The product side has the more positive charge (0 is more positive than −2), so add 2 electrons to the product side. The balanced half-reaction is

$$\mathrm{Pb}(s) + 2\,\mathrm{Cl^-}(aq) \rightarrow \mathrm{PbCl_2}(s) + 2\,\mathrm{e^-}$$

This is an oxidation.

c. $O_3(g) + 2\,H^+(aq) \rightarrow O_2 + H_2O(l)$
The total charge on the reactant side is +2 (O_3 is neutral; 2 H^+ ions, each with a charge of +1).
The total charge on the product side is zero (both O_2 and H_2O are neutral species).
The reactant side has the more positive charge (+2 vs. 0), so add 2 electrons to the reactant side. The balanced half-reaction is

$$\mathrm{O_3}(g) + 2\,\mathrm{H^+}(aq) + 2\,\mathrm{e^-} \rightarrow \mathrm{O_2}(g) + \mathrm{H_2O}(l)$$

This is a reduction.

d. $H_2S(g) \rightarrow S(s) + 2\,H^+(aq)$
The total charge on the reactant side is zero (H_2S is a neutral substance).
The total charge on the product side is +2 (S is neutral; 2 H^+ ions, each with a charge of +1).

The product side has the more positive charge so add 2 electrons to the product side. The balanced half-reaction is

$$H_2S(g) \rightarrow S(s) + 2\,H^+(aq) + 2\,e^-$$

This is an oxidation.

93. The unbalanced half-reaction is

$$Fe_3O_4(s) \rightarrow Fe_2O_3(s)$$

To balance an oxidation or a reduction half-reaction, use the following steps:

1. Balance the atoms. When balancing the atoms make sure you balance the hydrogen atoms last and the oxygen atoms next to last.
2. Determine the total charge on both sides.
3. Balance the charge by adding electrons to the side of the half-reaction with the more positive total charge.

In this question we need to first balance the iron atoms, then the oxygen atoms, and lastly the hydrogen atoms. Note there are currently no hydrogen atoms in this reaction. However, to balance oxygen atoms, we can add water molecules to either the reactant or product side as needed. Then to balance the hydrogen atoms we can add hydrogen ions to either side of the equation as needed.

To balance the iron atoms put the coefficient 2 in front of the Fe_3O_4 and the coefficient 3 in front of the Fe_2O_3. This will result in 6 iron atoms on both the reactant and product sides of the equation.

$$2\,Fe_3O_4(s) \rightarrow 3\,Fe_2O_3(s)$$

Next, balance the oxygen atoms. Currently, there are 8 oxygen atoms on the reactant side (from the 2 Fe_3O_4) and 9 oxygen atoms on the product side (from the 3 Fe_2O_3). To balance the oxygen atoms add 1 H_2O molecule to the reactant side. This will increase the number of oxygen atoms on the reactant side from 8 to 9.

$$2\,Fe_3O_4(s) + H_2O \rightarrow 3\,Fe_2O_3(s)$$

Lastly, balance the hydrogen atoms. Currently, there are 2 hydrogen atoms on the reactant side (from the single H_2O) and zero on the product side. To balance the hydrogen atoms, add 2 H^+ ions to the product side.

The atom-balanced half-reaction is

$$2\,Fe_3O_4(s) + H_2O(l) \rightarrow 3\,Fe_2O_3(s) + 2\,H^+(aq)$$

Next balance the total charge:

The total charge on the reactant side is zero (both Fe_3O_4 and H_2O are neutral compounds).

The total charge on the product side is +2 (Fe_2O_3 is neutral and there are 2 H^+ ions).

The product side has the more positive total charge (0 on the reactant side and +2 on the product side), so add 2 electrons to the product side. The balanced half-reaction is

$$2\,Fe_3O_4(s) + H_2O(l) \rightarrow 3\,Fe_2O_3(s) + 2\,H^+(aq) + 2\,e^-$$

Two electrons are lost in the oxidation process.

95. The method used to determine whether a reaction is an oxidation-reduction process is to determine the oxidation numbers (see Special Tips: Rules for Determining Oxidation Numbers before the answer to problem 89) of all the atoms in the reaction. If none of the atoms change their oxidation number in going from reactants to products, then the reaction is not an oxidation-reduction process. If elements do change their oxidation numbers then it is a redox process.

a. $3\,SiO_2 + 2\,Fe_3O_4 \rightarrow 3\,Fe_2SiO_4 + O_2$

Rule 1 applies only to O_2 since it is the only elemental form present here. The oxidation number for the oxygen atoms in O_2 is zero.

Rules 2–4 do not apply to any of the substances present in this reaction.

From rule 5, the oxidation number on oxygen in all compound forms (SiO_2, Fe_3O_4, and Fe_2SiO_4) in this reaction must be −2.

From rule 6, the sum of all the oxidation numbers for all the atoms in each substance must equal zero because they are all neutral substances.

Let's determine the oxidation numbers on the atoms in each of the remaining three substances in this reaction.

SiO_2

Rules 1–4 do not apply, so the oxidation number on each oxygen atom is −2.

$$1\text{ Si atom} + 2\text{ O atoms} = 0$$

$$1\,Si + 2(-2) = 0;\ 1\,Si - 4 = 0;\ Si = +4$$

In SiO_2 the oxidation numbers are Si = +4 and O = −2.

Fe_3O_4

Rules 1–4 do not apply, so the oxidation number on each oxygen atom is −2.

$$3\text{ Fe atoms} + 4\text{ O atoms} = 0$$

$$3\,Fe + 4(-2) = 0;\ 3\,Fe - 8 = 0;\ 3\,Fe = +8$$

$$Fe = 8/3 \text{ (notice the fractional oxidation number)}$$

In Fe_3O_4, the oxidation numbers are Fe = 8/3 and O = −2.

Fe_2SiO_4

Rules 1–4 do not apply, so the oxidation number of each oxygen atom is −2.

The problem here is that we do not know the oxidation numbers of two different elements (Fe and Si). The above rules provide no help in determining how to find the oxidation number either of iron or silicon. Once we know the oxidation of one of these two elements, we can then determine the other one.

Here is where knowing the charge on a polyatomic ion can be useful. From Chapter 4, we might recall that the silicate ion has a net charge of −4, that is, SiO_4^{4-}. From this we can determine the oxidation number on the silicon atom in this polyatomic ion species.

The oxidation number of each oxygen atom is –2 as previously stated.

$$1\ \text{Si atom} + 4\ \text{O atoms} = -4$$

$$1\ \text{Si} + 4(-2) = -4;\ 1\ \text{Si} - 8 = -4;\ 1\ \text{Si} = +4$$

The oxidation number on the Si atom is +4.

We can use this information to determine the oxidation number on the iron atom in Fe_2SiO_4.

$$2\ \text{Fe atoms} + 1\ \text{Si atom} + 4\ \text{O atoms} = 0$$

$$2\ \text{Fe} + 1(+4) + 4(2) = 0$$

$$2\ \text{Fe} + 4 - 8 = 0;\ 2\ \text{Fe} - 4 = 0;\ 2\ \text{Fe} = +4;$$

$$\text{Fe} = +2$$

In Fe_2SiO_4, the oxidation numbers are Fe = +2, Si = +4, and O = −2.

Summarizing the information for the reaction

$$3\ SiO_2 + 2\ Fe_3O_4 \rightarrow 3\ Fe_2SiO_4 + O_2$$

On the reactant side:

SiO_2: Si = +4; O = −2

Fe_3O_4: Fe = $+\frac{8}{3}$; O = −2

On the product side:

Fe_2SiO_4: Fe = +2; Si = +4; O = −2

O_2: O = 0

We see that during this reaction the oxidation number of oxygen in some of the atoms has changed from −2 to zero. This is a numerical increase in oxidation number so some of the oxygen atoms have been oxidized. Also the oxidation number on the Fe atoms has changed from 8/3 to +2. This is a numerical decrease in oxidation number, so the Fe atoms have been reduced.

This reaction is an oxidation-reduction process.

Fe is reduced and O is oxidized.

Note: We cannot determine which oxygen atoms have been oxidized (i.e., the oxygen atoms in SiO_2 or those in Fe_3O_4).

b. $SiO_2 + 2\ Fe + O_2 \rightarrow Fe_2SiO_4$

Since Fe and O_2 are elemental forms, the oxidation numbers are zero (rule 1).

In part *a*, we determined the oxidation numbers on the atoms in SiO_2 and Fe_2SiO_4.

The oxidation numbers are as follows:

SiO_2	Si = +4;	O = −2	(see part *a*)
Fe	Fe = 0		
O_2	O = 0		
Fe_2SiO_4	Fe = +2;	Si = +4; (see part *a*)	O = −2

Summarizing the information for the reaction

$$SiO_2 + 2\ Fe + O_2 \rightarrow Fe_2SiO_4$$

On the reactant side:

SiO_2: Si = +4; O = −2

Fe: Fe = 0

O_2: O = 0

On the product side:

Fe_2SiO_4: Fe = +2; Si = +4; O = −2

Since the oxidation numbers of Fe and O change, this is a redox process.

Since the oxidation number of Fe changes from 0 to +2, the iron atoms are oxidized.

There are four oxygen atoms on each side of this equation. On the reactant side two of the oxygen atoms have an oxidation state of −2 and the other two have an oxidation state of zero. On the product side, all of the oxygen atoms have an oxidation number of −2. In other words, only two of the oxygen atoms are reduced. Each oxygen atom gains 2 electrons. A total of 4 electrons are gained in the reduction process.

This is an oxidation-reduction reaction.

Fe is oxidized and O_2 is reduced.

c. $4\ FeO + O_2 + 6\ H_2O \rightarrow 4\ Fe(OH)_3$

Determine the oxidation numbers on each atom.

FeO

The oxidation number of the oxygen atom is −2.

1 Fe atom +1 O atom = 0

$$\text{Fe} + (-2) = 0$$

$$\text{Fe} - 2 = 0$$

$$\text{Fe} = +2$$

In FeO the oxidation numbers are Fe = +2 and O = −2.

H_2O

Each hydrogen atom has an oxidation number of +1 and the oxygen atom has an oxidation of −2.

O_2

This is elemental oxygen. According to rule 1, each oxygen atom has an oxidation number of zero.

$Fe(OH)_3$

Since the hydroxide ion has a charge of −1, we see that the oxidation number on the oxygen atoms must be -2 and the oxidation number on the hydrogen atom is $+1$.

$$\text{Fe atom} + 3\,\text{O atoms} + 3\,\text{H atoms} = 0$$

$$\text{Fe} + 3(-2) + 3(+1) = 0$$

$$\text{Fe} - 6 + 3 = 0$$

$$\text{Fe} - 3 = 0$$

$$\text{Fe} = +3$$

Alternatively, we could use the fact that since each hydroxide ion has a charge of -1, the total charge on 3 OH^- ions must be -3. That means that the charge (and thus the oxidation number) of the iron atom is $+3$.

Summarizing the information for the reaction

$$4\,FeO + O_2 + 6\,H_2O \rightarrow 4\,Fe(OH)_3$$

On the reactant side:

FeO	$Fe = +2;$	$O = -2$
O_2	$O = 0$	
H_2O	$H = +1;$	$O = -2$

On the product side:

$Fe(OH)_3$	$Fe = +3;$	$O = -2;$	$H = +1$

The oxidation number of Fe changes from $+2$ to $+3$, so Fe must be oxidized.

The oxidation number of the O atoms changes from zero to -2, so O must be reduced.

97. The total number of electrons transferred must be the same in the oxidation and reduction process. If the electrons do not cancel out, we must multiply each half-reaction by a coefficient such that they do.

Note: In an oxidation half-reaction (process) electrons are lost. In the balanced oxidation half-reaction the electrons are on the product (right) side.

In a reduction half-reaction electrons are gained. In the balanced reduction half-reaction the electrons are on the reactant (left) side.

a. In the following oxidation half-reaction 2 electrons are lost:

$$2\,FeCO_3 + H_2O \rightarrow Fe_2O_3 + 2\,CO_2 + 2\,H^+ + 2\,e^-$$

In the following reduction half-reaction 4 electrons are gained:

$$O_2 + 4\,H^+ + 4\,e^- \rightarrow 2\,H_2O$$

Multiply the oxidation half-reaction by 2 to equalize the number of electrons transferred in the two processes.

$$4\,FeCO_3 + 2\,H_2O \rightarrow 2\,Fe_2O_3 + 4\,CO_2 + 4\,H^+ + 4\,e^-$$

Now that we have 4 electrons in both half-reactions, we can add them together.

$$4\,FeCO_3 + 2\,H_2O \rightarrow 2\,Fe_2O_3 + 4\,CO_2 + 4\,H^+ + 4\,e^-$$

$$O_2 + 4\,H^+ + 4\,e^- \rightarrow 2\,H_2O$$

$$4\,FeCO_3 + 2\,H_2O + O_2 + 4\,H^+ + 4\,e^- \rightarrow 2\,Fe_2O_3 + 4\,CO_2 + 4\,H^+ + 2\,H_2O + 4\,e^-$$

Cancelling like out species ($4\,H^+$, $2\,H_2O$, and $4\,e^-$) yields

$$4\,FeCO_3 + O_2 \rightarrow 2\,Fe_2O_3 + 4\,CO_2$$

b. In the following oxidation half-reaction 2 electrons are lost:

$$3\,FeCO_3 + H_2O \rightarrow Fe_3O_4 + 3\,CO_2 + 2\,H^+ + 2\,e^-$$

In the reduction half-reaction 4 electrons are gained.

$$O_2 + 4\,H^+ + 4\,e^- \rightarrow 2\,H_2O$$

We need to multiply the oxidation half-reaction by 2, to give us 4 electrons lost. These will then cancel out the 4 electrons gained by the reduction.

$$6\,FeCO_3 + 2\,H_2O \rightarrow 2\,Fe_3O_4 + 6\,CO_2 + 4H^+ + 4e^-$$

$$O_2 + 4\,H^+ + 4\,e^- \rightarrow 2\,H_2O$$

$$6\,FeCO_3 + 2\,H_2O + O_2 + 4\,H^+ + 4\,e^- \rightarrow 2\,Fe_3O_4 + 6\,CO_2 + 4\,H^+ + 2\,H_2O + 4\,e^-$$

Cancelling out the common species ($2\,H_2O$, $4\,H^+$, and $4\,e^-$) yields

$$6\,FeCO_3 + O_2 \rightarrow 2\,Fe_3O_4 + 6\,CO_2$$

c. The oxidation half-reaction loses 2 electrons. The reduction half-reaction gains 4 electrons. We need to multiply the oxidation half-reaction by 2 to get a net transfer of 4 electrons in both half-reactions.

$$\text{Oxidation:} \quad 2\,Fe_3O_4 + H_2O \rightarrow 3\,Fe_2O_3 + 2\,H^+ + 2\,e^-$$

$$\text{Reduction:} \quad O_2 + 4\,H^+ + 4\,e^- \rightarrow 2\,H_2O$$

Multiplying the oxidation half-reaction by 2 yields

$$4\,Fe_3O_4 + 2\,H_2O \rightarrow 6\,Fe_2O_3 + 4\,H^+ + 4\,e^-$$

Now we can add the two half-reactions together.

$$4\,Fe_3O_4 + 2\,H_2O \rightarrow 6\,Fe_2O_3 + 4\,H^+ + 4\,e^-$$

$$O_2 + 4\,H^+ + 4\,e^- \rightarrow 2\,H_2O$$

$$4\,Fe_3O_4 + 2\,H_2O + O_2 + 4\,H^+ + 4\,e^- \rightarrow 6\,Fe_2O_3 + 4\,H^+ + 2\,H_2O + 4\,e^-$$

Cancelling out the like terms ($2\,H_2O$, $4\,H^+$, and $4\,e^-$) gives

$$4\,Fe_3O_4 + O_2 \rightarrow 6\,Fe_2O_3$$

SPECIAL TIPS: BALANCING OXIDATION-REDUCTION REACTIONS

There are two methods commonly used for balancing oxidation-reduction reactions. They are most often called the half-reaction method and the oxidation number method. In both methods, the idea is to split the redox process into two half-reactions: one that corresponds to the oxidation and the other corresponding to the reduction. We do not need to know which one is the oxidation or the reduction at this point.

The major difference in the two methods is the order of steps used to balance the half-reactions. In the half-reaction method, atoms are balanced first and then the net charge is balanced (this step determines the number of electrons lost or gained by that particular half-reaction). In the oxidation number method, electron transfer is determined first (by determining the oxidation numbers associated with each atom in the process), then net charge and lastly atoms are balanced. In general, the half-reaction method is easier to use.

Most oxidation-reduction reactions do not occur under neutral conditions. Often, the reaction occurs in aqueous solution and the medium must either be acidic or basic. When the redox reaction occurs in acidic solution, it means hydrogen ions (H^+) must be present along with water. In balancing a half-reaction in acidic solution we can add H_2O molecules and H^+ ions as needed to either side of the equation in order to balance the atoms.

In basic solution, hydroxide ions (OH^-) and H_2O molecules can be added as needed to balance the atoms. In practice, it is nearly impossible to balance a half-reaction in basic solution by adding OH^- ions and H_2O by inspection. This is because one gets into an endless loop in which each of the species added has one oxygen and one cannot go back and balance the hydrogen atoms without adding additional oxygens. The solution to this is to balance the half-reactions in acidic solution and then convert them to basic solution by adding an equal number of hydroxide ions to both sides of the equation. This conversion will be described later.

Using the oxidation number method, one can more easily balance a half-reaction in basic solution because one balances the net charge before balancing the atoms. However, in general, the half-reaction is still the easier and more popular method used to balance redox reactions.

Note: Many unbalanced redox reactions will not show H_2O, H^+, or OH^- even though these species are involved in the reaction.

Let's consider the following unbalanced redox reaction

$$Mn^{2+} + Cr_2O_7^{\,2-} \rightarrow MnO_2 + Cr^{3+}$$

Since this reaction takes place in acidic solutions, water molecules and H^+ ions are present in the solution; however, they are not listed anywhere in the preceding unbalanced equation.

Rules for Balancing Oxidation-Reduction Reactions by the Half-Reaction Method

We can see how these rules work by balancing the following unbalanced redox process

$$H^+ + Mn^{2+} + Cr_2O_7^{\,2-} \rightarrow MnO_2 + Cr^{3+} + H_2O$$

1. Write unbalanced half-reactions for both the oxidation process and the reduction process.

   We do not need to know which half-reaction is the oxidation and which is the reduction at this point. Also, the unbalanced half-reactions may not contain any H_2O molecules, H^+ ions, or OH^- ions. We can often determine the half-reactions by inspection.

   If we cannot, we need to determine the oxidation on each atom for both the reactants and products and use this information to determine which species are oxidized and which are reduced.

   The two half-reactions are

$$Cr_2O_7^{\,2-} \rightarrow Cr^{3+}$$

$$Mn^{2+} \rightarrow MnO_2$$

2. Balance each half-reaction for atoms.

   Apply steps *a–d* to one of the half-reactions and then repeat the procedure for the other half-reaction.

   a. First, balance all atoms other than hydrogen and oxygen.
   b. Balance the oxygen atoms by adding enough water molecules to the side of the reaction with less oxygen atoms until both sides have an equal number.
   c. Balance the hydrogen atoms by adding H^+ ions to the side of the half-reaction that needs more hydrogen atoms.
   d. Balance the charge by adding electrons to the side of the half-reaction that has the more positive total charge.

   Let's balance the half-reaction

$$Cr_2O_7^{\,2-} \rightarrow Cr^{3+}$$

Step *a* requires us to balance the Cr atoms. We can do this by putting the coefficient 2 in front of the species Cr^{3+}.

$$Cr_2O_7^{2-} \rightarrow 2\,Cr^{3+}$$

Step *b* requires us to balance the oxygen atoms. Currently, there are 7 oxygen atoms on the reactant (left) side and none on the product (right) side. To get 7 oxygen atoms on the product side, add 7 H_2O molecules to this side.

$$Cr_2O_7^{2-} \rightarrow 2\,Cr^{3+} + 7\,H_2O$$

Step *c* requires us to balance the hydrogen atoms by adding H^+ ions. There are currently 14 hydrogen atoms on the product side and none on the reactant side, so add 14 H^+ ions to the reactant side.

$$14\,H^+ + Cr_2O_7^{2-} \rightarrow 2\,Cr^{3+} + 7\,H_2O$$

This half-reaction is now balanced for atoms. However, it is not yet balanced for charge.

Step *d* requires us to balance the reaction for net charge. This means that the charge on both sides of the arrow needs to be the same. (This does not mean the charge must be zero, as you will see.) We can balance net charge by adding electrons to the side of the half-reaction with the more positive total charge.

First, we need to sum the charges on each side of the reaction. From step *c*, we see that on the reactant side the balanced half-reaction has 14 hydrogen ions (H^+) and 1 dichromate ion ($Cr_2O_7^{2-}$). Each H^+ ion has a net charge of +1, so the charge on 14 H^+ ions must be +14 (14 × 1). The $Cr_2O_7^{2-}$ ion has a net charge of −2. The total charge on the reactant side is +12 (14 + (−2)). On the product side, there are two chromium(III) ions (Cr^{3+}), each with a net charge of +3, and 7 water molecules. The water molecules are neutral so they have a net charge of zero. The total charge on the product side is + 6 ([2 × 3] + [7 × 0] = +6).

This can be written more simply as follows:

	$14\,H^+ + Cr_2O_7^{2-}$	$\rightarrow$	$2\,Cr^{3+} + 7\,H_2O$
	14(+1) + (−2)		2(+3) + 7(0)
	14 + (−2)		6 + 0
Total charge	+ 12		+ 6

The total charge is not equal (+12 does not equal +6). To equalize total charge, add electrons to the side with the greater total positive charge. In this half-reaction, we need to add electrons to the reactant side. To make the two sides equal in charge, add 6 electrons to the reactant side. Since each electron has a charge of −1, the total charge on 6 electrons is −6. Both sides now have the same charge.

$$+12 + 6(-1) = +6$$
$$+6 = +6$$

The balanced half-reaction (including electron transfer is)

$$6\,e^- + 14\,H^+ + Cr_2O_7^{2-} \rightarrow 2\,Cr^{3+} + 7\,H_2O$$

Since the electrons are on the reactant (left) side of the equation, this half-reaction corresponds to a reduction process.

Now balance the second half-reaction $Mn^{2+} \rightarrow MnO_2$.

Step *a*: The manganese atoms are already balanced.

Step *b*: Add 2 H_2O to the reactant (left) side to balance the 2 oxygen atoms in the MnO_2 species.

$$2\,H_2O + Mn^{2+} \rightarrow MnO_2$$

Step *c*: Add 4 H^+ions to the product (right) side to balance out the 4 hydrogen atoms from the 2 H_2O molecules.

$$2\,H_2O + Mn^{2+} \rightarrow MnO_2 + 4\,H^+$$

Step *d*: Determine the total charge on each side of the reaction, then balance the charge.

Left side:

H_2O is a neutral species so the total charge on 2 H_2O molecules is zero.

Mn^{2+} has a net charge of +2. There is one of these ions so the total charge on this species is +2.

Right side:

MnO_2 is a neutral substance, so the total charge on MnO_2 is zero.

H^+ has a net charge of +1. Since there are 4 H^+ ions, the total charge on the H^+ ions is +4.

$2\,H_2O$	$+\ Mn^{2+}$	$\rightarrow$	MnO_2	$+\ 4\,H^+$
2(0)	+ 1(+2)		1(0)	+ 4(+1)
0	+ 2		0	+ 4
Total charge	+2			+4

Add 2 electrons to the more positive side (product or right side).

The completely balanced half-reaction is

$$2\,H_2O + Mn^{2+} \rightarrow MnO_2 + 4\,H^+ + 2\,e^-$$

Since the electrons are on the product side, this corresponds to the oxidation half-reaction.

3. Multiply each half-reaction by a whole number so that the total number of electrons lost in the

oxidation equals the total number of electrons gained in the reduction.

Compare the number of electrons in each balanced half-reaction and determine the least common multiple for the number of electrons on the left and the right. Multiply each half-reaction through by a whole number so that the number of electrons on the left equals the number of electrons on the right.

For the two half-reactions in this example:

$$6\,e^- + 14\,H^+ + Cr_2O_7^{2-} \rightarrow 2\,Cr^{3+} + 7\,H_2O$$
$$2\,H_2O + Mn^{2+} \rightarrow MnO_2 + 4\,H^+ + 2\,e^-$$

the least common multiple is 6. If we multiply the top reaction by 1 (that is, we can leave it alone) and multiply the bottom reaction by 3, this will result in both half-reactions now having an equal number of electrons (6).

This yields

$$6\,e^- + 14\,H^+ + Cr_2O_7^{2-} \rightarrow 2\,Cr^{3+} + 7\,H_2O$$
$$6\,H_2O + 3\,Mn^{2+} \rightarrow 3\,MnO_2 + 12\,H^+ + 6\,e^-$$

4. Add the two half-reactions together and cancel out like species.

$$6\,e^- + 14\,H^+ + Cr_2O_7^{2-} \rightarrow 2\,Cr^{3+} + 7\,H_2O$$
$$6\,H_2O + 3\,Mn^{2+} \rightarrow 3\,MnO_2 + 12\,H^+ + 6\,e^-$$
$$6\,e^- + 6\,H_2O + 14\,H^+ + Cr_2O_7^{2-} + 3\,Mn^{2+} \rightarrow 2\,Cr^{3+} + 3\,MnO_2 + 7\,H_2O + 12\,H^+ + 6\,e^-$$

Canceling out $6\,e^-$, $6\,H_2O$, and $12\,H^+$ from both sides leaves us with

$$2\,H^+ + Cr_2O_7^{2-} + 3\,Mn^{2+} \rightarrow 2\,Cr^{3+} + 3\,MnO_2 + H_2O$$

This is the balanced equation.

It is a good idea to check the final equation to make sure everything is balanced.

Since both sides contain 2 hydrogen atoms, 2 chromium atoms, 3 manganese atoms, and 7 oxygen atoms, the atoms are balanced.

Now, check the charges.

Left side:	Right side:
$2(+1) + (-2) + 3(+2)$	$2(+3) + 3(0) + 1(0)$
$= +2 + (-2) + 6 = +6$	$+6 = +6$

Both sides have a charge of +6, so the charges are balanced.

5. To balance a half-reaction in basic solution (if required) add to both sides of the equation a number of OH^- ions equal to the number of H^+ ions in the overall balanced equation from step 4.

From the balanced equation in step 4, we see that there are 2 H^+ ions present, so we need to add 2 OH^- ions to both sides of the equation. Remember that adding equal amounts of anything to both sides of an equals sign does not change the overall equality.

Note: $H^+ + OH^-$ combine to make water, so adding $2\,OH^-$ to the reactant side below results in the formation of $2\,H_2O$ molecules.

$$2\,H^+ + Cr_2O_7^{2-} + 3\,Mn^{2+} \rightarrow 2\,Cr^{3+} + 3\,MnO_2 + H_2O$$
$$+2\,OH^- \qquad\qquad +2\,OH^-$$
$$2\,OH^- + 2\,H^+ + Cr_2O_7^{2-} + 3\,Mn^{2+} \rightarrow 2\,Cr^{3+} + 3\,MnO_2 + H_2O + 2\,OH^-$$

Since $2\,OH^- + 2\,H^+ = 2\,H_2O$, we can write the above equation as

$$2\,H_2O + Cr_2O_7^{2-} + 3\,Mn^{2+} \rightarrow 2\,Cr^{3+} + 3\,MnO_2 + H_2O + 2\,OH^-$$

Lastly, cancel out the common H_2O molecule to get the balanced equation.

$$H_2O + Cr_2O_7^{2-} + 3\,Mn^{2+} \rightarrow 2\,Cr^{3+} + 3\,MnO_2 + 2\,OH^-$$

This is the balanced equation in a basic solution.

Checking the totals for atoms and charges confirms that this reaction is balanced.

Rules for Balancing Redox Reactions Using Oxidation Number Method

We can balance the same redox reaction as previously, namely

$$H^+ + Mn^{2+} + Cr_2O_7^{2-} \rightarrow Cr^{3+} + MnO_2 + H_2O$$

1. Determine the oxidation numbers of all atoms in the reaction.

   H^+ has an oxidation number of +1
   Mn^{2+} has an oxidation number of +2
   In $Cr_2O_7^{2-}$
   oxygen has the oxidation number −2
   chromium has the oxidation number +6

   We can determine the oxidation state for chromium in $Cr_2O_7^{2-}$ from the equation

$$2\,\text{Cr atoms} + 7\,(\text{O atoms}) = -2$$
$$2\,Cr + 7(-2) = -2$$
$$2\,Cr - 14 = -2; \qquad 2\,Cr = +12; \qquad Cr = +6$$

Cr^{3+} has an oxidation number of +3.

In MnO_2
oxygen has the oxidation number −2
manganese has the oxidation number + 4
Since

$$\text{Mn atom} + 2\,\text{O atoms} = 0$$

$$\text{Mn} + 2(-2) = 0$$

$$\text{Mn} - 4 = 0$$

$$\text{Mn} = +4$$

In H_2O
oxygen has the oxidation number −2
hydrogen has the oxidation number +1

$$\overset{+1}{H^+} + \overset{+2}{Mn^{2+}} + \overset{+6-2}{Cr_2O_7^{2-}} \rightarrow \overset{+3}{Cr^{3+}} + \overset{+4-2}{MnO_2} + \overset{+1-2}{H_2O}$$

2. Determine the two half-reactions by finding out which species have changed oxidation numbers.

   From the preceding scheme, we can see that the only two elements whose oxidation numbers have changed are the manganese atoms (from +2 to +4) and the chromium atoms (from +6 to +3). The two half-reactions are

$$Mn^{2+} \rightarrow MnO_2$$

$$Cr_2O_7^{2-} \rightarrow Cr^{3+}$$

3. Balance all atoms other than hydrogen and oxygen. The manganese atoms are balanced as written but we need to get 2 chromium atoms on the product side of the second half-reaction by putting a coefficient of 2 in front of the Cr^{3+} species.

$$Mn^{2+} \rightarrow MnO_2$$

$$Cr_2O_7^{2-} \rightarrow 2\,Cr^{3+}$$

4. Determine the number of electrons lost by the atoms oxidized or gained by the atoms reduced.

   The manganese changes from a +2 to a +4 oxidation state, so the manganese has lost 2 electrons

   The chromium atoms change from a +6 to a +3 oxidation state, so each chromium atom has gained 3 electrons. Since 2 chromium atoms changed oxidation state, the total number of electrons gained by the chromium atoms is 6 (2 × 3).

5. Add the number of electrons determined in step 4 to the appropriate side of the half-reaction.

   In an oxidation process, electrons are lost and electrons are added to the product side. The manganese atom is oxidized, so add 2 electrons to the product side.

$$Mn^{2+} \rightarrow MnO_2 + 2\,e^-$$

   In a reduction process, electrons are gained and electrons are added to the reactant side. The chromium atoms are reduced and each chromium atom gains 3 electrons, so add 6 electrons to the reactant side.

$$6\,e^- + Cr_2O_7^{2-} \rightarrow 2\,Cr^{3+}$$

6. Balance charge by adding H^+ ions (if in acidic solution) or OH^- (if in basic solution).

   For the oxidation half-reaction, the total charges are as follows

$$\begin{array}{ccccc} Mn^{2+} & \rightarrow & MnO_2 & + & 2\,e^- \\ +2 & & 0 & + & (-2) \\ +2 & & & -2 & \end{array}$$

   To get the two sides equal in charge, add $4\,H^+$ to the product side in acidic solution or $4\,OH^-$ to the reactant side in basic solution.

   The half-reaction is now

$$Mn^{2+} \rightarrow MnO_2 + 4\,H^+ + 2\,e^- \quad \text{(in acidic solution), or}$$

$$Mn^{2+} + 4\,OH^- \rightarrow MnO_2 + 2\,e^- \quad \text{(in basic solution)}$$

   For the reduction half-reaction, the charges are

$$\begin{array}{ccccc} 6\,e^- & + & Cr_2O_7^{2-} & \rightarrow & 2\,Cr^{3+} \\ -6 & + & (-2) & & 2(+3) \\ & -8 & & & +6 \end{array}$$

   To get the two sides equal in charge add $14\,H^+$ to the reactant side in acidic solution or $14\,OH^-$ to the product side in basic solution.

   The half-reaction is now

$$6\,e^- + 14\,H^+ + Cr_2O_7^{2-} \rightarrow 2\,Cr^{3+} \quad \text{(in acidic solution), or}$$

$$6\,e^- + Cr_2O_7^{2-} \rightarrow 2\,Cr^{3+} + 14\,OH^- \quad \text{(in basic solution)}$$

7. Balance the hydrogen atoms by adding water molecules as needed.

   In the half-reaction involving the manganese atoms, there are 4 hydrogen atoms present, so we need to add 2 H_2O molecules. In acidic solution, the H_2O molecules are placed on the reactant side. In the basic solution, the H_2O molecules are needed on the product side.

$$Mn^{2+} + 2\,H_2O \rightarrow MnO_2 + 4\,H^+ + 2\,e^- \quad \text{(in acidic solution), or}$$

$$Mn^{2+} + 4\,OH^- \rightarrow MnO_2 + 2\,H_2O + 2\,e^- \quad \text{(in basic solution)}$$

   In the half-reactions involving the chromium atoms, there are 14 hydrogen atoms present, so we need to add 7 H_2O molecules to balance the half-reaction. This produces the following balanced half-reactions:

$$6\,e^- + 14\,H^+ + Cr_2O_7^{2-} \rightarrow 2\,Cr^{3+} + 7\,H_2O$$
(in acidic solution), or

$$6\,e^- + 7\,H_2O + Cr_2O_7^{2-} \rightarrow 2\,Cr^{3+} + 14\,OH^-$$
(in basic solution)

8. Multiply each half-reaction by a whole number so that the total number of electrons lost in the oxidation equals the total number of electrons gained in the reduction.

Compare the number of electrons in each balanced half-reaction and determine the least common multiple for the number of electrons on the left and the right. Multiply each half-reaction through by a whole number (including 1) so that each half-reaction contains the same number of electrons.

In these two half-reactions, the least common multiple is 6 (2 × 3), so we need to multiply the oxidation half-reaction (the one containing Mn^{2+}) by 3 to obtain

$$3\,Mn^{2+} + 6\,H_2O \rightarrow 3\,MnO_2 + 12\,H^+ + 6\,e^-$$
(in acidic solution)

Adding the two-half reactions together yields the following balanced redox reaction in acid solution

$$3\,Mn^{2+} + 6\,H_2O \rightarrow 3\,MnO_2 + 12\,H^+ + 6\,e^-$$
$$6\,e^- + 14\,H^+ + Cr_2O_7^{2-} \rightarrow 2\,Cr^{3+} + 7\,H_2O$$

$$6\,e^- + 6\,H_2O + 14\,H^+ + 3\,Mn^{2+} + Cr_2O_7^{2-}$$
$$\rightarrow 3\,MnO_2 + 2\,Cr^{3+} + 12\,H^+ + 7\,H_2O + 6\,e^-$$

Cancelling out like terms ($6\,e^-$, $6\,H_2O$, and $12\,H^+$) leaves

$$2\,H^+ + 3\,Mn^{2+} + Cr_2O_7^{2-}$$
$$\rightarrow 3\,MnO_2 + 2\,Cr^{3+} + H_2O$$

As expected, we get the same result as we did when using the half-reaction method.

In basic solution, multiplying the oxidation half-rection by 3 gives

$$3\,Mn^{2+} + 12\,OH^- \rightarrow 3\,MnO_2 + 6\,H_2O + 6\,e^-$$

Adding the two half-reactions together, yields

$$3\,Mn^{2+} + 12\,OH^- \rightarrow 3\,MnO_2 + 6\,H_2O + 6\,e^-$$
$$6\,e^- + 7\,H_2O + Cr_2O_7^{2-} \rightarrow 2\,Cr^{3+} + 14\,OH^-$$

$$6\,e^- + 7\,H_2O + 12\,OH^- + 3\,Mn^{2+} + Cr_2O_7^{2-}$$
$$\rightarrow 3\,MnO_2 + 2\,Cr^{3+} + 14\,OH^- + 6\,H_2O + 6\,e^-$$

Cancelling out like terms ($6\,e^-$, $6\,H_2O$, and $12\,OH^-$) leaves

$$H_2O + 3\,Mn^{2+} + Cr_2O_7^{2-}$$
$$\rightarrow 3\,MnO_2 + 2\,Cr^{3+} + 2\,OH^-$$

The result is identical to the balanced reaction in basic solution that we got when using the half-reaction method.

99. In order to answer this question, we need to determine the oxidation and reduction half-reactions from the unbalanced redox reaction. We also will need to then balance each of the half-reactions.

Remember that H_2O, H^+, and OH^- can be added as needed.

The starting reaction is

$$NH_4^+ + O_2 \rightarrow NO_3^-$$

It is likely that one of the half-reactions is

$$NH_4^+ \rightarrow NO_3^-$$

It is also likely that the other half-reaction involves oxygen, O_2. Since in any redox process we can add H_2O and H^+ ions (in acidic solution), it is always possible that the second half-reaction involves one of these species. In this case, it would most likely be the water because we need to have some oxygen atoms on the product side.

The second half reaction is

$$O_2 \rightarrow H_2O$$

Now we are ready to balance each half-reaction.

Let's begin by balancing the first half-reaction

$$NH_4^+ \rightarrow NO_3^-$$

The nitrogen atoms are balanced. There are 3 oxygen atoms on the product side. To get 3 oxygen atoms on the reactant side, add 3 H_2O molecules to the reactant side.

$$3\,H_2O + NH_4^+ \rightarrow NO_3^-$$

Now we can balance the hydrogen atoms. There are 10 hydrogen atoms on the reactant side and none on the product side. To balance the hydrogen atoms, add 10 H^+ ions to the product side. The half-reaction is now balanced for atoms.

$$3\,H_2O + NH_4^+ \rightarrow NO_3^- + 10\,H^+$$

Lastly, balance the half-reaction for total charge. The reactant side has a total charge of +1. (The 3 H_2O molecules are neutral and the 1 NH_4^+ ion has a charge of +1.) The product side has a net charge of +9. (The NO_3^- ion has a charge of −1 and the 10 H^+ ions have a total charge of +10.)

To make the charges on both sides equal, add electrons to the side that has the greater positive charge.

The reactant side has a charge of +1 and the product side +9. We need to add 8 electrons to the product side. This will result in both sides having a charge of +1. The balanced half-reaction is

$$3\,H_2O + NH_4^+ \rightarrow NO_3^- + 10\,H^+ + 8\,e^- \quad \text{(oxidation)}$$

Now balance the second half-reaction

$$O_2 \rightarrow H_2O$$

To balance the oxygen atoms requires us to put a coefficient of 2 in front of the H_2O.

$$O_2 \rightarrow 2\,H_2O$$

This results in 4 hydrogen atoms on the product side. We need to add 4 H^+ ions to the reactant side to balance the hydrogen atoms.

$$4\,H^+ + O_2 \rightarrow 2\,H_2O$$

Now we can balance the total charge.

The reactant side has a charge of +4, while the product side has zero net charge. We need to add 4 electrons to the reactant side (more positive side). This results in both sides having a charge of zero.

The balanced half-reaction is

$$4\,e^- + 4\,H^+ + O_2 \rightarrow 2\,H_2O \quad \text{(reduction)}$$

Both half-reactions are now balanced. To balance the overall process, we equalize the number of electrons transferred in each half-reaction. In these two half-reactions, the oxidation produces 8 electrons and the reduction requires 4 electrons. To get 8 electrons present in each half-reaction requires us to multiply the reduction half-reaction by 2.

Multiplying the reduction half-reaction by 2 and adding it to the oxidation half-reaction gives us

$$8\,e^- + 8\,H^+ + 2\,O_2 \rightarrow 4\,H_2O$$

$$3\,H_2O + NH_4^+ \rightarrow NO_3^- + 10\,H^+ + 8\,e^-$$

$$8\,e^- + 3\,H_2O + 8\,H^+ + NH_4^+ + 2\,O_2 \rightarrow NO_3^- + 10\,H^+ + 4\,H_2O + 8\,e^-$$

Cancelling out like terms ($8\,e^-$, $8\,H^+$, and $3\,H_2O$) leaves

$$NH_4^+ + 2\,O_2 \rightarrow NO_3^- + 2\,H^+ + H_2O$$

This is the balanced oxidation-reduction reaction for this process.

101. a. $Mn^{2+} + O_2 \rightarrow MnO_2$ (in basic solution)

The two half-reactions are

$$Mn^{2+} \rightarrow MnO_2$$

$$O_2 \rightarrow H_2O$$

The reaction $Mn^{2+} + O_2 \rightarrow MnO_2$ is balanced for atoms. However, it is not balanced for charges (the total charge on the reactant side is +2, while the total charge on the product side is zero). The conclusion we come to is that this reaction must be an oxidation-reduction reaction. It is easy to predict that one of the two half-reaction is

$$Mn^{2+} \rightarrow MnO_2$$

The other half-reaction is not readily apparent. It must involve oxygen (O_2) forming H_2O (this is typically what happens when O_2 is present in a reaction and there is no possible product listed in the unbalanced equation).

Also, the easiest way to balance a redox reaction in basic solution is to first balance it in acid solution and then convert it to a reaction in basic solution by adding OH^- ions to both sides until all the H^+ ions are converted to H_2O. It is not necessary to do this conversion for each half-reaction. Instead, wait until the final balanced redox equation is obtained and then convert it to a reaction in basic solution.

To balance the atoms in the half-reaction $Mn^{2+} \rightarrow MnO_2$, add 2 H_2O molecules to the reactant side and 4 H^+ ions to the product side.

$$2\,H_2O + Mn^{2+} \rightarrow MnO_2 + 4\,H^+$$

To balance the charge add 2 electrons to the product side. (The total charge on reactant side is +2 and total charge on product side is +4.)

$$2\,H_2O + Mn^{2+} \rightarrow MnO_2 + 4\,H^+ + 2e^- \quad \text{(oxidation)}$$

Now let's balance the second half-reaction

$$O_2 \rightarrow H_2O$$

To balance the atoms, put the coefficient 2 in front of the H_2O and then add 4 H^+ ions to the reactant side.

$$4\,H^+ + O_2 \rightarrow 2\,H_2O$$

To balance the charge, add 4 electrons to the reactant side.

$$4\,e^- + 4\,H^+ + O_2 \rightarrow 2\,H_2O \quad \text{(reduction)}$$

Next, balance the electron transfer. To do this we need to multiply the oxidation half-reaction by 2. Doing this and adding the half-reactions together yields

$$4\,H_2O + 2\,Mn^{2+} \rightarrow 2\,MnO_2 + 8\,H^+ + 4\,e^-$$

$$4\,e^- + 4\,H^+ + O_2 \rightarrow 2\,H_2O$$

$$4\,e^- + 4\,H^+ + 4\,H_2O + 2\,Mn^{2+} + O_2 \rightarrow 2\,MnO_2 + 2\,H_2O + 8\,H^+ + 4\,e^-$$

Cancelling out the like terms (4 e^-, 4 H^+, and 2 H_2O) leaves

$$2\,H_2O + 2\,Mn^{2+} + O_2 \rightarrow 2\,MnO_2 + 4\,H^+$$

In an acid solution, this is the balanced redox equation. To convert the reaction from acidic to a basic solution, we need to add 4 OH^- ions to both sides of the equation. (Remember that $H^+ + OH^- \rightarrow H_2O$.)

$$2\,H_2O + 2\,Mn^{2+} + O_2 \rightarrow 2\,MnO_2 + 4\,H^+$$

$$\underline{4\,OH^- \qquad\qquad\qquad\qquad 4\,OH^-}$$

$$4\,OH^- + 2\,H_2O + 2\,Mn^{2+} + O_2 \rightarrow 2\,MnO_2 + 4\,H_2O$$

Lastly, since there are 2 H_2O molecules on both sides, they can be cancelled out.

The correctly balanced redox equation (in a basic solution) is

$$4\,OH^- + 2\,Mn^{2+} + O_2 \rightarrow 2\,MnO_2 + 2\,H_2O$$

b. $MnO_2 + I^- \rightarrow Mn^{2+} + I_2$ (in acidic solution)
The two half-reactions are

$$MnO_2 \rightarrow Mn^{2+}$$

$$I^- \rightarrow I_2$$

The first half-reaction is simply the reverse of the oxidation in part *a* of this problem.

$$2\,e^- + 4\,H^+ + MnO_2 \rightarrow Mn^{2+} + 2\,H_2O \text{ (reduction)}$$

The second half-reaction $I^- \rightarrow I_2$ requires only that the coefficient 2 be placed in front of the I^- to balance the atoms.

$$2\,I^- \rightarrow I_2$$

To balance the charge, we need 2 electrons on the product side (total charge of –2 on reactant side and zero on the product side).

$$2\,I^- \rightarrow I_2 + 2\,e^- \qquad \text{(oxidation)}$$

Both half-reactions involve two electrons, so we just need to add the 2 half-reactions together to get the final balanced equation.

$$2\,e^- + 4\,H^+ + MnO_2 \rightarrow Mn^{2+} + 2\,H_2O$$

$$\underline{2\,I^- \rightarrow I_2 + 2\,e^-}$$

$$2\,e^- + 4\,H^+ + MnO_2 + 2\,I^- \rightarrow Mn^{2+} + I_2 + 2\,H_2O + 2\,e^-$$

The only species that cancels out are the 2 electrons. The final balanced equation is

$$4\,H^+ + MnO_2 + 2\,I^- \rightarrow Mn^{2+} + I_2 + 2\,H_2O$$

c. $I_2 + S_2O_3^{\,2-} \rightarrow S_4O_6^{\,2-} + I^-$
The two half-reactions are

$$I_2 \rightarrow I^-$$

$$S_2O_3^{\,2-} \rightarrow S_4O_6^{\,2-}$$

The first half-reaction is the reverse of the oxidation in part *b* of this problem.

$$2\,e^- + I_2 \rightarrow 2\,I^- \qquad \text{(reduction)}$$

To balance the half-reaction $S_2O_3^{\,2-} \rightarrow S_4O_6^{\,2-}$ for atoms, put the coefficient 2 in front of the $S_2O_3^{\,2-}$.

$$2\,S_2O_3^{\,2-} \rightarrow S_4O_6^{\,2-}$$

To balance this half-reaction for charge, add 2 electrons to the product side (−4 charge on reactant side and only –2 on the product side). The balanced half-reaction is

$$2\,S_2O_3^{\,2-} \rightarrow S_4O_6^{\,2-} + 2\,e^- \qquad \text{(oxidation)}$$

Note: To balance these half-reactions did not require any OH^- or H^+ ions, so we say this reaction occurs in a neutral solution.
Both half-reactions involve a transfer of 2 electrons. To get the balanced redox reaction we need only add the two balanced half-reactions together.

$$I_2 + 2\,e^- \rightarrow 2\,I^-$$

$$\underline{2\,S_2O_3^{\,2-} \rightarrow S_4O_6^{\,2-} + 2\,e^-}$$

$$2\,e^- + I_2 + 2\,S_2O_3^{\,2-} \rightarrow S_4O_6^{\,2-} + 2\,I^- + 2\,e^-$$

Cancelling out the 2 electrons gives

$$I_2 + 2\,S_2O_3^{\,2-} \rightarrow S_4O_6^{\,2-} + 2\,I^-$$

103. a. $MnO_4^{\,-} + S^{2-} \rightarrow MnS + S$
The two half-reactions are as follows:

$$MnO_4^{\,-} + S^{2-} \rightarrow MnS$$

$$S^{2-} \rightarrow S$$

We include S^{2-} as part of the first half-reaction because we need sulfur atoms on both sides of the equation. In this half-reaction, the oxidation number of the sulfur has not changed. It is −2 in both S^{2-} and MnS. The second half-reaction also contains S^{2-} because we need a source for elemental sulfur, S.

To balance the first half-reaction for atoms, add 4 H_2O molecules to the product side and 8 H^+ ions to the reactant side. Then to balance the charge add 5 electrons to the reactant side to give

$$5\,e^- + 8\,H^+ + MnO_4^{\,-} + S^{2-} \rightarrow MnS + 4\,H_2O \qquad \text{(reduction)}$$

The half-reaction $S^{2-} \rightarrow S$ is balanced for atoms as is written. It requires 2 electrons on the product side to balance the charge. The balanced half-reaction is

$$S^{2-} \rightarrow S + 2\,e^- \qquad \text{(oxidation)}$$

The half-reactions show that 5 electrons are required in the reduction and 2 electrons are produced in the oxidation. The lowest common multiple for the numbers 2 and 5 is 10. We need to multiply the reduction by 2 and the oxidation by 5.

Performing these multiplications and adding the half-reactions together yields

$$10\,e^- + 16\,H^+ + 2\,MnO_4^- + 2\,S^{2-} \rightarrow 2\,MnS + 8\,H_2O$$
$$5\,S^{2-} \rightarrow 5\,S + 10\,e^-$$
$$\overline{10\,e^- + 16\,H^+ + 2\,MnO_4^- + 7\,S^{2-} \rightarrow 2\,MnS + 5\,S + 8\,H_2O + 10\,e^-}$$

Cancelling out the electrons, we get

$$16\,H^+ + 2\,MnO_4^- + 7\,S^{2-} \rightarrow 2\,MnS + 5\,S + 8\,H_2O$$

Lastly, we add 16 OH^- ions to both sides to convert from an acidic to a basic solution.

$$16\,H^+ + 2\,MnO_4^- + 7\,S^{2-} \rightarrow 2\,MnS + 5\,S + 8\,H_2O$$
$$16\,OH^- \qquad\qquad 16\,OH^-$$
$$\overline{16\,H_2O + 2\,MnO_4^- + 7\,S^{2-} \rightarrow 2\,MnS + 5\,S + 8\,H_2O + 16\,OH^-}$$

We can cancel out 8 H_2O molecules to get the following balanced equation in a basic solution

$$8\,H_2O + 2\,MnO_4^-(aq) + 7\,S^{2-}(aq) \rightarrow 2\,MnS(s) + 5\,S(s) + 16\,OH^-(aq)$$

b. $MnO_4^- + CN^- \rightarrow CNO^- + MnO_2$
The two half-reactions are as follows:

$$MnO_4^- \rightarrow MnO_2$$
$$CN^- \rightarrow CNO^-$$

To balance the half-reaction $MnO_4^- \rightarrow MnO_2$, add 2 H_2O molecules to the product side and 4 H^+ ions to the reactant side. Then, to balance the charge, add 3 electrons to the reactant side to give the balanced half-reaction.

$$3\,e^- + 4\,H^+ + MnO_4^- \rightarrow MnO_2 + 2\,H_2O \qquad \text{(reduction)}$$

The half-reaction $CN^- \rightarrow CNO^-$ requires 1 H_2O molecule to be added to the reactant side and 2 H^+ ions to the product side. To balance the charge, add 2 electrons to the product side (-1 charge on the reactant side vs. a $+1$ charge on the product side).

$$H_2O + CN^- \rightarrow CNO^- + 2\,H^+ + 2\,e^- \qquad \text{(oxidation)}$$

There are 2 electrons produced in the oxidation half-reaction, but 3 electrons are needed in the reduction. The lowest common multiple for 2 and 3 is 6, so multiply the oxidation half-reaction by 3 and the reduction half-reaction by 2.

Multiplying and adding yields

$$6\,e^- + 8\,H^+ + 2\,MnO_4^- \rightarrow 2\,MnO_2 + 4\,H_2O$$
$$3\,H_2O + 3\,CN^- \rightarrow 3\,CNO^- + 6\,H^+ + 6\,e^-$$
$$\overline{6\,e^- + 8\,H^+ + 3\,H_2O + 2\,MnO_4^- + 3\,CN^- \rightarrow 3\,CNO^- + 2\,MnO_2 + 4\,H_2O + 6\,H^+ + 6\,e^-}$$

Cancelling out like terms (6 e^-, 6 H^+, and 3 H_2O) yields the balanced equation in acid solution

$$2\,H^+ + 2\,MnO_4^- + 3\,CN^- \rightarrow 3\,CNO^- + 2\,MnO_2 + H_2O$$

Lastly, convert this reaction to one in a basic solution by adding 2 OH^- ions to both sides and then cancelling out the appropriate number of water molecules.

$$2\,H^+ + 2\,MnO_4^- + 3\,CN^- \rightarrow 3\,CNO^- + 2\,MnO_2 + H_2O$$
$$2\,OH^- \qquad\qquad 2\,OH^-$$
$$\overline{2\,H_2O + 2\,MnO_4^- + 3\,CN^- \rightarrow 3\,CNO^- + 2\,MnO_2 + H_2O + 2\,OH^-}$$

The correctly balanced equation (in basic solution) is

$$H_2O + 2\,MnO_4^-(aq) + 3\,CN^-(aq) \rightarrow 3\,CNO^-(aq) + 2\,MnO_2(s) + 2\,OH^-(aq)$$

c. $MnO_4^- + SO_3^{2-} \rightarrow MnO_2 + SO_4^{2-}$
The two half-reactions are

$$MnO_4^- \rightarrow MnO_2$$
$$SO_3^{2-} \rightarrow SO_4^{2-}$$

The half-reaction $MnO_4^- \rightarrow MnO_2$ was balanced in part *b* of this problem. The balanced equation (in an acidic solution is)

$$3\,e^- + 4\,H^+ + MnO_4^- \rightarrow MnO_2 + 2\,H_2O \qquad \text{(reduction)}$$

To balance the other half-reaction, $SO_3^{2-} \rightarrow SO_4^{2-}$, add 1 H_2O molecule to the reactant side to balance the oxygen atoms and then add 2 H^+ ions to the product side to balance the hydrogen atoms. Lastly add 2 electrons to the product side to balance the charge.

$$H_2O + SO_3^{2-} \rightarrow SO_4^{2-} + 2\,H^+ + 2\,e^- \qquad \text{(oxidation)}$$

Next, equalize the total electron transfer at six. Multiply the reduction by 2 and the oxidation by 3.
Multiplying and adding yields

$$6e^- + 8H^+ + 2MnO_4^- \rightarrow 2MnO_2 + 4H_2O$$
$$3H_2O + 3SO_3^{2-} \rightarrow 3SO_4^{2-} + 6H^+ + 6e^-$$

$$6e^- + 3H_2O + 8H^+ + 2MnO_4^- + 3SO_3^{2-}$$
$$\rightarrow 3SO_4^{2-} + MnO_2 + 6H^+ + 4H_2O + 6e^-$$

Cancelling out $6e^-$, $3H_2O$, and $6H^+$ gives us the balanced equation in acid solution

$$2H^+(aq) + 2MnO_4^-(aq) + 3SO_3^{2-}(aq)$$
$$\rightarrow 3SO_4^{2-}(aq) + 2MnO_2(s) + H_2O$$

Lastly, add $2OH^-$ ions to each side to convert the process to one in basic solution.

$$2H^+ + 2MnO_4^- + 3SO_3^{2-} \rightarrow 3SO_4^{2-} + 2MnO_2 + H_2O$$
$$2OH^- \qquad\qquad 2OH^-$$

$$2H_2O + 2MnO_4^- + 3SO_3^{2-} \rightarrow 3SO_4^{2-} + 2MnO_2 + H_2O + 2OH^-$$

Cancel out $1H_2O$ molecule from each side to get the balanced equation in basic solution

$$H_2O + 2MnO_4^-(aq) + 3SO_3^{2-}(aq)$$
$$\rightarrow 3SO_4^{2-}(aq) + 2MnO_2(s) + 2OH^-(aq)$$

105. A proton donor is called an acid.

106. A strong acid completely dissociates in water. A weak acid only partially dissociates in water.

107. There are only a few strong acids:
HCl, hydrochloric acid; HNO_3, nitric acid
HBr, hydrobromic acid; $HClO_4$, perchloric acid
HI, hydroiodic acid; H_2SO_4, sulfuric acid
There are many weak acids. Some common weak acids are:
CH_3CO_2H, acetic acid; $H_3C_6H_5O_7$, citric acid
H_2CO_3, carbonic acid; H_3PO_4, phosphoric acid
HNO_2, nitrous acid; HCO_2H, formic acid

108. The more easily a substance donates a proton the stronger the acid. H_2SO_4 is a stronger acid than HSO_4^- because H_2SO_4 donates a proton more easily than does HSO_4^-.

109. A proton acceptor is called a base.

110. A strong base completely dissociates while a weak base only partially dissociates.

111. The soluble hydroxides are all strong bases.
LiOH NaOH KOH RbOH CsOH
$Ca(OH)_2$ $Sr(OH)_2$ $Ba(OH)_2$
There are many weak bases. Examples are
NH_3, ammonia; C_5H_5N, pyridine
Na_2CO_3, sodium carbonate

112. The net ionic equation when a strong acid reacts with a strong base is always

$$H^+(aq) + OH^-(aq) \rightarrow H_2O(l)$$

All of the other ions that are present in the solution are spectator ions; that is, they are present on both sides of the arrow in identical form.

113. Most acids have formulas that either start or end with a hydrogen atom. Many bases contain hydroxide ions in their formulas.

a. $H_2SO_4(aq) + Ca(OH)_2(aq) \rightarrow CaSO_4(s) + 2H_2O(l)$
Acid, H_2SO_4; sulfuric acid is a strong acid.
Base, $Ca(OH)_2$; calcium hydroxide is a strong base.
When a strong acid and strong base react, the net ionic equation is

$$H^+(aq) + OH^-(aq) \rightarrow H_2O(l)$$

b. $PbCO_3(s) + H_2SO_4(aq) \rightarrow PbSO_4(s) + CO_2(g) + H_2O(l)$
Acid, H_2SO_4; sulfuric acid is a strong acid.
Base, $PbCO_3$; $PbCO_3$ is a weak base (carbonate compounds are basic).
When writing total ionic equations, the following species are written as ions:
Soluble ionic compounds
Strong acids
Strong bases
The following species are written in molecular form:
Nonelectrolytes
Insoluble ionic compounds
Weak acids
Weak bases
For each species in the equation we need to determine whether it forms ions or remains molecular.
H_2SO_4, strong acid (forms ions)
$Pb(CO)_3(s)$, insoluble ionic compound (remains molecular)
$PbSO_4(s)$, insoluble ionic compound (remains molecular)
H_2O, nonelectrolyte (remains molecular)
$CO_2(g)$, nonelectrolyte (remains molecular)
In this problem, H_2SO_4 is the only species that forms ions. It dissociates to form $2H^+$ ions and a SO_4^{2-} ion.

The total molecular equation is

$$PbCO_3(s) + 2\ H^+(aq) + SO_4^{2-}(aq) \rightarrow PbSO_4(s) + CO_2(g) + H_2O(l)$$

There are no spectator ions to cancel out. This is the net ionic equation.

c. $Ca(OH)_2(s) + 2\ CH_3COOH(aq) \rightarrow Ca(CH_3COO)_2(aq) + 2\ H_2O(l)$

$Ca(OH)_2$ is a strong base. In aqueous solution this substance would form ions. In this case it reacts in solid form (not dissolved in water), so it is written in molecular form.

CH_3COOH is the formula for acetic acid, a weak acid. It remains in molecular form.

$Ca(CH_3COO)_2$ is a soluble ionic compound. It will split into a Ca^{2+} ion and 2 CH_3COO^- (acetate) ions.

H_2O is a nonelectrolyte. It remains in molecular form.

The only species that forms ions in this reaction is $Ca(CH_3COO)_2$. The total molecular equation is

$$Ca(OH)_2(s) + 2\ CH_3COOH(aq) \rightarrow Ca^{2+}(aq) + 2\ CH_3COO^-(aq) + 2\ H_2O(l)$$

There are no spectator ions to cancel out, so this is also the net ionic equation.

115. A saturated solution contains the maximum amount of solute that can be dissolved in a given quantity of solvent. In a saturated solution the dissolved solute is in equilibrium with the undissolved solute.

A supersaturated solution is not stable; that is, it is not at equilibrium. It contains more dissolved solute than a saturated solution, and the excess dissolved solute can very easily precipitate with minimal disturbance.

116. Some common units of solubility are
grams of dissolved solute per 100 mL of water (g/mL, the most common)
grams of dissolved solute per liter of water (g/L)
moles of dissolved solute per liter of solution (M, molarity)

117. In a precipitation reaction two solutions containing soluble substances are mixed to form one or more insoluble solid products.

118. When soluble substances are mixed together, the dissolved ions in the solutions can bounce into each other. If the positive and negative ions that collide form a soluble substance, they soon separate. If the ions that collide form an insoluble substance, then a precipitate forms, because the forces holding the positive and negative ions together cannot be broken apart by interactions with the water molecules.

119. The solubility rules predict the following:
Barium sulfate, limited solubility
Barium hydroxide, soluble
Lanthanum nitrate, soluble
Sodium acetate, soluble
Lead hydroxide, limited solubility
Calcium phosphate, limited solubility

121. The formula for lead(II) carbonate is $PbCO_3$ and for lead(II) hydroxide is $Pb(OH)_2$.

In general carbonates and bicarbonates dissolve in acid producing carbon dioxide (CO_2) gas and water molecules.

Hydroxides react with acids forming water molecules and dissolving the metal ions in water.

For $PbCO_3$ the reaction is

$$PbCO_3(s) + 2\ H^+(aq) \rightarrow Pb^{2+}(aq) + CO_2(g) + H_2O(l)$$

This reaction with H^+ results in the formation of Pb^{2+}. The net effect is that the insoluble compound dissolves.

For $Pb(OH)_2$ the reaction is

$$Pb(OH)_2(s) + 2\ H^+(aq) \rightarrow Pb^{2+}(aq) + 2\ H_2O(l)$$

In this reaction with H^+, $Pb(OH)_2$ dissolves with the formation of Pb^{2+} ions.

123. The reactions of metal ions with sodium hydroxide result in an ion exchange reaction in which the heavy metal ion replaces sodium ion. Since heavy metal hydroxides are insoluble, the precipitate can be filtered out to remove the heavy metals.

The balanced molecular equations are

$$Cr^{3+}(aq) + 3\ NaOH(aq) \rightarrow 3\ Na^+(aq) + Cr(OH)_3(s)$$

$$Cd^{2+}(aq) + 2\ NaOH(aq) \rightarrow 2\ Na^+(aq) + Cd(OH)_2(s)$$

In the total ionic equation, the sodium hydroxide is split into ions. In the net ionic equation the sodium ions are removed because they are spectator ions.

The total ionic equations are

$$Cr^{3+}(aq) + 3\ Na^+(aq) + 3\ OH^-(aq) \rightarrow 3\ Na^+(aq) + Cr(OH)_3(s)$$

$$Cd^{2+}(aq) + 2\ NaOH(s) \rightarrow 2\ Na^+(aq) + Cd(OH)_2(s)$$

The net ionic equations are

$$Cr^{3+}(aq) + 3\ OH^-(aq) \rightarrow Cr(OH)_3(s)$$

$$Cd^{2+}(aq) + 2\ OH^-(aq) \rightarrow Cd(OH)_2(s)$$

125. Since the solubility of AgCl in water is minimal, all of the Cl^- reacts with the Ag^+ ions in the silver nitrate to form an insoluble precipitate. Since $AgNO_3$ (silver nitrate) is in excess, the chloride ions are the limiting reactant.

From the given information we can calculate the number of moles of chloride (Cl^-) ions initially present in the solution.

From the formula of silver chloride (AgCl), we see that 1 mole of chloride ions is required to form 1 mole of silver chloride. That is, the ratio of chloride ions to silver chloride is 1 to 1.

Lastly, we can convert the moles of silver chloride to the mass of silver chloride in grams by multiplying the moles of silver chloride by the molar mass of silver chloride (143.5 g/mol).

$$\text{moles of } Cl^- = 5.0 \times 10^{-3}\frac{\text{mol}}{\text{L}} \times 0.010\text{ L}$$
$$= 5.0 \times 10^{-5}\text{ mol } Cl^-$$

Next, convert moles of Cl^- to moles of AgCl

$$\text{moles of AgCl} = 5.0 \times 10^{-3}\text{ mol } Cl^- \times \frac{1\text{ mol AgCl}}{1\text{ mol } Cl^-}$$
$$= 5.0 \times 10^{-5}\text{ mol AgCl}$$

Lastly, convert moles of AgCl to grams of AgCl

$$\text{mass of AgCl} = 5.0 \times 10^{-5}\text{ mol AgCl}$$
$$\times \frac{143.5\text{ g AgCl}}{\text{mol AgCl}} = 7.2 \times 10^{-3}\text{ g AgCl}$$

The maximum amount of AgCl that can precipitate out is 7.2×10^{-3} g, or 7.2 mg.

127. Deionized water can be produced using an ion exchange cartridge. In this process an ion exchange cartridge is filled with materials that have both anionic exchange sites (that is, an OH^- that can be replaced with another negative ion) and a cationic exchange site (that is, a H^+ that can be replaced by a different positive ion). The result is that H^+ and OH^- ions in the exchanger are replaced with different positive and negative ions (a possible set would be Na^+ and Cl^-). The free H^+ and OH^- ions then combine to form additional molecules of H_2O; that is, any ions that may have been present in the water are removed to give pure H_2O.

129. The starting point for this question is to write a balanced molecular equation. Since the reactants are an acid and a base in each case, we should expect a neutralization reaction.

a. $NaOH + HCl \rightarrow$

NaOH is a strong base and HCl is a strong acid. The ions that form when these species are dissolved in water are

$$Na^+(aq) + OH^-(aq) + H^+(aq) + Cl^-(aq)$$

Although the Na^+ ions and the Cl^- ions remain in solution, we can write the molecular reaction as

$$NaOH(aq) + HCl(aq) \rightarrow NaCl(aq) + H_2O(l)$$

We need only the molecular equation and not the net ionic equation to determine the volume of NaOH required.

Information given:

[NaOH] = 0.100 *M*

[HCl] = 0.0500 *M*

Volume of HCl = 10.0 mL (0.0100 L)

We need to find the volume of NaOH.

From the information given, we can calculate the number of moles of HCl, from the relationship moles = molarity × volume (in liters)

Next use the coefficients of the balanced equation to determine the moles of NaOH that will react with HCl.

Since we know know the moles of NaOH (calculated in the last step) and were given the molarity of NaOH, we can calculate the volume of NaOH required to neutralize the HCl solution.

Performing these calculations:

$$\text{moles of HCl} = \frac{0.0500\text{ mol HCl}}{1\text{ L}} \times 0.0100\text{ L}$$
$$= 5.00 \times 10^{-4}\text{ mol HCl}$$

$$\text{moles of NaOH}$$
$$= 5.00 \times 10^{-4}\text{ mol HCl} \times \frac{1\text{ mol NaOH}}{1\text{ mol HCl}}$$
$$= 5.00 \times 10^{-4}\text{ mol NaOH}$$

$$\text{volume of NaOH} = \frac{\text{moles NaOH}}{\text{molarity NaOH}}$$
$$= \frac{5.00 \times 10^{-4}\text{ mol NaOH}}{0.100\,M\text{ NaOH}}$$
$$= 5.00 \times 10^{-3}\text{ L NaOH,}$$
$$\text{or } 5.00\text{ mL}$$

The volume of 0.100 *M* NaOH required to completely react with 10.0 mL of 0.0500 *M* HCl is 5.00 mL.

b. The reactants are NaOH and HNO_3. The ions formed when these substances are in aqueous solution are Na^+, OH^-, H^+, and NO_3^-. The products of this reaction are $NaNO_3$ and H_2O. The balanced molecular equation is

$$NaOH(aq) + HNO_3(aq) \rightarrow NaNO_3(aq) + H_2O(l)$$

Information given:

[NaOH] = 0.100 *M*

[HNO_3] = 0.126 *M*

Volume of HNO_3 = 23.6 mL (0.0126 L)

We need to find the volume of NaOH.

From the information given, calculate the moles of HNO_3.

Use the moles of HNO_3 and the balanced equation to determine the moles of NaOH required to neutralize the HNO_3.

Use the moles of NaOH and the molarity of NaOH to determine the volume of NaOH required in this process.

$$\text{moles of } HNO_3 = \frac{0.126 \text{ mol } HNO_3}{1 \text{ L}} \times 0.0236 \text{ L}$$
$$= 2.97 \times 10^{-3} \text{ mol } HNO_3$$

moles of NaOH

$$= 2.97 \times 10^{-3} \text{ mol } HNO_3 \times \frac{1 \text{ mol NaOH}}{1 \text{ mol } HNO_3}$$
$$= 2.97 \times 10^{-3} \text{ mol NaOH}$$

$$\text{volume of NaOH} = \frac{\text{moles NaOH}}{\text{molarity NaOH}}$$
$$= \frac{2.97 \times 10^{-3} \text{ mol NaOH}}{0.100\, M \text{ NaOH}}$$
$$= 2.97 \times 10^{-2} \text{ L NaOH,}$$
$$\text{or } 29.7 \text{ mL}$$

It requires 29.7 mL of the 0.100 *M* NaOH solution for complete reaction.

c. The reactants are NaOH and H_2SO_4. The ions formed are Na^+, OH^-, H^+, and SO_4^{2-}. The products formed are H_2O and Na_2SO_4 (it requires 2 Na^+ ions to balance the –2 charge of the SO_4^{2-} ion).

The balanced molecular equation is

$$2\,NaOH(aq) + H_2SO_4(aq)$$
$$\rightarrow Na_2SO_4(aq) + 2\,H_2O(l)$$

Information given:
$[NaOH] = 0.100\, M$
$[H_2SO_4] = 0.215\, M$
Volume of H_2SO_4 = 15.8 mL (0.0158 L)

From the information given, we can calculate the moles of H_2SO_4.

From the moles of H_2SO_4 and the coefficients of the balanced equation, we can determine the moles of NaOH needed to neutralize the H_2SO_4.

From the moles of NaOH and the molarity of NaOH, we can calculate the volume of NaOH solution required to completely react with the H_2SO_4 solution.

$$\text{moles of } H_2SO_4 = \frac{0.215 \text{ mol } H_2SO_4}{1 \text{ L}} \times 0.0158 \text{ L}$$
$$= 3.40 \times 10^{-3} \text{ mol } H_2SO_4$$

moles of NaOH

$$= 3.40 \times 10^{-3} \text{ mol } H_2SO_4 \times \frac{2 \text{ mol NaOH}}{1 \text{ mol } H_2SO_4}$$
$$= 6.80 \times 10^{-3} \text{ mol NaOH}$$

$$\text{volume of NaOH} = \frac{\text{moles NaOH}}{\text{molarity NaOH}}$$
$$= \frac{6.8 \times 10^{-3} \text{ mol NaOH}}{0.100\, M \text{ NaOH}}$$
$$= 6.8 \times 10^{-2} \text{ L, or 68 mL NaOH}$$

It requires 68.0 mL of NaOH for complete reaction.

131. In this problem, we need to write a completely balanced molecular equation. The reactants are $Ca(OH)_2$ and HCl. The products formed are $CaCl_2$ and H_2O. The balanced equation is

$$Ca(OH)_2(aq) + 2\,HCl(aq) \rightarrow CaCl_2(aq) + 2\,H_2O(l)$$

Information given:
Solubility of $Ca(OH)_2$ = 0.148 g/100 mL
Volume of saturated $Ca(OH)_2$ solution = 10 mL
The molarity of the HCl solution = 0.00100 *M*
We need to find the volume of HCl needed to completely react with 10 mL of the $Ca(OH)_2$ solution.

To answer this problem, we first need to determine the moles of the $Ca(OH)_2$ in 10 mL of solution. We can expect the density of the $Ca(OH)_2$ solution to be very close to 1.00 g/mL because only 0.148 g of $Ca(OH)_2$ can be dissolved in 100 mL of water. To determine the number of moles of $Ca(OH)_2$, set up a ratio relating the mass of $Ca(OH)_2$ in 100 mL of solution to the mass of $Ca(OH)_2$ found in 10 mL of the solution.

$$\frac{0.148 \text{ g } Ca(OH)_2}{100 \text{ mL}} = \frac{x \text{ g } Ca(OH)_2}{10 \text{ mL}}$$

grams of $Ca(OH)_2$ in10 mL

$$= 10 \text{ mL} \times \frac{0.148 \text{ g } Ca(OH)_2}{100 \text{ mL}}$$
$$= 0.0148 \text{ g } Ca(OH)_2$$

Now, find the moles $Ca(OH)_2$ present in 10 mL of the solution. Next, find the moles of HCl required from the moles of $Ca(OH)_2$ and the coefficeients of the balanced equation. Lastly find the volume of HCl from the moles of HCl and the molarity of the HCl.

moles of $Ca(OH)_2$ in 10 mL

$$= 0.0148 \text{ g } Ca(OH)_2 \times \frac{1 \text{ mol } Ca(OH)_2}{74.1 \text{ g } Ca(OH)_2}$$
$$= 2.0 \times 10^{-4} \text{ mol } Ca(OH)_2$$

$$\text{moles of HCl} = 2.0 \times 10^{-4}\ \text{mol Ca(OH)}_2 \times \frac{2\ \text{mol HCl}}{1\ \text{mol Ca(OH)}_2} = 4.0 \times 10^{-4}\ \text{mol HCl}$$

$$\text{Volume of HCl} = \frac{\text{mol HCl}}{\text{molarity HCl}} = \frac{4.0 \times 10^{-4}\ \text{mol HCl}}{0.00100\ M\ \text{HCl}} = 4.0 \times 10^{-1}\ \text{L, or } 4.0 \times 10^{2}\ \text{mL HCl}$$

It requires 4.0×10^2 mL of the HCl solution for complete reaction.

133. Information given:
Volume of CrO_4^{2-} solution = 100 mL
$[CrO_4^{2-}] = 0.25\ M$
We need to find the mass of SO_2 in grams.
The balanced reaction is

$$2\,CrO_4^{-}(aq) + 3\,SO_2(g) + 4\,H^+(aq) \rightarrow Cr_2(SO_4)_3(s) + 2\,H_2O(l)$$

This is another solution stoichiometry problem.
In this problem, we use the molarity and volume of the chromate solution to find the number of moles of chromate ion (CrO_4^{2-}) given.
We can find the moles of SO_2 from the moles of chromate ion (CrO_4^{2-}) present and the coefficients of the balanced equation.
From the moles of SO_2 and its molar mass, we can find the mass of SO_2 in grams needed to precipitate out all of the chromium(VI) ions in the form of $Cr_2(SO_4)_3$.

$$\text{moles of } CrO_4^{2-} = 0.100\ \text{L} \times \frac{0.25\ \text{mol } CrO_4^{2-}}{1\ \text{L}} = 0.025\ \text{mol } CrO_4^{2-}$$

$$\text{moles of } SO_2 = 0.025\ \text{mol } CrO_4^{2-} \times \frac{3\ \text{mol } SO_2}{2\ \text{mol } CrO_4^{2-}} = 0.0375\ \text{mol } SO_2$$

$$\text{mass of } SO_2 = 0.0375\ \text{mol } SO_2 \times \frac{64.1\ \text{g } SO_2}{1\ \text{mol } SO_2} = 2.4\ \text{g } SO_2$$

It requires 2.4 g of SO_2 (sulfur dioxide) to remove all of the Cr(VI) ions from this solution.

135. Information given
$[OCl^-] = 0.125$ M
Volume of cyanide solution (wastewater) = 3.4×10^6 L
concentration of cyanide, $CN^- = 0.58$ mg/L
We need to find the volume of hypochlorite (OCl^-) solution.
The balanced equation is

$$2\,CN^-(aq) + 5\,OCl^-(aq) + H_2O \rightarrow N_2(g) + 2\,HCO_3^-(aq) + 5\,Cl^-(aq)$$

First, we can find the number of grams of CN^- present in 3.4×10^6 L of wastewater. This can be calculated by converting the concentration of the cyanide ions from mg/L to g/L. Then multiply this concentration by the volume of wastewater.

$$\text{grams of } CN^- \text{ in } 3.4 \times 10^6\ \text{L} = 3.4 \times 10^6\ \text{L} \times \frac{0.58\ \text{mg}}{\text{L}} \times \frac{1\ \text{g}}{1000\ \text{mg}} = 1972\ \text{g } CN^-$$

Note: The answer should have only 2 significant figures. However, I will carry extra significant figures into the next calculation. Now we can determine the number of moles of cyanide ions in the wastewater.

$$\text{moles of } CN^- = 1972\ \text{g } CN^- \times \frac{1\ \text{mol } CN^-}{26.0\ \text{g } CN^-} = 76\ \text{mol } CN^-$$

Then use the coefficients of the balanced equation to determine the moles of OCl^- required. Lastly find the volume of OCl^- solution needed from the moles and molarity of the OCl^- solution.

$$\text{moles of } OCl^- = 76\ \text{mol } CN^- \times \frac{5\ \text{mol } OCl^-}{2\ \text{mol } CN^-} = 190\ \text{mol } OCl^-$$

$$\text{volume of } OCl^- = \frac{\text{mol } OCl^-}{\text{molarity } OCl^-} = \frac{190\ \text{mol } OCl^-}{0.125\ M} = 1520\ \text{L of } OCl^- \text{ solution}$$

It requires 1.5×10^3 L (to 2 significant figures) of 0.125 *M* OCl^- solution to completely remove all the CN^- (cyanide) ions.

137. The problem here is to determine how much $PbCO_3$ will react with 1.00 L of 1.00 *M* H^+ ions.
To determine this quantity, we find the number of moles of H^+ ions present and then use the balanced equation to determine the moles of $PbCO_3$ that react. Lastly we convert the moles of $PbCO_3$ to mass of $PbCO_3$ in grams.

$$\text{moles of } H^+ = 1.00\ \text{L} \times \frac{1.00\ \text{mol } H^+}{1.00\ \text{L}} = 1.00\ \text{mol } H^+$$

$$\text{moles of } PbCO_3 = 1.00\,\text{mol } H^+ \times \frac{1\,\text{mol } PbCO_3}{2\,\text{mol } H^+}$$

$$= 0.500\,\text{mol } PbCO_3$$

$$\text{mass of } PbCO_3 = 0.500\,\text{mol } PbCO_3 \times \frac{267.2\,\text{g } PbCO_3}{1\,\text{mol } PbCO_3}$$

$$= 134\,\text{g } PbCO_3$$

The acid can dissolve 134 g of $PbCO_3$.
All 5.00 g will dissolve.

139. Colloidal particles scatter light. Homogeneous solutions do not (that is, light passes straight through them). When light strikes suspended particles it can be bent. Light scattering can be observed by the unaided eye or measured using a spectrometer designed for this purpose.

140. 1 nm to 100 nm

141. hydrophobic (water fearing) and hydrophilic (water loving)

142. adding electrolytes to a hydrophobic colloid; heating a hydrophilic colloid

CHAPTER 6 Chemical Bonding and Atmospheric Molecules

1. The electrons that are in the outermost energy level (have the highest energy) are considered valence electrons.

2. Yes, but only for H and He.

3. Yes, for main group elements this is true. This is not true for most transition metals.

4. Yes, for main group elements this is true. This is also true for *most* transition metals, as you have already learned in Chapter 3.

5. $Li\cdot$ $\quad \dot{Mg}\cdot$ $\quad \cdot\dot{Al}\cdot$

7. $[Na]^{+}$ $\quad [:\ddot{\underset{\cdot\cdot}{As}}:]^{3-}$ $\quad [Ca]^{2+}$ $\quad [:\ddot{\underset{\cdot\cdot}{S}}:]^{2-}$

9. B^{3+} has no valence electrons. It does not have a valence-shell octet.
   I^{-} has a valence-shell octet.
   Ca^{2+} has no valence electrons. It does not have a valence-shell octet.
   Pb^{2+} has 2 valence electrons. It does not have a valence-shell octet.

11. a. Group IA
    b. Group IVA
    c. Group VIA

13. a. Any +1 ions of the Group IIA elements will have one valence electron, for example, $[Mg\cdot]^{+}$
    b. Any Group IIIA element in the +3 state, $[Al]^{3+}$. All transition metals in the +3 state. Since they only have 1 or 2 valence electrons, they will definitely have none left if they are 3+ ions.

15. $[Mg]^{2+}$ represents the most stable ion of magnesium.

17. Covalent bonds are formed when atoms share electrons, while ionic bonds form when electrons are transferred from one atom to another to form cations and anions. These cations and anions are held together by electrostatic forces.

18. The basis for the criticism was the idea that an electron could be "owned" by more than one atom. When determining the number of valence electrons "owned" by each atom, the electrons involved in bonds are, in a sense, counted twice. But this is only true when looking for the number of "owned" electrons, not when counting the total number of electrons in a molecule. For example, one of these critics would say that there appears to be four electrons in H_2, H:H.

19. no

20. no

21. In a polar covalent bond, one of the atoms has a greater pull on the bonding electrons. The atom with the higher electronegativity will have the greater pull on the electrons.

22. Every element has a particular ability to attract electrons involved in a covalent bond. If atoms of two different elements are involved in a covalent bond, one will have a greater ability to attract the elements in the bond, so they are not shared equally. The atom with the higher electronegativity will have a greater attraction for the bonding electrons.

23. a. BN has 8 valence electrons.
    b. HF has 8 valence electrons.
    c. OH^{-} has 8 valence electrons.
    d. CN^{-} has 10 valence electrons.

25. a. $:C\equiv O:$
    b. $\ddot{\underset{\cdot\cdot}{O}}=\ddot{\underset{\cdot\cdot}{O}}$

c. $[:\ddot{\underset{..}{Cl}}-\ddot{\underset{..}{O}}:]^-$
d. $[:C{\equiv}N:]^-$

27. a. 3 pairs
 b. 2 pairs
 c. 1 pair
 d. 3 pairs

29. Ionization energy and electron affinity can be measured, but electronegativity is calculated using a number of experimentally determined values.

30. When the difference in electronegativities ($\Delta\chi$) of the two atoms involved in a bond is greater than 2.0, the bond is considered to be ionic in nature.

31. When moving down a group, electronegativity values generally decrease. When moving from left to right across a period, electronegativity values generally increase.

 When moving down a group, the value of n (the principle quantum number) of the valence electrons increases. As the principle quantum number increases, the distance of the valence orbitals from the nucleus increases. There is also increased shielding due to more inner-shell electrons as the principle quantum number increases. Both of these factors result in a decreased pull on the electrons by the nucleus, and hence a lower electronegativity.

 When moving across a period, the electrons added have the same value of n, the principle quantum number. As we learned in Chapter 3, Z_{eff} increases when moving across a period. A larger Z_{eff} means that the electrons around an atom will be pulled more tightly toward the nucleus. The increased pull on the electrons by the nucleus leads to the higher electronegativity.

32. This is a direct result of Z_{eff} (effective nuclear charge) increasing across a row and decreasing down a group. It makes sense that an electron feeling a larger Z_{eff}, more energy will be required to remove it from an atom.

33. Essentially they do follow the same trend, but it is the sign of the value that makes the trend seem different. If an element and an electron combine to form a more stable system than when they were apart, energy will be released and ΔH for this event will be negative. The more stable the anion formed, the more negative the value of the electron affinity. For example, as we move across a group, electron affinities will decrease (become more negative). Changes in the ionization energy across a row result from changes in Z_{eff}. Recall that Z_{eff} increases when moving across a row. As Z_{eff} increases, more energy is required to remove the electron from an atom. The result is that ionization energy increases (becomes more positive) when moving across a row.

34. This is a direct result of Z_{eff} (effective nuclear charge) increasing across a row and decreasing down a group. It makes sense that an electron feeling a larger Z_{eff} will be pulled closer to the nucleus and result in a smaller atomic size. Electrons feeling a smaller Z_{eff} will be pulled less by the nucleus, and this results in a larger atomic size.

35. Energy must be added in order to force the electron and the atom to combine to form an anion. A process that requires energy is endothermic.

36. No. The most electronegative element is F, but Cl actually has a more negative electron affinity. If you base your answer solely on the general trend, you would say that the most electronegative element also has the most negative electron affinity. Recall that the electron being added to F is going into a half-filled $2p$ orbital and into a half-filled $3p$ orbital for Cl. The $3p$ orbitals are larger in size than the $2p$ orbitals since orbital size increases as the principle quantum number increases. Adding an electron to the half-filled $3p$ orbital will be easier because there will be smaller electron–electron repulsions in the larger orbitals. The relatively compact size of the $2p$ orbitals of F results in larger electron–electron repulsions.

37.

C—S Polar	Determining which has the higher electronegativity is tricky in this case. If you look at the table, they both have electronegativity of 2.5. If you know the trend, you would have to say that S is more electronegative without the actual numbers in front of you.
C—O Polar	O is more electronegative.
Cl—Cl Nonpolar	
O=O Nonpolar	
N—H Polar	N is more electronegative.
C—H Polar	C is more electronegative.

39. In order to determine if the bonds are polar covalent or ionic, we need to calculate the difference in electronegativity, $\Delta\chi$. When $0 < \Delta\chi < 2.0$, a bond is considered polar covalent. When $\Delta\chi \geq 2.0$, the bond is considered ionic.

a. C—S $\Delta\chi = 0$	According to the calculation we would need to say that this bond is nonpolar. Based on the trend in electronegativities we would say that the bond is polar.
b. C—O $\Delta\chi = 1.0$	The bond is polar covalent.
c. Al—Cl $\Delta\chi = 1.5$	The bond is polar covalent.
d. Ca—O $\Delta\chi = 2.5$	The bond is ionic.

41. The valence shell in hydrogen has the principle quantum number 1, so hydrogen is capable of sharing at most two bonding electrons. In H—H—Ö:, a central H would have to share 4 electrons.

42. No, coordinate covalent bonds are a good example of unequal contribution of electrons. An example is the reaction of NH_3 and H^+ to form $NH_4{}^+$.

```
    H                      H    +
    |                  [   |   ]
 H—N:   +   H+  ——>    [ H—N—H ]
    |                  [   |   ]
    H                      H
```

43. a. CF_2Cl_2 b. Cl_2FCCF_2Cl c. C_2Cl_3F

```
    :F:           :F: :F:          :Cl: :F:
     |             |   |             |   |
 :F—C—Cl:     :Cl—C—C—Cl:     :Cl—C=C—Cl:
     |             |   |
    :Cl:          :Cl: :F:
```

45. $CH_3CH_2CH_2CH_2SH$

```
    H  H  H  H
    |  |  |  |
 H—C—C—C—C—S—H          H—S—H
    |  |  |  |
    H  H  H  H
```

47. Cl_2O $ClO_3{}^-$

```
                    [ :O—Cl—O: ] -1
 :Cl—O—Cl:          [     |    ]
                    [    :Cl:  ]
```

49. Resonance is a concept used to describe bonding when more than one valid Lewis structure can be written for a particular molecule or ion. The actual electronic structure of the molecule is a combination of the valid Lewis structures.

50. Yes, resonance stabilizes a molecule by lowering its total energy.

51. Generally, only compounds with one or more multiple bonds can have resonance. Any valid Lewis structure for H_2S will have two S—H single bonds. There is no other way to arrange the electrons and still have a valid Lewis structure. For SO_2, there are many valid Lewis structures and hence resonance is possible. Two valid resonance structures are

```
 O=S—O:  <—>  :O—S=O
```

52. All resonance forms must have the same skeletal arrangement of atoms and have the same number of valence electrons.

53. The concept of resonance helps us go beyond the limitations of Lewis structures.

54. Resonance forms are needed when experimental data, such as bond lengths and angles, do not match those expected for the lowest energy resonance form.

55. No. Although they all have the same number of valence electrons, the three structures all have different skeletal arrangements, so they are not resonance forms.

56. These two structures are not resonance structures since they have different skeletal arrangements.

57. C_6H_6

```
        H                      H
        |                      |
   H    C    H            H    C    H
     C=   C                 C   =C
     |    ||     <——>       ||    |
     C=   C                 C   =C
   H    C    H            H    C    H
        |                      |
        H                      H
```

59. S_2O_2

```
 O=S—S—O:  <—>  :O—S=S—O:
           <—>  :O—S—S=O
```

S_2O_3

```
    :O:              :O:
     |                |
 O=S—S—O:  <—>  :O—S=S—O:

       :O:                 :O:
        ||                  |
 <—> :O—S—S—O:  <—>  :O—S—S=O
```

61. $CO_3{}^{2-}$

```
 [ :O—C=O ] 2-        [ O=C—O: ] 2-
 [    |   ]   <——>    [   |    ]
 [   :O:  ]           [  :O:   ]

                      [ :O—C—O: ] 2-
              <——>    [    ||   ]
                      [   :O:   ]
```

63. Formal charges can help us determine the relative stability of molecular structures and resonance structures. The following rules are listed in order of importance. Rule 2 is generally used when structures are equivalent based on Rule 1.

   1. Forms with zero formal charges or those closest to zero are more stable.

2. Forms with negative formal charges on the most electronegative elements are more stable. It follows that forms with positive formal charges on the least electronegative elements are more stable.

64. Forms with negative formal charges on the most electronegative elements are more stable. It follows that forms with positive formal charges on the least electronegative elements are more stable.

65. The form with the negative formal charge on O, the more electronegative element, will contribute more to the bonding in the molecule.

66. The form with the positive formal charge on N, the less electronegative element, will contribute more to the bonding in the molecule.

67. N_2O as NON

$$\ddot{\underset{..}{N}}=O=\ddot{\underset{..}{N}} \longleftrightarrow :N\equiv O-\ddot{\underset{..}{N}}: \longleftrightarrow :\ddot{\underset{..}{N}}-O\equiv N:$$

(formal charges: −1 +2 −1; 0 +2 −2; −2 +2 0)

With the NON arrangement, the oxygen has a +2 formal charge in each resonance form, yet it is the most electronegative element in this molecule. Stable forms generally have a negative formal charge on the most electronegative element. Since there are no resonance forms of the NON arrangement where this is the case, the NON structure is not likely to be stable.

69. HCN HNC

$$\underset{0}{H}-\underset{0}{C}\equiv\underset{0}{N}: \qquad \underset{0}{H}-\underset{+1}{N}\equiv\underset{-1}{C}:$$

For HCN, the formal charge of each atom is zero in the only valid Lewis structure. For HNC, the formal charges are as follows: H = 0, N = +1, and C = −1.

71.

$$\text{(H}^0\text{ above N)}\quad {}^0H-\underset{0}{\underset{..}{N}}-\underset{0}{C}\equiv\underset{0}{N}: \qquad \text{(H}^0\text{ above N)}\quad {}^0H-\underset{+1}{N}=\underset{0}{C}=\underset{-1}{\ddot{\underset{..}{N}}}$$

The structure with the C≡N bond is preferred because all the formal charges are zero.

73. a.

$$H-\underset{H}{\overset{H}{C}}-\overset{:O:\,\|}{N}-\ddot{\underset{..}{O}}: \longleftrightarrow H-\underset{H}{\overset{H}{C}}-\overset{:\ddot{O}:\,|}{N}=\ddot{\underset{..}{O}}$$

b.

$$\underset{-1}{:C}\equiv\underset{+1}{N}-\underset{+1}{\overset{\overset{0}{:O:}}{\|}{N}}-\underset{-1}{\ddot{\underset{..}{O}}}: \longleftrightarrow \underset{-1}{:C}\equiv\underset{+1}{N}-\underset{+1}{\overset{\overset{-1}{:\ddot{O}:}}{|}{N}}=\underset{0}{\ddot{\underset{..}{O}}}$$

$$\underset{0}{:N}\equiv\underset{0}{C}-\underset{+1}{\overset{:O:\,0}{\|}{\underset{..}{N}}}-\underset{-1}{\ddot{\underset{..}{O}}}: \qquad \underset{0}{:N}\equiv\underset{0}{C}-\underset{+1}{\overset{:\ddot{O}:\,-1}{|}{N}}=\underset{0}{\ddot{\underset{..}{O}}}$$

The $CNNO_2$ arrangement is preferred because its resonance forms have lower magnitude formal charges.

c. No. Resonance forms must have the same skeletal arrangement of atoms.

75. CNO^-

$$[\ddot{\underset{..}{C}}=N=\ddot{\underset{..}{O}}]^{-1} \longleftrightarrow [:C\equiv N-\ddot{\underset{..}{O}}:]^{-1} \longleftrightarrow [:\ddot{\underset{..}{C}}-N\equiv O:]^{-1}$$

(formal charges: −2 +1 0; −1 +1 −1; −3 +1 +1)

Based on formal charges, $:C\equiv N-\ddot{\underset{..}{O}}:$ is the most stable resonance of CNO^-.

NCO^-

$$[\ddot{\underset{..}{N}}=C=\ddot{\underset{..}{O}}]^{-1} \longleftrightarrow [:N\equiv C-\ddot{\underset{..}{O}}:]^{-1} \longleftrightarrow [:\ddot{\underset{..}{N}}-C\equiv O:]^{-1}$$

(formal charges: −1 0 0; 0 0 −1; −2 0 +1)

Based on formal charges, $:N\equiv C-\ddot{\underset{..}{O}}:$ is the most stable resonance form of NCO^-.

CON^-

$$[\ddot{\underset{..}{C}}=O=\ddot{\underset{..}{N}}]^{-1} \longleftrightarrow [:C\equiv O-\ddot{\underset{..}{N}}:]^{-1} \longleftrightarrow [:\ddot{\underset{..}{C}}-O\equiv N:]^{-1}$$

(formal charges: −2 +2 −1; −1 +2 −2; −3 +2 0)

Based on formal charges, $:C\equiv O-\ddot{\underset{..}{N}}:$ is the most stable resonance form of CON^-. None of the resonance forms of CON^- are particularly stable.

77. NO_2^-

$$[:\ddot{\underset{..}{O}}-\ddot{N}=\ddot{\underset{..}{O}}]^{-1} \longleftrightarrow [\ddot{\underset{..}{O}}=\ddot{N}-\ddot{\underset{..}{O}}:]^{-1}$$

NO_3^-

$$\left[:\ddot{\underset{..}{O}}-\underset{\underset{:\ddot{O}:}{|}}{N}=\ddot{\underset{..}{O}}\right]^{-1} \longleftrightarrow \left[:\ddot{\underset{..}{O}}-\underset{\underset{:O:}{\|}}{N}-\ddot{\underset{..}{O}}:\right]^{-1} \longleftrightarrow \left[\ddot{\underset{..}{O}}=\underset{\underset{:\ddot{O}:}{|}}{N}-\ddot{\underset{..}{O}}:\right]^{-1}$$

No, we expect the bond distances for NO_2^- to be slightly shorter than those for NO_3^-. For NO_2^- both resonance forms have three NO bonds and they must be positioned in two locations (3/2 = 1.5) and we expect the average bond distance to be intermediate between those of a double and a single bond. For NO_3^- we have four NO bonds and they must be placed in three locations (4/3 = 1.333), so we expect the NO bonds in NO_3^- to be a bit longer than those expected for NO_2^-.

78. $\ddot{O}=\ddot{O}$ $\quad \ddot{O}=\ddot{O}-\ddot{O}: \longleftrightarrow :\ddot{O}-\ddot{O}=\ddot{O}$

The resonance in O_3 leads us to believe that the average bond distance in O_3 is somewhere between a single and a double bond. Therefore, the O=O double bond in O_2 will be shorter than the O−O bonds in O_3. Also, there is no valid Lewis structure for O_3 where both O−O bonds are double bonds.

79.

```
     0   -1                      0   -1
    :O: :O:                     :O: :Ö:
     ||  ||  0              -1   ||  |   0
:Ö - N - N - Ö:  <-->  :Ö - N - N = Ö
-1  +1  +1                     +1  +1

       -1   0                    -1  -1
       :Ö: :O:                   :Ö: :Ö:
     0  |   ||  -1             0  |   |   0
<-->  Ö = N - N - Ö:  <-->  Ö = N - N = Ö
         +1  +1                  +1  +1

 0  +1  -1           -1  +1  0          -2  +1  +1
:N≡N-Ö:    <-->     N = N = Ö   <-->   :N - N≡O:
```

In the four resonance forms of N_2O_4 there are 6 NO bonds in 4 possible locations: $6/4 = 1.5$. The average NO order in N_2O_4 is 1.5, and we expect these NO bonds to be a length somewhere between a single NO bond and a double NO bond.

In the two most important resonance forms of N_2O, :N≡N−Ö: and N=N=Ö, there are 3 NO bonds in the 2 structures: $3/2 = 1.5$

These simple calculations yield an average bond order of 1.5 for N_2O_4 and N_2O. We expect their bond lengths to be similar.

80. No. Based on the Lewis structure of SO_3^{2-}, we expect all three SO bonds to be single bonds. For any of the Lewis structures of SO_4^{2-}, we find some single S−O bonds and some double S=O bonds. Due to resonance, we expect the average bond distance to be somewhere between those of a single S−O bond and a double S=O bond.

81. For NO_2^- we expect the NO bond distance to be intermediate between those of a single bond and a double bond and we expect the NO bond distance in NO_3^- to be slightly longer (see the answer to Problem 77). For NO^+, we expect a triple bond.

We expect the NO bond lengths to increase in the order $NO^+ < NO_2^- < NO_3^-$

83. C_2O_3

```
          :O:                       :Ö:
           ||                        |
:O≡C - C - Ö:   <-->   :O≡C - C = Ö
```

We expect two different bond lengths based on the preceding Lewis structures. One CO bond will be shorter (a triple bond), while the other two will be longer (between a single and a double bond because of resonance).

85. No. Two p atomic orbitals can overlap head-to-head (along the internuclear axis) to form two molecular orbitals, a σ and a σ^*.

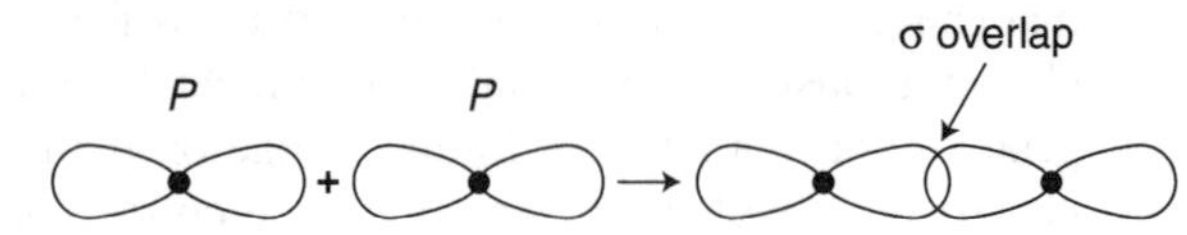

86. No, π molecular orbitals can be formed when d orbitals overlap and an example is shown below. For the vast majority of M.O. theory questions in this text, you need not be concerned with d orbitals.

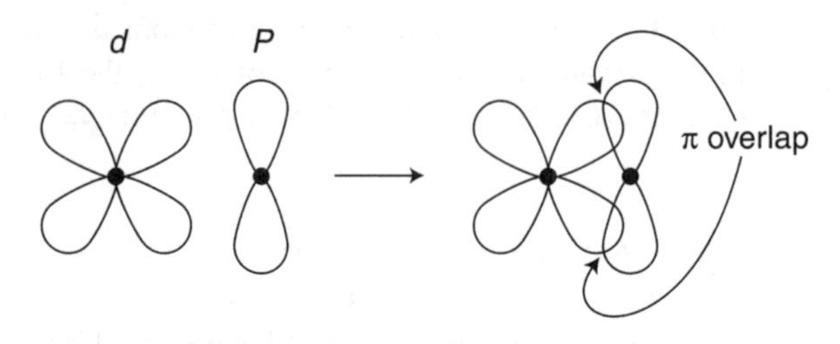

87. No. Orbitals with similar energies are more likely to mix.

88. No. Orbitals with similar shapes are more likely to mix.

89. 1s + 1s

σ overlap

Anti bonding, σ*, orbital.

Bonding, σ, orbital.

91.

$\overline{\sigma^*(2p_x)}$

$\overline{\pi^*(2p_y)}$ $\overline{\pi^*(2p_z)}$

↑

$\sigma(2p_x)$

↑↓ ↑↓

$\pi(2p_y)$ $\pi(2p_z)$

↑↓

$\sigma^*(2s)$

↑↓

$\sigma(2s)$

↑↓

$\sigma^*(1s)$

↑↓

$\sigma(1s)$

For N_2^+: Bond order $= \frac{1}{2}(9 - 4) = 2.5$

Expected to exist.

$\overline{\sigma^*(2p_x)}$

↑

$\pi^*(2p_y)$ $\overline{\pi^*(2p_z)}$

↑↓ ↑↓

$\pi(2p_y)$ $\pi(2p_z)$

↑↓

$\sigma(2p_x)$

↑↓

$\sigma^*(2s)$

↑↓

$\sigma(2s)$

↑↓

$\sigma^*(1s)$

↑↓

$\sigma(1s)$

For O_2^+: Bond order $= \frac{1}{2}(10 - 5) = 2.5$

Expected to exist.

$\overline{\sigma^*(2p_x)}$

$\overline{\pi^*(2p_y)}$ $\overline{\pi^*(2p_z)}$

$\overline{\sigma(2p_x)}$

↑ ↑

$\pi(2p_y)$ $\pi(2p_z)$

↑↓

$\sigma^*(2s)$

↑↓

$\sigma(2s)$

↑↓

$\sigma^*(1s)$

↑↓

$\sigma(1s)$

For C_2^{2+}: Bond order $= \frac{1}{2}(6 - 4) = 1$

Expected to exist.

↑↓

$\sigma^*(4p_x)$

↑↓ ↑↓

$\pi^*(4p_y)$ $\pi^*(4p_z)$

↑↓ ↑↓

$\pi(4p_y)$ $\pi(4p_z)$

↑↓

$\sigma(4p_x)$

For Br_2^{2-}: Bond order $= \frac{1}{2}(36 - 36) = 0$

Not expected to exist.

93.

$\overline{\sigma^*(2p_x)}$

$\overline{\pi^*(2p_y)}$ $\overline{\pi^*(2p_z)}$

↑

$\sigma(2p_x)$

↑↓ ↑↓

$\pi(2p_y)$ $\pi(2p_z)$

↑↓

$\sigma^*(2s)$

↑↓

$\sigma(2s)$

↑↓

$\sigma^*(1s)$

↑↓

$\sigma(1s)$

N_2^+ will have 1 unpaired electron.

$\overline{\sigma^*(2p_x)}$

↑

$\pi^*(2p_y)$ $\overline{\pi^*(2p_z)}$

↑↓ ↑↓

$\pi(2p_y)$ $\pi(2p_z)$

↑↓

$\sigma(2p_x)$

↑↓

$\sigma^*(2s)$

↑↓

$\sigma(2s)$

↑↓

$\sigma^*(1s)$

↑↓

$\sigma(1s)$

O_2^+ will have 1 unpaired electron.

$\overline{\sigma^*(2p_x)}$

$\overline{\pi^*(2p_y)}$ $\overline{\pi^*(2p_z)}$

$\overline{\sigma(2p_x)}$

↑ ↑

$\pi(2p_y)$ $\pi(2p_z)$

↑↓

$\sigma^*(2s)$

↑↓

$\sigma(2s)$

↑↓

$\sigma^*(1s)$

↑↓

$\sigma(1s)$

C_2^{2+} will have 2 unpaired electrons.

Even though we do not know how to draw the complete MO diagram for Br_2^{2-}, we can reason out the number of unpaired electrons. Begin by imagining Br_2^{2-} as being formed by 2 Br^- ions coming together. Each of the bromide ions will have 36 electrons and have the electron configuration $1s^22s^22p^63s^23p^64s^23d^{10}4p^6$. Each of these ions will have 18 atomic orbitals, four s orbitals, nine *p* orbitals, and five *d* orbitals. When the 36 total atomic orbitals mix, they will form 36 molecular orbitals. Half will be bonding and half will be antibonding. There are a total of 72 electrons in Br_2^{2-}; 36 will be in bonding orbitals and 36 will be in antibonding orbitals. Therefore, Br_2^{2-} will have zero unpaired electrons; all 36 molecular orbitals will be completely filled.

Another way to look at it is to look only at the highest energy set of molecular orbitals. We can do this because the lower energy molecular orbitals will be completely filled. When a set of molecular orbitals is completely filled, there are equal numbers of electrons in the bonding and antibonding molecular orbitals. In this filled set, the bonding and antibonding orbitals will, in a sense, cancel each other out. For example, the lowest energy set in Br_2^{2-} is $\sigma(1s)$ and $\sigma^*(1s)$. If this set is filled, the molecular orbitals made by the mixing of the $1s$ orbitals will not contribute to the bonding in Br_2^{2-}. The highest energy set in Br_2^{2-} will be the molecular orbitals made by mixing the $4p$ orbitals. There will be three bonding molecular orbitals and three antibonding molecular orbitals and the electron configurations tell us that we have a total of 12 electrons to place in these six molecular orbitals. All electrons will be paired up.

95.

—
$\sigma^*(2p_x)$
— —
$\pi^*(2p_y)$ $\pi^*(2p_z)$
⇅
$\sigma(2p_x)$
⇅ ⇅
$\pi(2p_y)$ $\pi(2p_z)$
⇅
$\sigma^*(2s)$
⇅
$\sigma(2s)$
⇅
$\sigma^*(1s)$
⇅
$\sigma(1s)$
C_2^{2-} has no electrons in π^* orbitals.

—
$\sigma^*(2p_x)$
↑ ↑
$\pi^*(2p_y)$ $\pi^*(2p_z)$
⇅
$\sigma(2p_x)$
⇅ ⇅
$\pi(2p_y)$ $\pi(2p_z)$
⇅
$\sigma^*(2s)$
⇅
$\sigma(2s)$
⇅
$\sigma^*(1s)$
⇅
$\sigma(1s)$
N_2^{2-} has 2 electrons in π^* orbitals.

—
$\sigma^*(2p_x)$
⇅ ⇅
$\pi^*(2p_y)$ $\pi^*(2p_z)$
⇅ ⇅
$\pi(2p_y)$ $\pi(2p_z)$
⇅
$\sigma(2p_x)$
⇅
$\sigma^*(2s)$
⇅
$\sigma(2s)$
⇅
$\sigma^*(1s)$
⇅
$\sigma(1s)$
O_2^{2-} has 4 electrons in π^* orbitals.

⇅
$\sigma^*(4p_x)$
⇅ ⇅
$\pi^*(4p_y)$ $\pi^*(4p_z)$
⇅ ⇅
$\pi(4p_y)$ $\pi(4p_z)$
⇅
$\sigma(4p_x)$

Br_2^{2-} has 12 electrons in π^* orbitals.

97. a. The bond order will increase because these electrons will be placed in bonding molecular orbitals.
    b. The bond order will increase because these electrons will be placed in bonding molecular orbitals.
    c. The bond order will decrease because these electrons will be placed in antibonding molecular orbitals.
    d. The bond order will decrease because these electrons will be placed in antibonding molecular orbitals.

99. No. The bond length will increase if the electron is removed from an antibonding molecular orbital and the bond order will decrease if the electron is removed from a bonding molecular orbital.

100. For B: The bond order will increase in both anions because electrons will be added to bonding molecular orbitals.
     For C: The bond order will increase in both anions because electrons will be added to bonding molecular orbitals.
     For N: The bond order will decrease in both anions because electrons will be added to antibonding molecular orbitals.
     For O: The bond order will decrease in both anions because electrons will be added to antibonding molecular orbitals.
     For F: The bond order will decrease in both anions because electrons will be added to antibonding molecular orbitals.

CHAPTER 7 Molecular Shape and the Greenhouse Effect

1. One should be able to determine if the compound has a C—N or a C=N bond by comparing the IR absorption bands in the unknown compound to those in compounds known to have C—N and C=N bonds. One could also check a reference table of IR frequencies to see if the unknown has bands that fall within the range typical for C—N or C=N bonds.

2. Yes, CO_2 and CO have different bond lengths and different bond strengths; therefore, they have different IR spectra.

3. UV radiation has shorter wavelengths than IR radiation. The shorter the wavelength, the more energetic the radiation. IR radiation must not be energetic enough to break the bonds of the species that absorb it, while UV radiation sometimes contains sufficient energy.

4. Greenhouse gases are gases that absorb IR radiation. In order for a species to absorb IR radiation, it must undergo a change in its dipole moment when it is hit by the radiation. Ar is not IR-active because it has zero dipole moment and there can be no change in its dipole moment.

5. Yes, we expect them to be similar. Looking at Lewis structures for the two compounds we see that the most important resonance forms of the two compounds each have a N with one N—O and one N=O bond. We expect them both to have similar N—O stretching frequencies because they have similar bond distances and bond strengths because of resonance.

6. If the structure were S—S—O, we would expect to see an absorption in the region where S=S bonds are commonly found. We would expect no such absorptions in this region if the connectivity were S—O—S.

   S̤̈=S̈—Ö̤:   S̤̈=Ö—S̤̈:

7. The stronger SO bonds will have absorptions at higher frequencies, so we need to draw the Lewis structures for the three species.

   S̤̈=Ö̤   :Ö̤—S̈=Ö̤   (:Ö̤:)₂S=Ö̤

   SO is expected to have a bond order of 2.0, SO_2 will have a bond order of 1.5, and SO_3 will have a bond order of 1.333. The higher the bond order, the stronger the bond; the stronger the bond strength, the greater the stretching frequency.

   Based on the Lewis structures for these species, we obtain the following order for bond strength and stretching frequency:

SO	<	SO_2	<	SO_3
highest				lowest

9. The stronger NO bonds will have absorptions at higher frequencies, so we need to draw the Lewis structures for the three species.

   $[:N{\equiv}O:]^+$ $[:\ddot{O}-\ddot{N}=\ddot{O}]^-$ $[:\ddot{O}-N(:\ddot{O}:)_2-\ddot{O}:]^{+}$

   Looking at the Lewis structures, we see that NO^+ will have the highest NO stretching frequency and NO_4^{3-} will have the lowest stretching frequency.

11. All of these elements are in the second period of the periodic table. Atoms in the second row are not able to expand their octets. The 3*s* orbitals are too high in energy compared to the 2*p* orbitals to be used for bonding in second period elements.

12. Any odd electron species must violate the octet rule. The odd number of total electrons means that one or more atoms in the species cannot have an octet, an even number of electrons.

13. a. The S in SF_6 has an expanded octet.

```
 :F: :F:
   \ /
:F—S—F:
   / \
 :F: :F:
```

b. The S in SF_5 has an expanded octet.

```
 :F: :F:
   \ /
:F—S—F:
    |
   :F:
```

c. The S in SF_4 has an expanded octet.

```
   :F:
    |
:F—S—F:
    |
   :F:
```

d. The S in SF_2 obeys the octet rule.

```
:F—S—F:
```

15. a. Each Cl in Cl_2 obeys the octet rule.

```
:Cl—Cl:
```

b. The Cl in ClF_3 has an expanded octet.

```
   :F:
    |
:F—Cl—F:
```

c. The Cl in ClF_2^+ obeys the octet rule.

```
[:F—Cl—F:]+
```

d. The Cl in ClO^- obeys the octet rule.

```
[:Cl—O:]−
```

17. a. In this resonance structure of SF_4O, the S shares 12 valence electrons.

```
 :F: :F:
   \ /
:F—S—F:
    ||
   :O:
```

b. In this resonance structure of SOF_2, the S shares 10 valence electrons.

```
   :F:
    |
:F—S=O
```

c. In this resonance structure of SO_3, the S shares 8 valence electrons.

```
:O:
   \
    S=O
   /
:O:
```

d. The S in SF_5^- shares 12 valence electrons.

```
[ :F: :F:   ]−
[   \ /     ]
[:F—S—F:    ]
[     \     ]
[     :F:   ]
```

19. a. The Xe in XeF_2 shares 2 pairs of valence electrons.

```
:F—Xe—F:
```

b. In this resonance structure of $XeOF_2$, the Xe shares 4 pairs of valence electrons.

```
   :F:
    |
:F—Xe=O
```

c. The Xe in XeF^+ shares 1 pair of valence electrons.

```
[:Xe—F:]+
```

d. The Xe in XeF_5^+ shares 5 pairs of valence electrons.

```
[ :F: :F:    ]+
[   \ /      ]
[:F—Xe—F:    ]
[     \      ]
[     :F:    ]
```

e. In this resonance structure of XeO_4, the Xe shares 8 pairs of valence electrons. Other resonance structures are possible.

```
    :O:
     ||
O=Xe=O
     ||
    :O:
```

21.

```
   :F:          :F:
    |            |
:F—S—F:      :F—O—F:
    |            |
   :F:          :F:
```

Looking at the two Lewis structures shown we can see why SF_4 exists and OF_4 does not. In order for OF_4 to exist, the central oxygen must have an expanded octet. Since oxygen is in the second period, it cannot expand its octet. Basically, we cannot write a reasonable Lewis structure for OF_4 without using expanded valence shells. The fact that the central S of SF_4 can use *d* orbitals to expand its octet allows it to be a stable gas.

23. POF_3: :O: double-bonded to P; P bonded to :F:, :F:, :F: NOF_4: :O: single-bonded to N; N bonded to :F:, :F:, :F:

The differences in the types of bonding are easily seen from the Lewis structures. Since the central P of POF_3 can have an expanded octet, we can draw a Lewis structure with a P$=$O bond and all formal charges equal to zero. The N in NOF_4 cannot expand its octet and the best Lewis structure we can draw for the molecule has an N$-$O single bond, a N with a $+1$ formal charge, and an O with a –1 formal charge.

25. $:\ddot{F}-\dot{S}e-\ddot{F}:$ (with :F: above and below Se) $[:\ddot{F}-Se-\ddot{F}:]^-$ (with two F above and :F: below Se)

The central Se has an expanded octet in both species.

27. In this resonance structure of Cl_2O_2, the central Cl has an expanded octet.

$:\ddot{Cl}-\ddot{Cl}$ with one single-bonded O and one double-bonded O

29. $:\ddot{Cl}-Al-\ddot{Cl}:$ (Al$=$F) $\longleftrightarrow$ $\ddot{Cl}=Al-\ddot{Cl}:$ (Al$-$F) $\longleftrightarrow$ $:\ddot{Cl}-Al=\ddot{Cl}:$ (Al$-$F)

$\longleftrightarrow$ $:\ddot{Cl}-Al-\ddot{Cl}:$ (Al$-$F)

31. a. Cl_2O_7 has a total of 56 valence electrons. It is not an odd-electron molecule.
    b. Cl_2O_6 has a total of 50 valence electrons. It is not an odd-electron molecule.
    c. ClO_4 has a total of 31 valence electrons. It is an odd-electron molecule.
    d. ClO_3 has a total of 25 valence electrons. It is an odd-electron molecule.
    e. ClO_2 has a total of 19 valence electrons. It is an odd-electron molecule.

33. a. $[\dot{S}=\ddot{O}]^+$

    In SO^+ we expect the unpaired electron to be found on S, the less electronegative element.

    b. $\dot{N}=\ddot{O}$

    In NO we expect the unpaired electron to be found on N, the less electronegative element.

    c. $\cdot C\equiv N:$

    In CN we expect the unpaired electron to be found on C, the less electronegative element.

    d. $\cdot\ddot{O}-H$

    In OH we expect the unpaired electron to be found on O. We do not expect it to be found on H because this would require it to expand its valence shell beyond two, or require a bond containing only 1 electron. In order to have 2 electrons in the OH bond and to have only 2 electrons on H, the unpaired electron should go on the oxygen.

35. Resonance structure *d* is expected to contribute most to the bonding description of CNO. It has the lowest formal charges (one 0, one $+1$, and one -1). Based on the magnitude of formal charges, *b*, *c*, and *d* are comparable and also better than *a*. Resonance structure *d* is the best of these three because the most electronegative element, O, has the negative formal charge.

37. Here is one resonance structure for Cl_2O_6 with a Cl$-$Cl bond.

$:\ddot{O}-Cl-Cl-\ddot{O}:$ (each Cl also bonded to one :O: above and one :O: below)

Here is one resonance structure for Cl_2O_6 with a Cl$-$O$-$Cl linkage.

$:\ddot{O}-Cl-\ddot{O}-Cl-\ddot{O}:$ (first Cl bonded to :O: above; second Cl bonded to :O: above and :O: below)

Here is one resonance structure for ClO_2.

$\ddot{O}=\dot{Cl}-\ddot{O}:$

39. $\cdot C\equiv N:$

The unpaired electron on CN is expected to be found on the carbon atom since it is less electronegative than nitrogen. When the CN molecules collide, it is likely that a bond will form between the two carbon atoms. This is expected because the carbons of CN have incomplete octets and bonds can be made when unpaired electrons in orbitals of similar shape and energy combine.

NCCN $:N\equiv C-C\equiv N:$ (formal charges 0 0 0 0)

CNNC $:C\equiv N-N\equiv C:$ (formal charges –1 +1 +1 –1)

The NCCN structure also has all zero formal charges while the CNNC structure has no zero formal charges.

41. In this Lewis structure, each atom has a formal charge of zero.

$:\ddot{F}-\dot{S}-C\equiv N:$ (S also bonded to :F: above and :F: below)

43. $TeOF_6^{2-}$ has a total of 56 valence electrons. Since the tellurium atom is the only one in this compound that can have an expanded octet, it must be the central atom in this pentagonal bipyramidal structure. In order for Te to share 14 valence electrons, it must be using *d* orbitals to expand its octet.

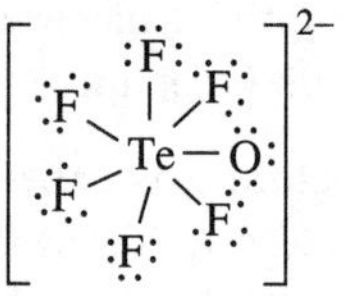

There are two possible positions for the oxygen—axial or equitorial.

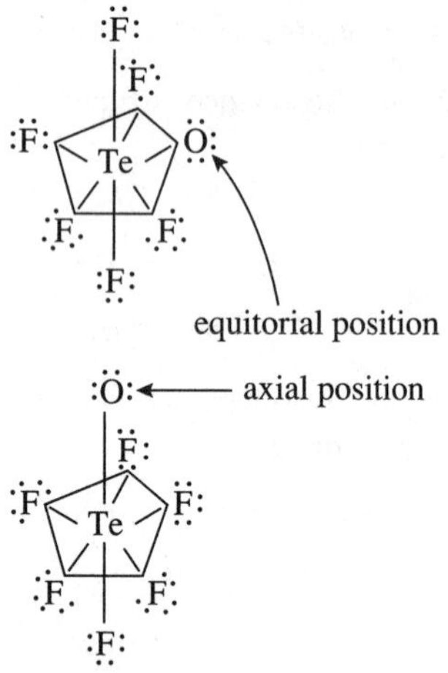

equitorial position

axial position

45. With increasing nuclear charge, the energy of the atomic orbitals within a series of atoms decreases. As nuclear charge increases, electrons are drawn closer to the nucleus and this makes the orbital energies more negative. So, the $2p$ orbitals of oxygen are lower in energy than those of nitrogen.

46. The MO diagram for CO should be more distorted because the energies of the atomic orbitals in the C and O atoms are further apart than those in the N and O atoms.

47.

— $\sigma^*(2p_x)$

— $\pi^*(2p_y)$ — $\pi^*(2p_z)$

$\uparrow$ $\sigma(2p_x)$

$\uparrow\downarrow$ $\pi(2p_y)$ $\uparrow\downarrow$ $\pi(2p_z)$

$\uparrow\downarrow$ $\sigma^*(2s)$

$\uparrow\downarrow$ $\sigma(2s)$

The above MO diagram is for the valence shell of CN. There is no need to draw in the molecular orbitals below the valence shell since they will not contribute to the bonding. CN has a total of 9 valence electrons. The bond order for CN $= \frac{1}{2}(7 - 2) = 2.5$. We have no way of drawing this bond. Our rules for drawing odd-electron species would yield the structure below.

$\cdot C \equiv N:$

49. In order to construct a MO diagram for PO, we need to determine which orbitals will mix. In the valence shell of P we have the following configuration: $3s^2 3p^3 (3p_x{}^1, 3p_y{}^1, 3p_z{}^1)$. In the valence shell of oxygen we expect the following configuration: $2s^2 2p^4 (2p_x{}^1, 2p_y{}^1, 2p_z{}^2)$. Since the valence shells of each atom are mixing, we expect the $3s$ of P to mix with the $2s$ of O since they are of similar energy and symmetry. We expect the $2p_x$ of O to mix with the $3p_x$ of P. Similarly, we expect the $2p_y$ and $2p_z$ of O to mix with $3p_y$ and $3p_x$ of P. When the $2s$ and $3s$ mix, a σ and a σ^* will form. When $2p_x$ and $3p_x$ mix, a σ and a σ^* will form. When the $2p_y$ and $3p_y$ mix, a π and a π^* will form. When the $2p_z$ and $3p_z$ mix, a π and a π^* will form.

— $\sigma^*(2p_x, 3p_x)$

$\uparrow$ $\pi^*(2p_z, 3p_z)$ — $\pi^*(2p_y, 3p_y)$

$\uparrow\downarrow$ $\pi(2p_z, 3p_z)$ $\uparrow\downarrow$ $\pi(2p_y, 3p_y)$

$\uparrow\downarrow$ $\sigma(2p_x, 3p_x)$

$\uparrow\downarrow$ $\sigma^*(2s, 3s)$

$\uparrow\downarrow$ $\sigma(2s, 3s)$

Bond order $= \frac{1}{2}$(Bonding electrons − antibonding electrons) $= \frac{1}{2}(8 - 3) = 2.5$.

With this diagram we obtain a bond order of 2.5. This is close to the double or triple bond we obtain from two resonance structures, but we know that Lewis structures have such limitations.

$P = O \longleftrightarrow P \equiv O:$

51. Xe_2^+ has a total of 15 valence electrons. The MO diagram below yields a bond order of 0.5.

$\uparrow$ $\sigma^*(5p_x)$

$\uparrow\downarrow$ $\pi^*(5p_y)$ $\uparrow\downarrow$ $\pi^*(5p_z)$

$\uparrow\downarrow$ $\pi(5p_y)$ $\uparrow\downarrow$ $\pi(5p_z)$

$\uparrow\downarrow$ $\sigma(5p_x)$

$\uparrow\downarrow$ $\sigma^*(5s)$

$\uparrow\downarrow$ $\sigma(5s)$

$[:Xe - Xe:]^+$

53. When a species has an unpaired electron, the electron's magnetic field is not cancelled out by an electron of opposite spin is the same orbital, and there is a small net magnetic field. When an external magnetic field is applied to an odd-electron molecule, the small magnetic field of the unpaired electron can align with the applied field or opposed to it. It is the difference in energy of these two states that makes an ESR spectrum possible.

54. The splitting in the spectrum allows one to determine where the unpaired electron resides. Splittings in ESR spectra arise from interactions between the spins of unpaired electrons and the spins of the protons in the nucleus. Oxygen has no unpaired proton spins and a one-line spectrum would result if the unpaired electron was on the oxygen. Nitrogen has one unpaired proton spin and a three-line spectrum is expected if the unpaired electron resides on the nitrogen. A three-line spectrum is observed and hence ESR leads us to believe that the unpaired electron is on the N.

55. Since a three-line spectrum is observed the unpaired electron is on an atom with an odd number of protons. The only atom with an odd number of protons in this structure is N. One possible Lewis structure is as follows:

```
       :O: :O: :O:        2-
        |   |   |
  [ :O—S —N—S —O: ]
        |       |
       :O:     :O:
```

57. The shape of molecules is determined by the repulsions of the bonded and nonbonded valence electrons. The theory we use to determine shapes is called: **V**alence **S**hell **E**lectron **P**air **R**epulsion theory.

58. Electron pairs involved in bonding are attracted by two nuclei while lone-pair electrons are attracted by only one nucleus. The result is that the lone-pair electrons have greater electron density further from the nucleus compared to bonding electrons. In general:

    lone pair–lone pair repulsions > lone pair–bond pair repulsions > bond pair–bond pair repulsions

    This results in a H—N—H bond angle in NH_3 that is less than 109.5°. This is because the lone pair–bond pair repulsions are greater than the bond pair–bond pair repulsions.

59. The Lewis structure is needed to determine the number of lone pairs and bonded groups around the central atom in a molecule or ion.

60. In the vast majority of cases this is true; use CO_2 as an example. No matter which resonance structure you use, there are two bonded groups coming off the central atom. All three resonance structures have a linear geometry.

```
:O—C≡O: ⟷ O=C=O ⟷ :O≡C—O:
```

61. The bond angle in a triatomic species depends on the number of lone pairs of electrons on the central atom. If the central atom has either zero or three lone pairs, the molecule will have a linear structure (180° bond angle). If the central atom has one lone pair, the molecule will have an angular geometry, with a bond angle between 109.5° and 120°. If there are two lone pairs, the molecule will have an angular geometry, with a bond angle between 90° and 109.5°.

62. In each of these molecules, the central atom has three bonded groups and no lone pairs. VSEPR theory predicts geometry based on the number of bonded groups and lone pairs attached to a central atom. Single, double, triple, and intermediate bonds (because of resonance) all count as one bonded group. Both SO_3 and BF_3 will have a trigonal planar geometry. Note that the different resonance structures of these two species all have the same geometry.

```
   :F:            :F:            :F:
    |              |              ||
:F—B—F:  ⟷   F=B—F:  ⟷   :F—B—F:

                  :F:
                   |
             ⟷ :F—B=F

:O               O              :O
   \              \\               \
    S=O  ⟷        S—O:  ⟷        S—O:
   /              /               //
:O              :O               O
```

63. a. GeH_4 is tetrahedral. b. PH_3 is trigonal pyramidal.

```
     H
     |                H—P—H
  H—Ge—H                |
     |                  H
     H
```

c. H_2S is angular. d. $CHCl_3$ is tetrahedral.

```
                         :Cl:
  H—S—H                    |
                      :Cl—C—H
                           |
                         :Cl:
```

65. a. NH_4^+ is tetrahedral. b. CO_3^{2-} is trigonal planar.

```
  [    H   ]+      [ :O        ]2-
  [    |   ]       [    \      ]
  [ H—N—H  ]       [     C=O   ]
  [    |   ]       [    /      ]
  [    H   ]       [ :O        ]
```

c. NO_2^- is angular. d. XeF_5^+ is square pyramidal.

```
                    [  :F:  :F:   ]+
[:O—N=O]−           [     \ /     ]
                    [ :F—Xe—F:    ]
                    [      \      ]
                    [      :F:    ]
```

67. a. $S_2O_3^{2-}$ is tetrahedral. One possible resonance form is

$$\left[\begin{array}{c} :\ddot{O}: \\ | \\ \ddot{S}=S-\ddot{O}: \\ \| \\ :O: \end{array}\right]^{2-}$$

b. PO_4^{3-} is tetrahedral. One possible resonance form is

$$\left[\begin{array}{c} :O: \\ \| \\ :\ddot{O}-P-\ddot{O}: \\ | \\ :\ddot{O}: \end{array}\right]^{3-}$$

c. NO_3^- is trigonal planar.

$$\left[\begin{array}{c} :\ddot{O}\cdot \\ \diagdown \\ N=\ddot{O} \\ \diagup \\ :\ddot{O}. \end{array}\right]^{-}$$

d. NCO is linear.

$$\cdot N\equiv C-\ddot{O}:$$

69. O_3 and SO_2 both have angular geometries. CO_2 has a linear geometry.

$$\ddot{O}=\ddot{O}-\ddot{O}: \quad :\ddot{O}-\ddot{S}=\ddot{O} \quad \ddot{O}=C=\ddot{O}$$

71. SCN^- and CNO^- both have linear geometries. NO_2^- has an angular geometry.

$$[\ddot{S}=C=\ddot{N}]^- \quad [:C\equiv N-\ddot{O}:]^- \quad [:\ddot{O}-\ddot{N}=\ddot{O}]^-$$

73. S_2O is angular.

$$\ddot{S}=\ddot{S}-\ddot{O}:$$

S_2O_2 is trigonal planar.

$$\begin{array}{c} :\ddot{O}: \\ | \\ \ddot{S}=S-\ddot{O}: \end{array}$$

75. XeF_4 is square planar.

$$\begin{array}{c} :\ddot{F}: \\ | \\ :\ddot{F}-\ddot{X}e-\ddot{F}: \\ | \\ :\ddot{F}: \end{array}$$

Lewis structure and sketch of XeF_5^-.

$$\begin{array}{c} :\ddot{F}: \\ | \\ :\ddot{F}-\ddot{X}e-\ddot{F}: \\ \diagup \quad \diagdown \\ :\ddot{F}. \quad .\ddot{F}: \end{array}$$

77. In the Lewis structure shown all atoms have formal charge of zero. The geometry about the P is tetrahedral.

$$\begin{array}{c} :O: \qquad\quad CH_3 \\ \| \qquad\quad \diagup \\ CH_3-P-\ddot{O}-C-H \\ | \qquad\quad \diagdown \\ :\ddot{F}: \qquad\quad CH_3 \end{array}$$

79. Orbitals that are sp^3 hybridized are not capable of forming π bonds because of their shape. π bonds are most commonly formed by an "above the plane" and a "below-the-plane" overlap of two p orbitals. The shape of an sp^3 hybridized orbital is such that most of the electron density is along the axis of the bond. An sp^3 orbital would not be able to overlap with another orbital effectively in an "above the plane" and a "below the plane" fashion. sp^3 hybridized orbitals can overlap to form σ bonds.

80. Valance bond theory better accounts for molecular geometry. This theory, along with the idea of hybridized orbitals, generally explains the shapes we see via X-ray diffraction quite well.

81. Molecular orbital theory is able to determine if a species is paramagnetic or diamagnetic, regardless of the number of total electrons in the species.

82. d^2sp^3 hybrid orbitals are constructed of orbitals with different principal quantum numbers. For example, sulfur can mix two $3d$ orbitals, one $4s$, and three $4p$ orbitals to make six d^2sp^3 hybrid orbitals.

83. Free atoms are generally regarded as not having hybrid orbitals. Normally, the central atom mixes its orbitals into hybrids (this requires energy) in order to bond with other atoms (this gives off energy). Without the benefit of bonding, a free atom has no need to hybridize and will remain in its ground state.

84. Yes. The central nitrogen has sp hybridized orbitals in each of the three resonance structures of N_2O.

$$:N\equiv N-\ddot{O}: \longleftrightarrow \ddot{N}=N=\ddot{O} \longleftrightarrow :\ddot{N}-N\equiv O:$$

85. a. The N in NO_2^+ is sp hybridized.

$$[\ddot{O}=N=\ddot{O}]^+$$

b. The N in NO_2^- is sp^2 hybridized.

$$[:\ddot{O}-\ddot{N}=\ddot{O}]^-$$

c. The N in N_2O is sp hybridized.

$$:N\equiv N-\ddot{O}:$$

d. Both nitrogens in N_2O_5 are sp^2 hybrids.

$$\begin{array}{c} :O: \qquad :O: \\ \| \qquad\quad \| \\ :\ddot{O}-N-\ddot{O}-N-\ddot{O}: \end{array}$$

e. Both nitrogens in N_2O_3 are sp^2 hybrids.

$$\begin{array}{c} :O: \\ \| \\ \ddot{O}=\ddot{N}-N-\ddot{O} \end{array}$$

87.

$$\begin{array}{ccc} [\ddot{N}= & N= & \ddot{N}]^- \\ \uparrow & \uparrow & \uparrow \\ sp^2 & sp & sp^2 \end{array} \qquad \begin{array}{cccc} \ddot{N}= & N= & \ddot{N}- & \ddot{F}: \\ \uparrow & \uparrow & \uparrow & \\ sp^2 & sp & sp^2 & \end{array}$$

The nitrogens of N_3^- show similar hybridization to those in N_3F. The difference between the two species is that in N_3F one of the terminal nitrogens has two bonded groups and one lone pair (sp^2) while in N_3^- the terminal nitrogens are also sp^2, but they both contain one bonded group and two lone pairs.

89. The S in SF_2 is sp^3 hybridized.

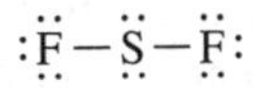

:F–S–F:

The S in SF_4 is dsp^3 hybridized.

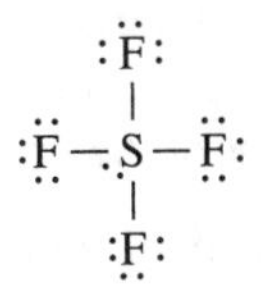

```
    :F:
     |
:F – S – F:
     |
    :F:
```

The S in SF_6 is d^2sp^3 hybridized.

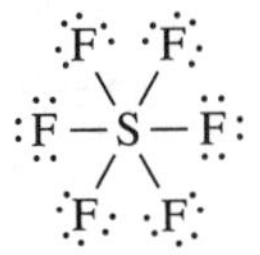

```
  :F:  :F:
     \ /
:F – S – F:
     / \
  :F:  :F:
```

91.

```
      :O:       :O:
       |         |
[ :O – S – O – S – O: ]2–
       |         |
      :O:       :O:
```

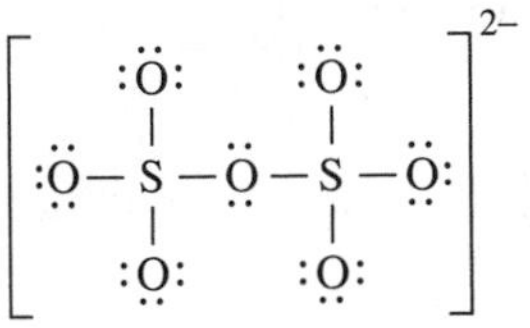

The central oxygen in $S_2O_7^{2-}$ is sp^3 hybridized.

93. No, ClO_2^- is angular and Cl is sp^3 hybridized, while SO_2 is angular and S is sp^2 hybridized.

[:O–Cl–O:]$^-$:O–S=O

95. Cl_3^+ is angular and the central Cl is sp^3 hybridized.

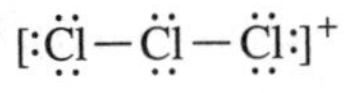

[:Cl–Cl–Cl:]$^+$

97. In N_4O, there are two sp^2 hybridized nitrogens and two sp hybridized nitrogens.

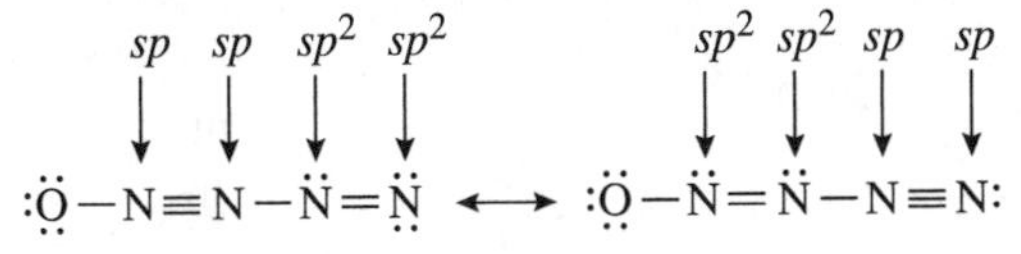

sp sp sp^2 *sp*^{2} *sp*^{2} *sp*^{2} *sp sp*

:O–N≡N–N=N ⟷ :O–N=N–N≡N:

99.

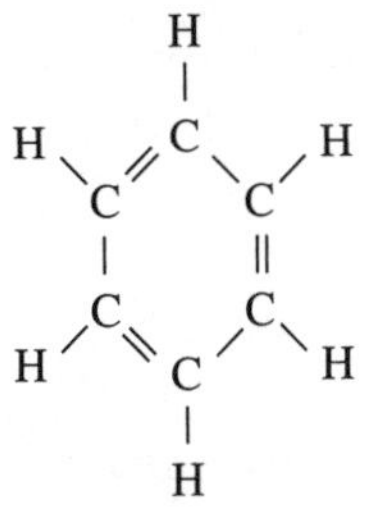

All carbons are sp^2 hybridized.

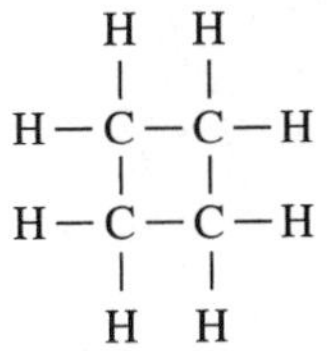

All carbons are sp^3 hybridized.

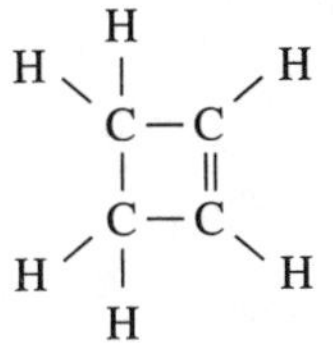

Two carbons are sp^3 hybridized and two carbons sp^2 hybridized.

101. A polar bond is one with an uneven or unsymmetrical distribution of electron density between the two bonding atoms. The atom with the greater electronegativity will have a greater share of the electron density between the two atoms. In a polar molecule there is overall an uneven charge distribution, which results in a permanent dipole moment. To determine if a molecule is polar, we must look at the polarity of the individual bonds and their spatial distribution.

102. Yes, a polar molecule must have overall an uneven charge distribution. In order for this to be the case, the molecule must contain polar bonds.

103. Yes, there are many examples of nonpolar molecules that contain polar bonds. The polarity of the individual bonds can cancel out if they are aligned and spatially distributed in the correct manner.

104. A dipole moment is a measure of the magnitude of a molecule's polarity.

105. CCl_4 is a nonpolar molecule.

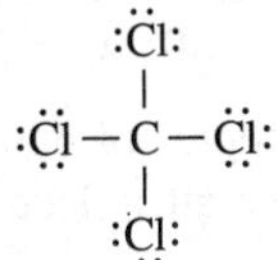

```
      :Cl:
       |
:Cl – C – Cl:
       |
      :Cl:
```

$CHCl_3$ is a polar molecule.

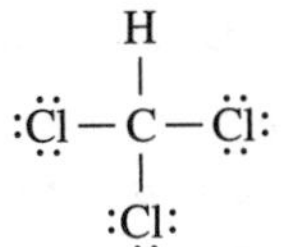

```
       H
       |
:Cl – C – Cl:
       |
      :Cl:
```

CO_2 is a nonpolar molecule.

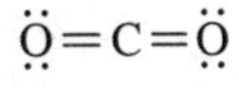

O=C=O

H_2S is a polar molecule.

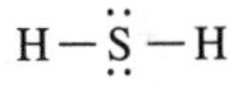

H–S–H

SO_2 is a polar molecule.

:O–S=O

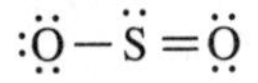

107. a. $CFCl_3$ is a polar molecule.

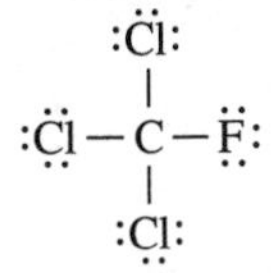

```
      :Cl:
       |
:Cl – C – F:
       |
      :Cl:
```

b. CF_2Cl_2 is a polar molecule.

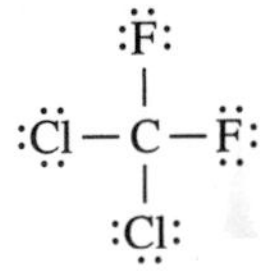

```
      :F:
       |
:Cl – C – F:
       |
      :Cl:
```

c. Cl_2FCCF_2Cl is a polar molecule.

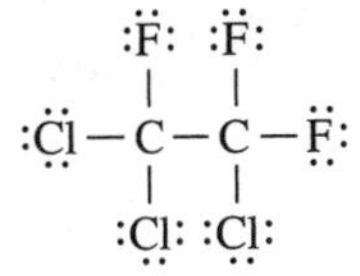

```
      :F:  :F:
       |    |
:Cl – C  – C – F:
       |    |
      :Cl: :Cl:
```

109. a. $CBrF_3$ is more polar than $CClF_3$.

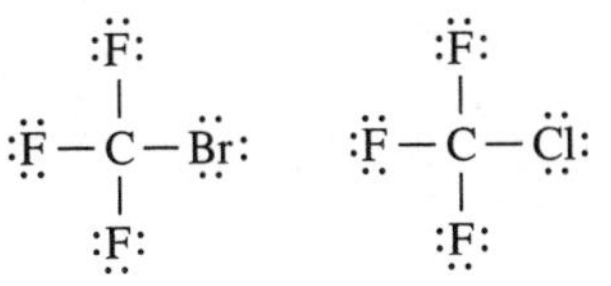

```
     :F:              :F:
      |                |
:F – C – Br:     :F – C – Cl:
      |                |
     :F:              :F:
```

b. CHF_2Cl is more polar than CF_2Cl_2.

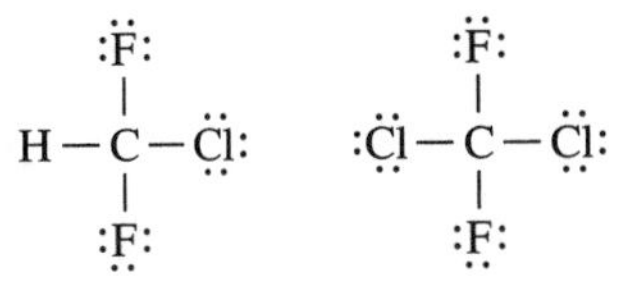

```
     :F:              :F:
      |                |
H – C – Cl:     :Cl – C – Cl:
      |                |
     :F:              :F:
```

c. Cl_2FCCF_2Cl is more polar than ClF_2CCF_2Cl.

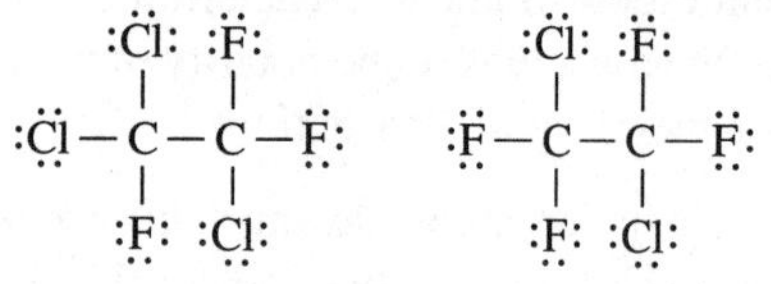

```
      :Cl: :F:              :Cl: :F:
       |    |                |    |
  :Cl—C—C—F:           :F—C—C—F:
       |    |                |    |
      :F: :Cl:              :F: :Cl:
```

111.

```
:Cl               :Br               :I
    \                 \                 \
     C=O               C=O               C=O
    /                 /                 /
:Cl               :Br               :I
```

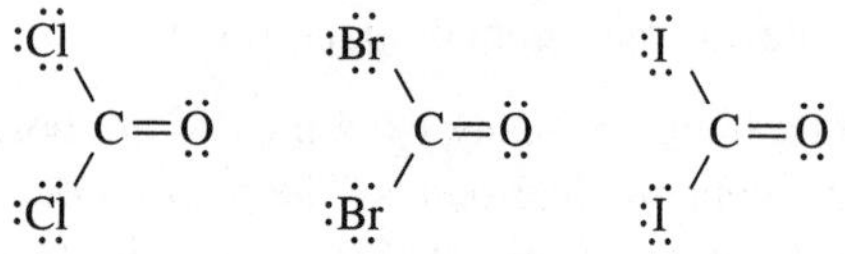

All three of the above molecules are trigonal planar. In order to determine their relative polarities, we must perform vector analysis of the polar bonds. We will use the difference in electronegativity (Δx) to determine the degree of polarity in each band.

Cl—C $\Delta x = 0.5$ Br—C $\Delta x = 0.3$ I—C $\Delta x = 0.0$
C=O $\Delta x = 1.0$ C=O $\Delta x = 1.0$ C=O $\Delta x = 1.0$

Since Δx is a measure of the degree of bond polarity, we can use it to determine the overall molecule polarity.

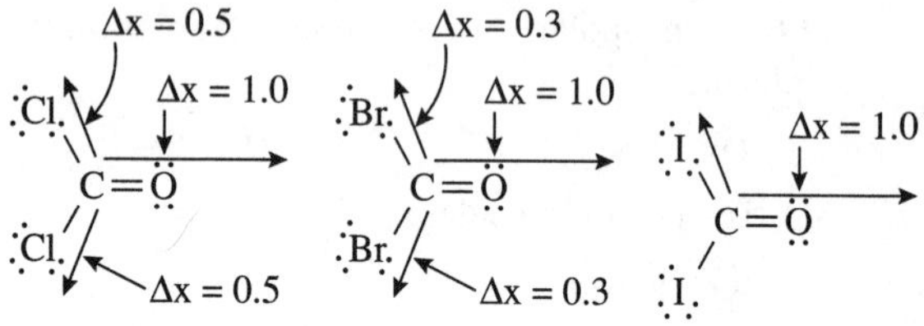

Δx = 0.5
Δx = 1.0
Cl
C=O
Cl
Δx = 0.5

Δx = 0.3
Δx = 1.0
Br
C=O
Br
Δx = 0.3

Δx = 1.0
I
C=O
I

COI_2 will be the most polar molecule since the C—I bonds it contains are essentially nonpolar and do not cancel out with the vector representing the C—O bond polarity. Similarly we can see that Cl_2CO is the least polar of the three molecules.

113. The two components will definitely have different dipole moments, which can be used to distinguish them. We expect the structure on the left to be polar and the structure on the right to be nonpolar.

115. :C≡O:

$\mu = Qd$

$$\frac{(0.112\,\text{D})(3.336 \times 10^{-30}\,\text{C}\cdot\text{m})}{1\,\text{D}} = 3.74 \times 10^{-31}\,\text{C}\cdot\text{m}$$

$$\frac{(113\,\text{pm})\,(\text{m})}{1 \times 10^{12}\,\text{pm}} = 1.13 \times 10^{-10}\,\text{m}$$

$$3.74 \times 10^{-31}\,\text{C}\cdot\text{m} = Q(1.13 \times 10^{-10}\,\text{m})$$

$$3.31 \times 10^{-21}\,\text{C} = Q$$

$$\frac{(3.3 \times 10^{-21}\,\text{C})(1\,e^-)}{1.602 \times 10^{-19}\,\text{C}} = 0.0206\,e^-$$

The carbon atom will have a partial positive charge equivalent to the charge on 0.0206 electrons.

CHAPTER 8 Properties of Gases and the Air That We Breathe

1. We can consider a force to be a push or a pull and we define it as mass times acceleration. Pressure is force per unit area (the total force divided by the area (l × w).

$$\text{Force} = \text{m} \times \text{a}$$

$$\text{Pressure} = \frac{\text{force}}{\text{area}}$$

2. The earliest barometers consisted of a glass tube that was about a meter in length. The tube was filled with liquid mercury and was quickly inverted into a container that also contained liquid mercury. Some of the mercury in the tube was pulled into the mercury pool due to the force of gravity. In such an apparatus, the empty space above the mercury in the tube is effectively a vacuum. The air surrounding the mercury pool is also applying a force due to the air molecules pushing against the mercury pool. This is called the air pressure or atmospheric pressure. This force pushes the mercury in the pool back up the glass tube. The net effect of these opposing forces is indicated by the height of the mercury in the column. We can easily measure the height of this column, which provides a measure of atmospheric pressure.

   If the atmospheric pressure decreases, the force acting on the mercury pool decreases and the height of the mercury column is lower.

3. A torr is 1/760 of an atmosphere.
   1 atmosphere = 760 mm Hg = 760 torr

4. 10 millibars = 1 kilopascal = 1000 pascals
   1 millibar = 100 pascals

5. The three barometers are identical in every way except they are filled with different liquids. In order to exert the same pressure, they must contain the same mass of liquid because both the cross-sectional area of the column and the acceleration associated with gravity will be same in each barometer.

$$\text{pressure} = \frac{\text{force}}{\text{area}} = \frac{(\text{mass})(\text{acceleration})}{\text{area}}$$

   The volume of the liquids in the barometer must be different because each of the three liquids has a different density. Since the area of each column is the same, the height of the columns must differ in order for their volumes to differ.

$$V = \frac{m}{d}$$

   The liquid with the smallest density needs the largest volume and so the height of the column must be the highest.

   Ethanol is the least dense of the three liquids, so the height of the ethanol column must be the tallest.

6. The denser the liquid, the shorter the column height needed (see Problem 5). For example, if the liquid chosen is mercury, the column at standard atmospheric pressure of 1 atmosphere is 760 mm in height. If water is chosen, the column must contain the same mass as the mercury column. Water has a density of 1 g/cm^3 while mercury has a density of about 13.6 g/cm^3. This means that the water column must be 13.6 times higher than the mercury column or about 13.6 × 760 mm = 10,336 mm water (407 inches of water or about 34 feet of water).

For mercury, a very dense liquid, the column height is easily accommodated in a normal room. We do not need a column longer than about 1 meter in height. If the liquid in the barometer were water, the column length would need to be close to 40 feet in length. Since this is higher than many houses, the size of the mercury barometer is much more practical.

7. Pressure is defined as force per unit area. A sharpened blade has a smaller area than a dull blade, so dividing the same force by a smaller area results in a larger pressure.

8. The increased area of the snowshoe results in a lower pressure being exerted on the snow. This should make it easier to walk because one is not likely to sink into the snow as deeply.

9. Pressure is defined as force per unit area, that is, $P = F/A$.

   Force is defined as mass times acceleration, $F = ma$. For objects on the surface of the earth, the acceleration is due to the force of gravity. This value for acceleration (a) is 9.8 m/s^2.

   We need to determine the force exerted by the cube of iron. The mass of the iron cube is 1.00 kg.

$$m = 1.00 \text{ kg}; \qquad a = 9.8 \text{ m/s}^2$$

$$F = (1.00 \text{ kg})(9.8 \text{ m/s}^2) = 9.8 \text{ kg} \cdot \text{m/s}^2 \text{ or N (newtons)}$$

Note: The newton is the SI unit for force. It has units of kg $\cdot$ m/s^2.

To determine the pressure, we need to divide the force by the area of the cube. Also, since the distance unit in a newton is the meter, we need to convert the length of each side of the cube from centimeters to meters. (1 m = 100 cm)

$$5 \text{ cm} = 0.05 \text{ m}$$

$$A = l \times w = 0.05 \text{ m} \times 0.05 \text{ m} = 2.5 \times 10^{-3} \text{ m}^2$$

Now we can calculate the pressure exerted by the iron cube.

$$P = \frac{F}{A} = \frac{9.8 \text{ kg} \cdot \text{m/s}^2}{2.50 \times 10^{-3} \text{ m}^2} = 3.92 \times 10^3 \frac{\text{kg}}{\text{m} \cdot \text{s}^2},$$

$$\text{or} \quad 3.92 \times 10^3 \frac{\text{N}}{\text{m}^2} \; (3.92 \times 10^3 \text{ Pa})$$

11. Some conversion factors for pressure:

   1 atm = 760 mm Hg = 760 torr = 30 in Hg
   1 atm = 14.7 lbs/in^2
   1 atm = 1.013×10^5 Pa = 1.013×10^2 kPa
   1 atm = 1013.25 mb (millibars)
   1 Pa = 1 N/m^2 (N is newton)

   a. $2.0 \text{ kPa} \times \dfrac{1.00 \text{ atm}}{1.013 \times 10^2 \text{ kPa}} = 0.01974 \text{ atm},$

   or 0.020 atm (to 2 significant figures)

   b. $562 \text{ mm Hg} \times \dfrac{1.00 \text{ atm}}{760 \text{ mm Hg}} = 0.739 \text{ atm}$

   c. $4.19 \times 10^5 \dfrac{\text{N}}{\text{m}^2} \times \dfrac{1 \text{ Pa}}{1 \text{ N/m}^2} \times \dfrac{1.00 \text{ atm}}{1.013 \times 10^2 \text{ Pa}}$

   $= 4.14 \text{ atm}$

13. The density of air decreases as the altitude increases. Pressure is related to the total mass of molecules of gas above a given altitude. As altitude increases, the mass of the gases still above it must decrease. Less mass means less pressure.

   Alternatively, we know that at higher altitude there are fewer molecules per unit of volume. If we measure pressure as the number of collisions between gas particles and any other object, then fewer molecules result in many fewer collisions between gas particles themselves and/or gas particles and any sensor. The end result is lower pressure.

14. STP stands for standard temperature and pressure, where standard pressure is 1.00 atmosphere and standard temperature is 0.0°C or 273 K.

   The volume of 1.00 mole of an ideal gas at STP is 22.414 L. Usually this quantity is expressed to only three significant figures (i.e., 22.4 L/mole).

15. If we compress the gas to half its volume, the pressure should double. If we change the temperature from 20°C to 10°C (293 K to 283 K), the pressure should decrease only slightly because the fraction 283/293 is quite close to 1.

   In this example, the change in volume has a much greater effect on the pressure than does the change in temperature. The pressure of the gas will increase.

16. According to Boyle's law, pressure and volume are inversely proportional, or pressure times volume must equal a constant (temperature and sample size are constant).

$$P_1V_1 = P_2V_2 \quad \text{or} \quad \frac{P_1}{P_2} = \frac{V_2}{V_1}$$

   If the volume is reduced by one tenth then the pressure must increase by a factor of 10.

17. Information given:

   Amount of ammonia = 1.00 mol
   P_1 = 1.00 atm
   V_1 = 78 mL
   V_2 = 39 mL

   We want to find the final pressure of ammonia.

Since the quantity of ammonia is always 1 mole and the temperature is constant, this problem reduces to a Boyle's law problem, $P_1V_1 = P_2V_2$.

Note: We do not need to convert the volumes into liters when using Boyle's law because there are always two volume quantities and two pressure quantities so the units will cancel. The only requirement is that the two volumes are in the same unit (both in mL, or gallons, etc.) and the two pressures have the same unit.

We can substitute the values and solve for the final pressures as follows:

$$P_1V_1 = P_2V_2$$

$$(1.00 \text{ atm})(78. \text{ mL}) = (P_2)(39. \text{ mL})$$

$$P_2 = \frac{P_1V_1}{V_2} = \frac{(1.00 \text{ atm})(78. \text{ mL})}{39. \text{ m}L} = 2.0 \text{ atm}$$

The final pressure of the ammonia gas is 2.0 atm.

If we decrease the volume, then the pressure should increase. The answer seems reasonable.

19. Information given:

$V_1 = 153$ L
$V_2 = 352$ L
$P_2 = 1.00$ atm

The pressure on the diver increases by 1 atm for every 10 m of water.

We want to find the pressure at the underwater site.

Before we can answer the second question (the depth of the diver), we need to calculate the pressure where the diver released the balloon.

From the information given, we can use Boyle's law to determine this pressure. From the pressure, we can then determine the depth of the diver.

$$P_1V_1 = P_2V_2$$

$$(P_1)(153 \text{ L}) = (1.00 \text{ atm})(352 \text{ L})$$

Solving for the initial pressure:

$$P_1 = \frac{P_2V_2}{V_1} = \frac{(1.00 \text{ atm})(352 \text{ L})}{153 \text{ L}} = 2.30 \text{ atm}$$

The pressure at the underwater site is 2.33 atm.

If the pressure increases by 1 atm for every 10 m and the pressure at the surface is 1.00 atm, then the increase in pressure due to the water is 1.33 atm. To find the depth that corresponds to an additional 1.33 atm, multiply this value by 10 m per atm.

$$1.33 \text{ atm} \times \frac{10 \text{ m}}{1 \text{ atm}} = 13.3 \text{ m}$$

The diver is at a depth of 13.3 m (42.9 ft).

21. Let's make a plot of volume vs. $1/P$ and see if we get a straight line going through the origin (the point on the graph when both the x and y values are zero). If we do, then the data obeys Boyle's law.

As you can see from the plot, the data does obey Boyle's law.

According to the ideal gas law, if the number of moles and temperature of a gas are identical then the volume occupied by any gas at identical pressure should be the same. For example, 1 mol of any ideal gas at 1.00 atm and 273 K will occupy a volume of 22.4 L. The identity of the gas is not important.

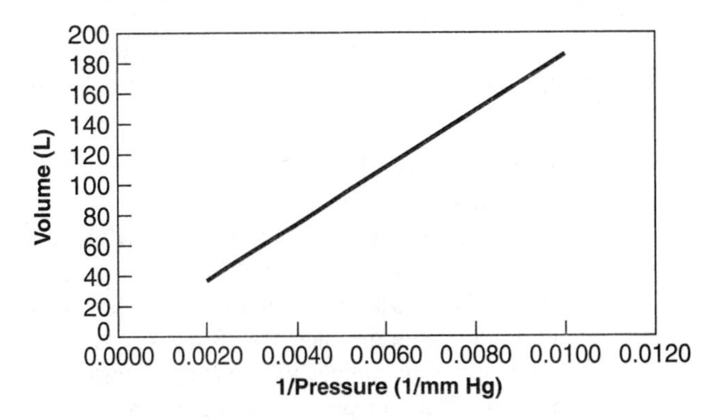

The graph would not change if the identity of the gas changed from H_2 to Ar as long as it behaved ideally.

23. The temperature inside the cylinder doubles from 298 K (25°C) to 596 K (323°C).

If the volume of the cylinder doubles ($\frac{V_2}{V_1} = 2$), the temperature (in kelvins) must also double, that is, $\frac{T_2}{T_1} = 2$.

The temperature inside the cylinder doubles.

25. Information given:

$T_1 = 250$ K
$T_2 = 398$ K
$V_1 = 2.68$ L

Pressure and sample size are constant.

We want to find the volume of the gas when the temperature is changed (V_2).

We have been given two temperatures and one volume and are asked to find the final volume, so we can use Charles's law.

$$\frac{V_2}{V_1} = \frac{T_2}{T_1}$$

Substituting in the equation gives

$$\frac{V_2}{2.68 \text{ L}} = \frac{398 \text{ K}}{250 \text{ K}}$$

Solving for V_2 yields

$$V_2 = \frac{(398 \text{ K})(2.68 \text{ L})}{250 \text{ K}} = 4.27 \text{ L}$$

We expect that if the temperature of a gas increases, its volume should also increase. This answer makes sense.

27. Plot 1 is the correct plot of volume versus temperature for an ideal gas. According to Charles's law the volume of a gas is directly proportional to its Kelvin temperature. When a plot of volume versus temperature is drawn, the result should be a straight line with a positive slope.

   Note: A straight line is obtained even if the temperature axis is in Centigrade (Celsius) degrees and not Kelvin degrees. If the Kelvin scale is used, the line will go through the origin. If the Centigrade scale is used, the line will intersect the x axis at $-273°C$.

   Analysis of plot 2 reveals that the volume decreases as temperature increases. This is not consistent with Charles's law.

29. The change that has the largest fractional ratio will have the greater effect on the volume.

   Decreasing the pressure from 760 mm Hg to 720 mm Hg increases the volume occupied by the gas sample (according to Boyle's law) by a factor of 760/720 (1.06).

   When the temperature increases from 0°C to 40°C the volume of the gas sample will increase according to Charles's law. The increase in volume is directly proportional to the change in Kelvin temperature, so the volume will increase by a factor of 313/283 (1.11).

   Raising the temperature increases the volume of the container by a factor of 1.11. Decreasing the pressure only increases the volume by a factor of 1.06, so raising the temperature, in this case, produces the greater increase in volume.

31. a. In this problem the temperature and pressure of a gas sample are doubled.

      According to the ideal-gas law, we find that the product of the pressure and volume is directly proportional to the Kelvin temperature. Mathematically this is represented as

$$PV \propto T$$

      Doubling the pressure results in the volume being halved, while doubling the temperature results in the volume doubling. These two factors counteract each other and the volume remains the same.

   b. The temperature is halved and the pressure is doubled:

      Halving the temperature should halve the volume, and doubling the pressure also halves the volume. Each change halves the volume, so the overall effect is to reduce the volume to one quarter of its original level.

   c. The temperature is increased by 75% and the pressure is increased by 50%, so

$$T_2 = 1.75\ T_1; \qquad P_2 = 1.50\ P_1$$

      We can use the expression $\frac{P_1 V_1}{T_1} = \frac{P_2 V_2}{T_2}$ to determine what happens to the volume when both temperature and pressure are allowed to change.

      Solving for V_2/V_1 gives

$$\frac{V_2}{V_1} = \frac{T_2 P_1}{T_1 P_2} = (1.75)(0.666) = 1.17$$

      The volume increases by a factor of 1.17 or about 17%.

33. We can assume that the sample size (number of moles) is constant and that the gas sample in the container is not changed.

   Volume is directly proportional to temperature and inversely proportional to pressure. Using the expression $PV \propto T$ we can deduce that a larger volume will correspond to a higher temperature and lower pressure. At a given pressure (or a given value of $1/P$) the sample represented by line 2 has a larger volume than the sample represented by line 1. The remaining variable, the temperature, must be greater for line 2 than line 1.

   Line 2 corresponds to the sample of gas at the higher temperature.

35. Information given:

   $V_1 = 200$ L
   $T_1 = 20°C = 293$ K
   $T_2 = 210$ K
   $P_1 = 1.00$ atm $= 760$ mm Hg
   $P_2 = 63$ mm Hg

   We need to find the volume of the balloon at 20,000 m.

   The formula to use is the one that relates the three variables, pressure, volume, and temperature, to one another.

$$\frac{P_2 V_2}{T_2} = \frac{P_1 V_1}{T_1}, \quad \text{or} \quad \frac{P_2 V_2}{P_1 V_1} = \frac{T_2}{T_1}$$

   Before we can substitute values into this formula, we need to have the two pressure values in terms of the same unit. We can convert 1 atm to 760 mm Hg. Now we can perform the calculation. Solving for V_2 yields

$$V_2 = \frac{T_2 P_1 V_1}{T_1 P_2} = \frac{(210\text{ K})(760\text{ mm Hg})(200\text{ L})}{(293\text{ K})(63\text{ mm Hg})}$$

$$= 1729\text{ L}, \quad \text{or} \quad 1.73 \times 10^3\text{ L}$$

   The weather balloon expands to a volume of 1.73×10^3 L at 20,000 m.

37. Information given:

   $V_1 = 2.0$ L
   $T_1 = 22°C = 295$ K
   $T_2 = -22°C = 251$ K

   We want to find the final volume of the balloon, V_2.

In this problem the only variables are volume and temperature. The pressure and the sample size remain constant, so we can use Charles's law.

$$\frac{V_1}{T_1} = \frac{V_2}{T_2}, \quad \text{or} \quad \frac{V_2}{V_1} = \frac{T_2}{T_1}$$

The temperature must be in kelvins.

Solving for V_2

$$V_2 = \frac{T_2 T_1}{T_1} = \frac{(251 \text{ K})(2.0 \text{ L})}{(295 \text{ K})} = 1.70 \text{ L}$$

The volume of the balloon outside the window is 1.70 L.

This answer makes sense because the volume occupied by an ideal gas is directly proportional to the Kelvin temperature. Decreasing the temperature must decrease the volume of the balloon.

39. Information given:

$V = 2.36$ L
$T = 17°\text{C} = 290$ K
$P = 6.8$ atm

We want to find the number of moles.

From this information, we see that the formula we need is the ideal-gas law, $PV = nRT$, where P is the pressure in atmosphere, V is the volume in liters, n is the number of moles, and T is the temperature in kelvins. R is the ideal-gas law constant. This constant has several different values depending on its units. The most common value of R is 0.0821 L · atm/mol · K.

One can either substitute the data into the ideal-gas law and then solve for n, or rearrange the gas law to find n.

$$n = \frac{PV}{RT} = \frac{(6.8 \text{ atm})(2.36 \text{ L})}{(0.0821)(290 \text{ K})}$$

$$= 0.674 \text{ mol of air}$$

There are 0.67 (2 significant figures) mol of air in the tire.

41. Information given:

$V = 500$ mL $= 0.500$ L
$T = 298$ K
Mass of deuterium = 1.00 g

We want to find the pressure exerted by the D_2 (deuterium) gas.

This appears to be an ideal-gas law problem; however we do not know the number of moles of D_2.

Deuterium is an isotope of hydrogen that contains 1 proton, 1 neutron, and 1 electron. It has an atomic mass of 2.02 g/mol (or 2.02 amu). From this information we can find the number of moles of D_2 in the bottle.

The molar mass of D_2 = 4.04 g/mol

$$\text{moles of } D_2 = 1.00 \text{ g } D_2 \times \frac{1 \text{ mol } D_2}{4.04 \text{ g } D_2} = 0.248 \text{ mol } D_2$$

Using the ideal-gas equation, solving for P, and substituting yields

$$P = \frac{nRT}{V}$$

$$= \frac{(0.248 \text{ mol})(0.0821 \text{ L} \cdot \text{atm/mol} \cdot \text{K})(298 \text{ K})}{0.500 \text{ L}}$$

$$= 12.1 \text{ atm}$$

The pressure of D_2 in the container is 12.1 atm (3 significant figures).

43. Information given:

$V = 50.0$ L
$T = 20°\text{C} = 293$ K
$P = 2850$ pounds/in (psi)

We want to find the mass of hydrogen in the container.

In this problem, we know the pressure, volume, and temperature. From this information we can determine the number of moles of hydrogen (H_2) gas in the container. To find the mass of H_2, multiply the moles calculated by the molar mass of H_2 (2.02 g/mol).

We can use $PV = nRT$ and solve for n (moles). However, before we can substitute the data into the ideal-gas equation, we need to convert the pressure from units of psi to atmospheres.

$$1 \text{ atm} = 14.7 \text{ psi}$$

$$2850 \text{ psi} \times \frac{1 \text{ atm}}{14.7 \text{ psi}} = 194 \text{ atm}$$

Now we can use the ideal-gas equation to find the number of moles of H_2.

$$n = \frac{PV}{RT} = \frac{(194 \text{ atm})(50.0 \text{ L})}{(0.0821 \text{ L} \cdot \text{atm/mol} \cdot \text{K})(293 \text{ K})}$$

$$= 403 \text{ mol } H_2$$

Converting 403 mol to mass of H_2 in grams:

$$403 \text{ mol } H_2 \times \frac{2.02 \text{ g } H_2}{\text{mol } H_2} = 814 \text{ g } H_2$$

There are 814 g of H_2 in the tank.

45. Use the information given to determine the number of moles of acetylene, C_2H_2. Then use the balanced equation to determine the moles of calcium carbide (CaC_2). Lastly, convert the moles of calcium carbide to grams of calcium carbide.

Note: The process described in the problem deals with the formation of acetylene. We do not need to concern ourselves with the next step in this process, the burning of the acetylene.

a. Information given:

Acetylene consumption = 1.0 L/hr
Time = 4.0 hr
$P = 1.00$ atm
$T = 18°C = 291$ K

We know that 1.0 L of acetylene is consumed per hour. We can convert this quantity to moles of acetylene using the ideal-gas equation. Moles of C_2H_2 in 1.0 L

$$n = \frac{PV}{RT} = \frac{(1.00 \text{ atm})(1.0 \text{ L})}{(0.0821 \text{ L} \cdot \text{atm/mol} \cdot \text{K})(291 \text{ K})}$$

$$= 0.0419 \text{ mol } C_2H_2$$

It requires 0.0419 mol (3 significant figures) of C_2H_2 to keep the lamp burning for an hour.

b. Since the lamp needs to burn for 4.0 hr and it requires 0.042 mol of C_2H_2 per hour, the total mole of acetylene required is 4 hr × 0.042 mol/hr = 0.168 mol C_2H_2.

Next, use the balanced equation to find how many moles of calcium carbide (CaC_2) are needed. From the balanced equation we get the conversion factor 1 mole CaC_2 = 1 mole C_2H_2.

Remember to use the 0.168 mol of C_2H_2 (that is needed to light the lamp for 4 hr and not the 0.042 mol C_2H_2 needed to light the lamp for only 1 hr).

$$\text{moles of } CaC_2 = 0.168 \text{ mol } C_2H_2 \times \frac{1 \text{ mol } CaC_2}{1 \text{ mol } C_2H_2}$$

$$= 0.168 \text{ mol } CaC_2$$

Lastly, determine the mass of calcium carbide needed.

$$\text{mass of } CaC_2 = 0.168 \text{ mol } CaC_2 \times \frac{64.1 \text{ g } CaC_2}{1 \text{ mol } CaC_2}$$

$$= 10.8 \text{ g } CaC_2, \quad \text{or} \quad 11 \text{ g}$$

Note: Since the time, 4.0 hr, contains only two significant figures the answer should be reported to two significant figures.

The lamp must contain a minimum of 11 g of calcium carbide in order to burn for at least 4 hr.

47. Information given:

$P = 0.85$ atm
$T = 273$ K
$V = 200.$ L

We want to find the mass of $KClO_3$ required for this process.

From the information given, we can calculate how many moles of oxygen (O_2) gas are required. We then use the coefficients of the balanced equation to determine the moles of potassium chlorate ($KClO_3$) required, and convert this to mass of $KClO_3$.

First, find how many moles of O_2 are consumed using $PV = nRT$.

$$n = \frac{PV}{RT} = \frac{(0.85 \text{ atm})(200. \text{ L})}{(0.0821 \text{ L} \cdot \text{atm/mol} \cdot \text{K})(273 \text{ K})}$$

$$= 7.6 \text{ mol } O_2$$

Next, use the balanced equation to determine the moles of $KClO_3$ required.

$$2\, KClO_3(s) \rightarrow 2\, KCl(s) + 3\, O_2(g)$$

$$3 \text{ moles } O_2 = 2 \text{ moles } KClO_3$$

$$\text{moles of } KClO_3 = 7.6 \text{ mol } O_2 \times \frac{2 \text{ mol } KClO_3}{3 \text{ mol } O_2}$$

$$= 5.1 \text{ mol } KClO_3$$

Lastly, convert moles of $KClO_3$ to mass of $KClO_3$ in grams (molar mass = 122.6 g/mol).

$$\text{mass of } KClO_3 = 5.1 \text{ mol } KClO_3 \times \frac{122.6 \text{ g } KClO_3}{1 \text{ mol } KClO_3}$$

$$= 625 \text{ g } KClO_3, \text{ or } 6.3 \times 10^2 \text{ g } KClO_3$$

The mass of $KClO_3$ needed is 6.3×10^2 g.

49. In this problem, we need to determine the total volume of carbon dioxide (CO_2) exhaled in 24.0 hours. We can then use this volume of CO_2, along with the other information given, to determine how many moles of CO_2 are exhaled by the sailor (using the ideal-gas equation). Next, we can use the balanced equation to convert moles of CO_2 to moles of Na_2O_2. Lastly, we convert moles of Na_2O_2 to mass of Na_2O_2 in grams. The molar mass of Na_2O_2 is 78.0 g/mol.

Information given:

Rate of CO_2 exhalation = 150 mL/min
Time = 24 h
$T = 20°C = 293$ K
$P = 1.02$ atm

We want to find the mass of Na_2O_2 needed to remove the CO_2 exhaled by the sailor:

Total volume of carbon dioxide exhaled by the sailor:

$$\frac{150 \text{ mL}}{1 \text{ min}} \times \frac{60 \text{ min}}{1 \text{ hr}} \times \frac{24 \text{ hr}}{1 \text{ day}}$$

$$= 2.16 \times 10^5 \text{ mL } CO_2\text{/day, or } 216 \text{ L/day}$$

The sailor exhales 216 L of CO_2 per day. Next, use the ideal-gas law to find how many moles of CO_2 are exhaled per day.

$$\text{moles of } CO_2 = \frac{PV}{RT}$$
$$= \frac{(1.02 \text{ atm})(216 \text{ L})}{(0.0821 \text{ L} \cdot \text{atm/mol} \cdot \text{K})(293 \text{ K})}$$
$$= 9.16 \text{ mol } CO_2$$

Lastly, use the balanced equation to find the mass of Na_2O_2 required per sailor. The balanced equation is

$$2\, Na_2O_2(s) + 2\, CO_2(g) \rightarrow 2\, Na_2CO_3(s) + O_2(g)$$

The needed conversion factor is 2 moles Na_2O_2 = 2 moles Na_2CO_3.

$$\text{mass of } Na_2O_2 = 9.16 \text{ mol } CO_2 \times \frac{2 \text{ mol } Na_2O_2}{2 \text{ mol } CO_2}$$
$$\times \frac{78.0 \text{ g } Na_2O_2}{1 \text{ mol } Na_2O_2} = 714 \text{ g } Na_2O_2$$

Each sailor requires 714 g of sodium peroxide (Na_2O_2) per day.

51. No. The density of a gas is given by the formula $d = \frac{P\mathcal{M}}{RT}$, where P is the pressure, $\mathcal{M}$ is the molar mass, and T is the temperature of the gas in kelvins.

    At constant temperature and pressure, the density of a gas is directly proportional to the molar mass of the gas. In other words, the density of a gas increases as the molar mass of the gas increases. The larger the molar mass of the gas, the greater its density.

52. Warm air rises because it is less dense than colder air. If the pressure of a specified gas remains the same, the density of a gas is inversely proportional to the temperature. In other words, hotter gases have lower densities and rise and colder gases have larger densities and sink.

53. a. The density of a gas (at constant temperature) is directly proportional to the pressure of the gas. Increasing the pressure of a gas at constant temperature will increase the density of the gas.
    b. The density of a gas (at constant pressure) is inversely proportional to its Kelvin temperature. Decreasing the temperature at constant pressure will increase the density of the gas.

54. One method of determining the density of a gas uses a special type of glass bulb of known volume. The bulb is evacuated (that is, a pump is used to remove any air from the bulb, creating a vacuum). The glass bulb is weighed empty. It is then filled with a gas and the bulb is reweighed. The difference between the two mass measurements is the mass of the gas in the glass bulb. Lastly, the mass of the gas is divided by the known volume of the bulb to determine the density of the gas.

55. a. Information given:

       $T = 298$ K
       $P = 1.00$ atm

       We want to find the density of radon gas under these conditions.

       The formula for gas density is $d = \frac{P\mathcal{M}}{RT}$

       When using this formula, the molar mass (or atomic weight for atomic species) must be in grams/mole, the pressure in atmospheres, and the temperature in kelvins. The ideal gas constant, R, is 0.0821 L · atm/mol · K. The resulting units for the gas density are grams/liter.

       Solving for the density of radon gas yields:

$$d = \frac{P\mathcal{M}}{RT} = \frac{(1.00 \text{ atm})(222 \text{ g/mol})}{(0.0821 \text{ L} \cdot \text{atm/mol} \cdot \text{K})(298 \text{ K})}$$
$$= 9.07 \text{ g/L}$$

       The density of radon gas is 9.07 g/L.
    b. Radon concentrations are likely to be greater in the basement because the density of radon gas is much greater than the density of air and denser gases sink.

57. Under conditions of constant temperature and pressure, the density of a gas is directly proportional to the molar mass of the gas; that is, the gas that has the greater molar mass has the greater density. Since the molar mass of nitrogen (N_2) is 28.0 g/mol and the molar mass of methane (CH_4) is 16.0 g/mol, the density of nitrogen is greater than the density of methane if the temperature and pressure are the same. For any given pressure the density on line 2 is greater than the density on line 1, so line 1 should be labeled *methane*.

59. If we can determine the molar mass of the unknown gas, we can compare it to the molar masses of SO_2 and SO_3.

    To determine the molar mass of a gas using the ideal-gas equation we can use the formula.

$$d = \frac{P\mathcal{M}}{RT}$$

Solving for molar mass yields

$$\mathcal{M} = \frac{dRT}{P}$$

Since density $= \frac{\text{mass}}{\text{volume}}$ $(d = m/V)$, we can rewrite the preceding equation as

$$M = \frac{mRT}{PV}$$

Using this equation we can easily determine the molar mass of the gas in this question.

Information given:

Mass = 0.391 g
$P = 750$ mm Hg = 0.987 atm
$V = 150$ mL = 0.150 L
$T = 22°\text{C} = 295$ K

$$\begin{aligned}\mathcal{M} &= \frac{mRT}{PV} \\ &= \frac{(0.391\text{ g})(0.0821\text{ L}\cdot\text{atm/mol}\cdot\text{K})(295\text{ K})}{(0.987\text{ atm})(0.150\text{ L})} \\ &= 64.0\text{ g/mol}\end{aligned}$$

The molar mass of the unknown gas is 64.0 g/mol.

Since the molar mass of SO_2 is about 64 g/mol and the molar mass of SO_3 is about 80 g/mol, the unknown gas is SO_2.

61. We can use the gas density formula

$$\mathcal{M} = \frac{dRT}{P}$$

The molar mass of CO is 28.0 g/mol and that of CO_2 is 44.0 g/mol.

Information given:

$d = 1.107$ g/L
$T = 300$ K
$P = 740$ mm Hg = 0.974 atm (740/760)

$$\begin{aligned}\mathcal{M} &= \frac{dRT}{P} \\ &= \frac{(1.107\text{ g/L})(0.0821\text{ L}\cdot\text{atm/mol}\cdot\text{K})(300\text{ K})}{(0.974\text{ atm})} \\ &= 28.0\text{ g/mol}\end{aligned}$$

The unknown gas is CO (carbon monoxide).

63. The partial pressure of a gas is the pressure of one gas in a mixture of gases. The sum of the partial pressures equals the total pressure of all gases in the mixture.

64. No. A barometer measures the pressure of all gases that are present in the atmosphere.

65. The volume occupied by a gas can be calculated using the ideal-gas law.

$$V = \frac{nRT}{P}$$

If temperature and pressure are constant (as they are in this question) then the volume of a gas will be directly proportional to the number of moles of gas present in the sample. In each case we have 0.500 mol of a gas being collected. However, in part *c* some water vapor is present in the gas mixture, so the total number of moles of gas is increased. The result is that the gases (H_2 and H_2O) in part *c* occupy the largest volume.

66. Using the ideal-gas equation, $PV = nRT$, we see that if P, V, and T are identical, then the number of moles in each balloon must be the same. If each balloon contains the same number of moles, then they must also contain the same number of particles (atoms or molecules), so both balloons contain the same number of particles.

67. The formula for mole fraction is

$$\text{mole fraction of A} = \frac{\text{moles of A}}{\text{total moles}}, \quad \text{or}$$

$$X_A = \frac{\text{moles A}}{\text{total mole}}$$

There are 0.2 mol of H_2 and the total number of moles (moles N_2 + moles H_2 + moles CH_4) is equal to 1.0 (0.70 + 0.20 + 0.10).

The mole fraction of H_2:

$$XH_2 = \frac{\text{moles } H_2}{\text{total moles}} = \frac{0.20}{1.0} = 0.2$$

69. In this problem we need to use the ideal-gas law to find the pressure of each gas and to find the total pressure.

Using the ideal-gas law, we can calculate the pressure of each gas and the total pressure as follows:

$$P_{\text{gas}} = \frac{n_{\text{gas}}RT}{V}, \quad \text{and} \quad P_{\text{total}} = \frac{n_{\text{total}}RT}{V}$$

Information given:

$N_2 = 0.70$ mol
$CH_4 = 0.10$ mol
$H_2 = 0.20$ mol
$V = 10.0$ L
$T = 27°\text{C} = 300$ K

Total pressure:

The total amount of gas present in the sample is 1.0 mol.

$$\begin{aligned}P_{\text{total}} &= \frac{n_{\text{total}}RT}{V} \\ &= \frac{(1.0\text{ mol})(0.0821\text{ L}\cdot\text{atm/mol}\cdot\text{K})(300\text{ K})}{10.0\text{ L}} \\ &= 2.463\text{ atm}, \quad \text{or} \quad 2.5\text{ atm}\end{aligned}$$

Next find the pressure of each gas

$$\begin{aligned}P_{N_2} &= \frac{n_{N_2}RT}{V} \\ &= \frac{(0.70\text{ mol})(0.0821\text{ L}\cdot\text{atm/mol}\cdot\text{K})(300\text{ K})}{10.0\text{ L}} \\ &= 1.72\text{ atm}, \quad \text{or} \quad 1.7\text{ atm}\end{aligned}$$

$$P_{H_2} = \frac{n_{H_2}RT}{V}$$

$$= \frac{(0.20 \text{ mol})(0.0821 \text{ L} \cdot \text{atm/mol} \cdot \text{K})(300 \text{ K})}{10.0 \text{ L}}$$

$$= 0.4926 \text{ atm}, \quad \text{or} \quad 0.49 \text{ atm}$$

$$P_{CH_4} = \frac{n_{CH_4}RT}{V}$$

$$= \frac{(0.10 \text{ mol})(0.0821 \text{ L} \cdot \text{atm/mol} \cdot \text{K})(300 \text{ K})}{10.0 \text{ L}}$$

$$= 0.2463 \text{ atm}, \quad \text{or} \quad 0.25 \text{ atm}$$

The total pressure is $1.72 + 0.49 + 0.25 = 2.46$, or 2.5 atm (to 2 significant figures).

Note: From the total pressure and the mole fraction, one could also have calculated the partial pressure of each gas using the equation $P_{gas} = X_{gas}P_{total}$.

71. When one collects a gas over water, the total pressure is the sum of the vapor pressure of water and the pressure of the gas collected. In other words,

$$P_{total} = P_{gas} + P_{water}$$

The total pressure in this case is the atmospheric pressure (1.00 atm). The vapor pressure of water at 25°C is 23.8 torr.

Since 1.00 atm = 760. torr, the pressure of O_2 gas is

$$P_{oxygen} = 760. \text{ torr} - 23.8 \text{ torr} = 736.2 \text{ torr}$$

We now have the information needed to determine how many moles of O_2 are collected in this process.

Information given:

$P_{oxygen} = 736.2$ torr
$T = 25°C = 298$ K
$V = 0.480$ L

We can calculate the number of moles of O_2 using the ideal-gas law. However, we need to convert the pressure from torr to atmospheres.

$$P_{O_2} = 736.2 \text{ torr} \times \frac{1.00 \text{ atm}}{760 \text{ torr}} = 0.969 \text{ atm}$$

Now we can solve the ideal gas equation for moles of O_2 collected.

$$n_{O_2} = \frac{P_{O_2}V}{RT}$$

$$= \frac{(0.969 \text{ atm})(0.480 \text{ L})}{(0.0821 \text{ L} \cdot \text{atm/mol} \cdot \text{K})(298 \text{ K})}$$

$$= 0.0190 \text{ mol } O_2 \text{ collected}$$

73. Since the volume and temperature are constant, the pressure will depend only on the number of moles of gas. If there are more moles of gas after the reaction, the pressure is greater, and vice versa.

a. $N_2O_5(g) \rightarrow NO_2(g) + NO(g) + O_2(g)$

1 mole of gaseous reactants (1 mole of N_2O_5)
3 moles of gaseous products (1 mole of NO_2, 1 mole of NO, and 1 mole of O_2)

The total pressure is greater after the reaction takes place.

b. $2\ SO_2(g) + O_2(g) \rightarrow 2\ SO_3(g)$

3 moles of gaseous reactants (2 moles of SO_2 and 1 mole of O_2)
2 moles of gaseous products (2 moles of SO_3)

The total pressure is less after the reaction takes place.

c. $C_3H_8(g) + 5\ O_2(g) \rightarrow 3\ CO_2(g) + 4\ H_2O(g)$

6 moles of gaseous reactants (1 mole of C_3H_8 and 5 moles of O_2)
7 moles of gaseous products (3 moles of CO_2 and 4 moles of H_2O)

The total pressure is greater after the reaction takes place.

75. The mole fraction of oxygen in air is 0.209 (see textbook Section 8.5) and it does not change significantly with increasing altitude.

The normal atmospheric pressure is 1 atm at sea level. Using the equation $P_{O_2} = X_{O_2}P_{total}$, we can see that the pressure of O_2 gas in air on a typical day is 0.209 atm. At 8000 meters, the atmospheric pressure is 0.35 atm. If we inhale pure oxygen at this pressure, the pressure of O_2 would be 0.35 atm.

If all other factors are the same (which they are not), then the ratio of the pressures will correspond to the ratio of the moles of oxygen gas. That is,

$$\frac{P_2}{P_1} = \frac{n_2}{n_1}$$

Substituting the pressure values into the equation yields

$$\frac{P_2}{P_1} = \frac{n_2}{n_1} = \frac{0.35 \text{ atm}}{0.209 \text{ atm}} = 1.67$$

There should be 1.67 times more moles of O_2 in a lung full of pure O_2 at 8000 meters than in a lung full of air at sea level.

77. The volume remains constant, and since no change in temperature is mentioned, we can assume that the temperature remains constant also. Since the pressure of the CO (680 mm Hg) is twice that of the O_2 (340 mm Hg), there must be twice as many moles of CO as of O_2.

The balanced equation for the reaction described is

$$2\ CO(g) + O_2(g) \rightarrow 2\ CO_2(g)$$

From the balanced equation the number of moles of CO_2 gas produced is equal to the number of moles of CO reacting.

Since there are twice as many moles of CO as of O_2, these two gases are present in a stoichiometric ratio. In other words, all the CO and O_2 react and there are no gaseous reactants left unreacted.

Since the number of moles of CO_2 formed is the same as the number of moles of CO reacting, the pressure of the CO_2 gas formed will be the same as the initial pressure of CO.

The pressure of the CO_2 gas formed is 680 mm Hg. This is the only gas present in the container after the reaction has occurred, so the total pressure after the reaction takes place is 680 mm Hg.

79. Let's start by the writing a balanced equation for the process described.

$$3\,H_2(g) + N_2(g) \rightarrow 2\,NH_3(g)$$

After mixing of the gases and before reaction starts to occur we can assume that the temperature and the volume are constant. This means that the pressure inside the container is directly proportional to the number of moles present before the reaction takes place. There are 4.8×10^3 moles initially present (1.2×10^3 moles of N_2 and 3.6×10^3 moles of H_2).

After reaction occurs we can also assume that the temperature and volume have not changed (sealed container). The pressure inside the container after reaction will be directly proportional to the number of moles present after reaction.

The trick in this question is that not all of the reactants are used. So at the time of the pressure reading there will be gaseous molecules of all three gases (N_2, H_2, and NH_3).

The total pressure depends on the total number of moles present after the reaction takes place.

Before reaction occurs, there are 1.2×10^3 moles of N_2 gas and 3.6×10^3 moles of H_2. Not all of the N_2 reacts. The problem tells us that only half of the N_2 reacts. If we have 1.2×10^3 moles of N_2, then only 6.0×10^2 (600) moles of N_2 react. Let's determine how many moles of H_2 react and how many moles of NH_3 are formed (using the coefficients of the balanced equation).

moles of H_2 reacted

$$= 6.0 \times 10^2 \text{ mol } N_2 \times \frac{3 \text{ mol } H_2}{1 \text{ mol } N_2}$$

$$= 1.8 \times 10^3 \text{ mol } H_2$$

moles of NH_3 formed

$$= 6.0 \times 10^2 \text{ mol } N_2 \times \frac{2 \text{ mol } NH_3}{1 \text{ mol } N_2}$$

$$= 1.2 \times 10^3 \text{ mol } NH_3$$

Let's make a table summarizing this information.

	N_2	H_2	NH_3
initial moles present	1.2×10^3	3.6×10^3	0.0
number of moles reacted	6.0×10^2	1.8×10^3	
number of moles formed			1.2×10^3
moles present after reaction	6.0×10^2	1.8×10^3	1.2×10^3

The total number of moles present after reaction is 3.6×10^3 (6.0×10^2 moles of N_2, 1.8×10^3 moles of H_2, and 1.2×10^3 moles of NH_3).

At constant temperature and volume, the pressure of gases is directly proportional to the number of moles present. In a formula this is

$$\frac{P_2}{P_1} = \frac{n_2}{n_1}$$

Substituting in the 3.6×10^3 moles as n_2 and 4.8×10^3 moles as n_1, we get

$$\frac{P_2}{P_1} = \frac{3.6 \times 10^3 \text{ mol}}{4.8 \times 10^3 \text{ mol}} = 0.75$$

The total moles present after reaction is only $\frac{3}{4}$ of what was initially present, so the final pressure should be only $\frac{3}{4}$ as great as the initial pressure. That is, the pressure has decreased by $\frac{1}{4}$ of its original value.

The pressure must decrease by $\frac{1}{4}$ or 25%.

81. Let's say we have a closed system with a movable piston. The container is half-filled with water. We know that solubility is an equilibrium process and that at equilibrium the rate of gas molecules hitting the water surface and being trapped (or dissolved) is equal to the rate at which dissolved gas molecules escape from the water. Now, if we push down on the piston, decreasing the open space above the liquid, the gas molecules will have less room to move. The result is gas molecules hit the water surface more often and are trapped (dissolved in the water). That is, the rate of condensation is now greater than the rate of evaporation. When equilibrium is re-established more molecules have become dissolved. That is, the solubility of the gas has increased.

82. At the high pressures associated with deep-sea diving, the solubility of gases increases markedly. As a diver rises to the surface, the pressure decreases and the dissolved gases come out of blood. These gas bubbles can block capillaries and can cause severe pain. The excess oxygen that comes out of solution does not pose a problem because it is used in the metabolism process. If one surfaces slowly the excess nitrogen can be moved to the lungs where it is removed from the body by breathing. Mixtures that use helium gas are often used because the solubility of helium is much lower than the solubility of nitrogen.

83. No, it should be a weighted average of the two.

84. The formula for the solubility of a gas in a liquid is

$$C_{gas} = k_H P_{gas}$$

The partial pressure of oxygen in the atmosphere remains the same, so if the solubility changes with temperature then the constant k_H has to change with temperature.

85. Since CO_2 reacts with H_2O to form water-soluble HCO_3^-, CO_3^{2-}, and H^+ ions, more CO_2 is dissolved in the water and k_H is larger for CO_2 than for N_2 or O_2.

86. The bubbles are dissolved air, which is expelled as the water is heated because air is less soluble at higher temperatures.

87. In this problem, the solubility is given in grams per liter and the question asks us to find the solubility of O_2 in mol/L · atm.

We are given the solubility of O_2 gas in water at 37°C as 0.25 g/L. We need to convert this solubility into mol/L. All we need to do is to convert 0.25 g of O_2 to moles of O_2 (molar mass O_2 = 32.0 g/mol).

$$\text{moles } O_2 = 0.25 \text{ g } O_2 \times \frac{1 \text{ mol } O_2}{32.0 \text{ g } O_2}$$

$$= 7.8 \times 10^{-3} \text{ mol } O_2$$

The solubility of O_2 is 7.8×10^{-3} mol/L (M).

Lastly we need to add the pressure unit to this solubility.

According to Henry's law, the solubility of a gas is directly proportional to tne pressure of that gas above the solution. We need to know the pressure of oxygen gas in the body. We can assume that the partial pressure of O_2 will be the same as in the atmosphere. The textbook tells us that the partial pressure of O_2 in the atmosphere is 0.209 atm. We need to divide the solubility in mol/L by this pressure to determine the solubility in mol/L · atm.

$$\frac{7.8 \times 10^{-3} \text{ mol/L}}{0.209 \text{ atm}} = 3.7 \times 10^{-2} \text{ mol/L} \cdot \text{atm}$$

This is actually the Henry's law constant for O_2 at 37°C.

89. a. On Mt. Everest the atmospheric pressure is 0.35 atm. We assume that oxygen makes up 21% of the atmosphere. We need to determine the partial pressure of oxygen at this atmospheric pressure.

$$\text{partial pressure of } O_2 = (0.35 \text{ atm})(0.21)$$

$$= 0.0735 \text{ atm}$$

Now, we are ready to use Henry's law. From Problem 87, k_H for blood is 3.7×10^{-2} M/atm.

$$C_{gas} = k_H P_{gas} = (3.7 \times 10^{-2} M/\text{atm})(0.0735 \text{ atm})$$

$$= 2.7 \times 10^{-3} M$$

Lastly convert the solubility from M (mol/L) to g/L. Simply multiply the molarity (M) by the molar mass of O_2 (32.0 g/mol).

$$\text{solubility of } O_2 \text{ in g/L} = \frac{2.7 \times 10^{-3} \text{mol}}{\text{L}} \times \frac{32.0 \text{ g}}{\text{mol}}$$

$$= 8.6 \times 10^{-2} \text{ g/L}$$

The solubility of O_2 in blood at the top of Mt. Everest is 8.6×10^{-2} g/L.

b. At a depth of 100 ft underwater, the pressure exerted on a diver is 3.0 atm.

$$\text{partial pressure of } O_2 \text{ (at 100 ft)} = 0.21 \times 3.0 \text{ atm}$$

$$= 0.63 \text{ atm}$$

Again use Henry's law, except the partial pressure of oxygen is now 0.63 atm.

$$C_{gas} = (3.7 \times 10^{-2} M/\text{atm})(0.63 \text{ atm})$$

$$= 2.3 \times 10^{-2} M$$

Lastly, convert the solubility from M to g/L by multiplying the molarity by the molar mass of O_2.

$$\text{solubility of } O_2 \text{ in g/L} = \frac{2.3 \times 10^{-2} \text{ mol}}{\text{L}} \times \frac{32.0 \text{ g}}{\text{mol}}$$

$$= 0.74 \text{ g/L}$$

The solubility of oxygen in blood at a depth of 100 ft underwater is 0.74 g/L.

91. The graph is a plot of the solubility of oxygen gas (liters of O_2 gas that can be dissolved per liter of water) versus temperature. We need to convert the solubility of O_2 gas from this unit to molarity, M (moles/liter).

Determine the solubility of O_2 from the graph at 10°C, 20°C, and 30°C.

The solubility of O_2 at 10°C is about 0.038 L of O_2 per liter of water.

The solubility of O_2 at 20°C is about 0.031 L of O_2 per liter of water.

The solubility of O_2 at 30°C is about 0.026 L of O_2 per liter of water.

Next convert this unit to moles of O_2 per liter. We can use the ideal-gas equation to determine how many moles of O_2 gas dissolved in a liter of water.

$$n = \frac{PV}{RT}$$

where,
$P = 1.00$ atm; $V =$ solubility of O_2 in liters; $T =$ temperature of the solution

At 10°C the number of moles of O_2 that dissolve is

$$n_{O_2} = \frac{(1.00\,\text{atm})(0.038\,\text{L})}{(0.0821\,\text{L}\cdot\text{atm/mol}\cdot\text{K})(283\text{K})}$$

$$= 1.6 \times 10^{-3}\,\text{mol O}_2$$

The solubility of O_2 in units of molarity (M) is $1.6 \times 10^{-3} M$.

Henry's law constant can be calculated using

$$k_H = \frac{C_{gas}}{P_{gas}}.$$

The Henry's law constant for O_2 at 10°C is

$$k_H = \frac{1.6 \times 10^{-3} M}{1\,\text{atm}} = 1.6 \times 10^{-3} M/\text{atm}$$

We can do the same set of calculations to find Henry's law constants for O_2 at 20°C and 30°C.

At 20°C the number of moles of O_2 that dissolve is

$$n_{O_2} = \frac{(1.00\,\text{atm})(0.031\,\text{L})}{(0.0821\,\text{L}\cdot\text{atm/mol}\cdot\text{K})(293\,\text{K})}$$

$$= 1.3 \times 10^{-3}\,\text{mol O}_2$$

The solubility of O_2 in units of molarity (M) is $1.3 \times 10^{-3} M$.

The Henry's law constant for O_2 at 20°C is

$$k_H = \frac{1.3 \times 10^{-3} M}{1\,\text{atm}} = 1.3 \times 10^{-3} M/\text{atm}$$

At 30°C the number of moles of O_2 that dissolve is

$$n_{O_2} = \frac{(1.00\,\text{atm})(0.026\,\text{L})}{(0.0821\,\text{L}\cdot\text{atm/mol}\cdot\text{K})(303\,\text{K})}$$

$$= 1.0 \times 10^{-3}\,\text{mol O}_2$$

The solubility of O_2 in units of molarity (M) is $1.0 \times 10^{-3} M$.

The Henry's law constant for O_2 at 30°C is

$$k_H = \frac{1.0 \times 10^{-3} M}{1\,\text{atm}} = 1.0 \times 10^{-3} M/\text{atm}$$

93. The root-mean-square speed (μ_{rms}) of a gas molecule is the square root of the average of the squared speeds of a collection of gas molecules (or atoms). It is calculated using the formula $(\mu_{rms}) = \sqrt{\frac{3RT}{\mathcal{M}}}$.

94. Gas particles are in constant motion so there are many collisions occurring between gas particles at any instant in any gas sample. Each collision is elastic; that is, the total energy of the colliding particles is the same. However, the kinetic energy of each particle in the collision can change. This results in particles with many different speeds.

95. a. The formula for root-mean-square speed is

$$(\mu_{rms}) = \sqrt{\frac{3\,RT}{\mathcal{M}}}.$$

The root-mean-square speed is inversely proportional to the square root of the molar mass.

b. The root-mean-square speed is directly proportional to the square root of the Kelvin temperature.

96. No. In an ideal gas, the pressure does not directly affect the root-mean-square speed of a gas molecule.

97. Graham's law is given by the expression

$$\frac{\text{rate known gas}}{\text{rate unknown gas}} = \sqrt{\frac{\mathcal{M}_{unknown}}{\mathcal{M}_{known}}}$$

If one has an unknown gas and a known gas, the relative rates of effusion can be experimentally measured (or the exact rate of effusion of each gas can be measured). Since we know the rate of effusion of both the known and unknown gas and the molar mass of the known gas, we can use Graham's law to find the molar mass of the unknown gas. (We know three of the four quantities in the Graham's law equation and we solve for the missing quantity, the molar mass of the unknown gas.)

98. Yes, the ratio of the rates of effusion of two gases is the same as the ratio of their root-mean-square speeds (see Equations 8.17 and 8.18 in the text).

99. Diffusion is the spreading out of particles through another substance, such as a solution (e.g., adding cream to coffee). In this example, we can speed up the diffusion process by stirring the solution. Effusion is the escape of gas particles through a small opening (such as a pinhole).

100. Yes, Graham's law applies to both the diffusion and effusion of gases.

101. The root-mean-square speed of a gas species is inversely proportional to the square root of the molar mass.

Lighter gas particles move quicker than heavier gas particles. The mass of a gas molecule is related to its molar mass. The smaller the molar mass, the faster the average velocity (root-mean-square velocity) of the molecule.

For the gases in this problem, the molar masses increase in the order

CO_2 (44.0 g/mol) < NO_2 (46.0 g/mol) < SO_2 (64.0 g/mol)

The order of increasing root-mean-square speed is $SO_2 < NO_2 < CO_2$.

103. We need to calculate the root-mean-square velocity for O_2 using the formula

$$\mu_{rms} = \sqrt{\frac{3\,RT}{\mathcal{M}}}.$$

In this formula, R is the gas law constant, T is the Kelvin temperature, and $\mathcal{M}$ is the molar mass of the gas.

To get units of m/s, the R value we use is 8.314 J/mol · K and the molar mass must have units of kg/mol (instead of g/mol).

The molar mass of O_2 is 32.0 g/mol or 0.032 kg/mol.

$$\mu_{rms}\sqrt{\frac{3\,RT}{\mathcal{M}}} = \sqrt{\frac{(3)(8.314\text{ J/mol}\cdot\text{K})(286\text{ K})}{0.032\text{ kg/mol}}}$$

$$= \sqrt{2.229 \times 10^5 \frac{\text{m}^2}{\text{s}^2}} = 472\text{ m/s}$$

Oxygen corresponds to gas C.

Note: A joule has units of $\frac{\text{kg}\cdot\text{m}^2}{\text{s}^2}$.

The cancellation of units in our calculation is as follows:

$$\sqrt{\frac{(\text{kg}\cdot\text{m}^2/\text{s}^2)\,(1/\text{mol}\cdot\text{K})\,(\text{K})}{\text{kg/mol}}} = \sqrt{\frac{\text{m}^2}{\text{s}^2}} = \frac{\text{m}}{\text{s}}$$

105. The average kinetic energy of neon atoms is given to be 5.18 kJ/mole. We are asked to find the root-mean-square velocity of neon atoms. We can use the equation

$$\text{K.E.} = \tfrac{1}{2}m\mu^2$$

In this equation, the m refers to the mass of an individual atom or molecule. If we are given the kinetic energy of a mole of atoms or molecules we can write the equation using the either the atomic weight or the molar mass for m.

$$\text{K.E.} = \tfrac{1}{2}\,M\mu^2$$

In this problem, we use the atomic weight of neon.

Solving this equation for μ (the root-mean-square velocity) gives us

$$\mu = \sqrt{\frac{2\text{ K.E.}}{M}}$$

where K.E. is the average kinetic energy.

Before solving, let's look at units. We want the root-mean-square velocity to have units of meters per second (m/s). The unit of energy joule are

$$1\text{ J} = \frac{\text{kg}\cdot\text{m}^2}{\text{s}^2}$$

Examining this unit, we see that m/s are included, so we need to convert the 5.18 kJ/mol to J/mol. The other unit here is kilograms (kg). For any mass unit to cancel, it must also have units of kilogram. We need to convert the atomic weight (or molar mass) from units of g/mol to kg/mol. Let's see how these units cancel by plugging only the units back into the kinetic energy formula.

$$\mu = \sqrt{\frac{2\text{K.E.}}{m}} = \sqrt{\frac{\text{J/mol}}{\text{kg/mol}}} = \sqrt{\frac{\frac{\text{kg}\cdot\text{m}^2}{\text{s}^2/(\text{mol}\cdot\text{K})}}{\frac{\text{kg}}{\text{mol}}}}$$

$$= \sqrt{\frac{\frac{\text{kg}\cdot\text{m}^2}{\text{s}^2\cdot\text{mol}}}{\frac{\text{kg}}{\text{mol}}}}\sqrt{\frac{\text{kg}\cdot\text{m}^2\cdot\text{mol}}{\text{s}^2\cdot\text{mol}\cdot\text{kg}}} = \sqrt{\frac{\text{m}^2}{\text{s}^2}} = \text{m/s}$$

To use this formula, we only need to convert 5.18 kJ/mole to J/mole, and the atomic weight of neon from 20.2 g/mol to kg/mol.

$$5.18\frac{\text{kJ}}{\text{mol}} \times \frac{1000\text{ J}}{\text{kJ}} = 5.18 \times 10^3\frac{\text{J}}{\text{mol}}$$

Converting the molar mass, we get 0.0202 kg/mol.

Now we find the root-mean-square speed using the formula

$$\mu = \sqrt{\frac{2\text{ K.E.}}{m}} = \sqrt{\frac{2\left(\frac{5.18\times10^3\,\text{kg}\cdot\text{m}^2}{\text{s}^2/\text{mol}\cdot\text{K}}\right)}{0.0202\frac{\text{kg}}{\text{mol}}}}$$

$$= \sqrt{5.13 \times 10^5\,\frac{\text{m}^2}{\text{s}^2}} = 716\text{ m/s}$$

107. We can determine the ratio of the root-mean-square speeds by using Equation 8.17.

D_2 is deuterium, an isotope of hydrogen that contains 1 proton, 1 neutron, and 1 electron.

$$\frac{\mu_{D_2}}{\mu_{H_2}} = \sqrt{\frac{\mathcal{M}_{H_2}}{\mathcal{M}_{D_2}}} \qquad (8.17)$$

The molar mass of D_2 is 4.0 g/mol and the molar mass of H_2 is 2.0 g/mol. Substituting these values into Equation 8.17 yields the following:

$$\frac{\mu_{D_2}}{\mu_{H_2}} = \sqrt{\frac{2.0\text{ g/mol}}{4.0\text{ g/mol}}} = \sqrt{0.5} = 0.707$$

Note: Since we are using a comparison version of the root-mean-square speed equation, the molar masses are required only to have the same unit. They do not need to have units of kg/mol. You will get the same answer if you do convert them to kg/mol.

This means that the average D_2 molecule is moving at only 0.707 the speed of the average H_2 molecule.

109. The average kinetic energy of gas molecules is dependent only on the temperature of the gas molecules. At a given temperature, all gas molecules have the same average kinetic energy. Since kinetic energy $= \frac{1}{2}m\mu^2_{rms}$,

we can deduce that on average, lighter molecules must be moving faster than heavier molecules. (The effect of mass on the root-mean-square speed is shown in Figure 8.15. From this figure, we notice that the flatter the curve, the lower the molar mass of the gas.)

In the figure in this problem, gas *b* is flatter (lower peak maximum); it must be the lighter gas.

The molar mass of SO_2 is 64.1 g/mol and the molar mass of CO_2 is 44.0 g/mol, so the CO_2 curve should be flatter (gas *b*).

C_3H_8 should have a curve very similar to that of CO_2 because both C_3H_8 and CO_2 have a molar mass of 44.0 g/mol.

111. This problem requires us to use Graham's law of effusion. The formula is

$$\frac{\text{rate of effusion of gas A}}{\text{rate of effusion of gas B}} = \sqrt{\frac{\text{molar mass of gas B}}{\text{molar mass of gas A}}}$$

or

$$\frac{\text{rate A}}{\text{rate B}} = \sqrt{\frac{\mathcal{M}_B}{\mathcal{M}_A}}$$

It is usually easier to solve this equation if the lighter gas or the gas that has the greater rate of effusion is designated as gas A. We know that H_2 effuses 4 times as fast as gas X. This means that

$$\frac{\text{rate of effusion of } H_2}{\text{rate of effusion of gas X}} = 4$$

We can use this value for the left-hand side of the Graham's law formula.

$$\frac{\text{rate of effusion of } H_2}{\text{rate of effusion of gas X}} = 4 = \sqrt{\frac{\mathcal{M}_X}{\mathcal{M}_{H_2}}} = \sqrt{\frac{\mathcal{M}_X}{2.0 \text{ g/mol}}}$$

Next, square both sides of the equation.

$$4^2 = \left(\sqrt{\frac{\mathcal{M}_X}{2.0 \text{ g/mol}}}\right)^2, \quad \text{so} \quad 16 = \frac{\mathcal{M}_X}{2.0 \text{ g/mol}}$$

Solving for $\mathcal{M}_X$, we get

$$\mathcal{M}_X = (16)(2.0 \text{ g/mol}) = 32.0 \text{ g/mol}$$

The molar mass of gas X is 32.0 g/mol.

113. We can solve this problem in two ways.

Method 1

This is the longer of the two methods but is probably easier to understand.

First, we find the root-mean-square speed of H_2 molecules at 300 K.

Since the molecule of interest has only one-third the root-mean-square speed of H_2, we divide the root-mean-square of H_2 by three.

Lastly we use this root-mean-square speed and a temperature of 300 K to find the molar mass of this unknown gas species, using the formula $\mu = \sqrt{\frac{3RT}{\mathcal{M}}}$.

The root-mean-square speed of H_2:

(Remember when using the formula $\mu = \sqrt{\frac{3RT}{\mathcal{M}}}$ the molar mass must be in units of kg/mole if we use R = 8.314 J/mol · K. Also the units of a J are kg · m^2/s^2.)

$$\mu = \sqrt{\frac{3RT}{\mathcal{M}}} = \sqrt{\frac{3(8.314 \text{ J/mol} \cdot \text{K})(300 \text{ K})}{0.002 \text{ kg/mol}}}$$

$$= \sqrt{3.74 \times 10^6 \frac{\text{m}^2}{\text{s}^2}} = 1.93 \times 10^3 \text{ m/s}$$

The root-mean-square speed of H_2 is 1.93×10^3 m/s, so root-mean-square speed of the unknown substance is

$$(1/3)(1.93 \times 10^3 \text{ m/s}) = 6.45 \times 10^2 \text{ m/s}$$

Now we can use the root-mean-square formula to solve for the molar mass of the unknown gas.

To solve this equation for molar mass, we need to square both sides and then solve for $\mathcal{M}$.

$$\mathcal{M} = \frac{3RT}{\mu^2} = \frac{3(8.314 \text{ J/mol} \cdot \text{K})(300 \text{ K})}{(6.45 \times 10^2 \text{ m/s})^2}$$

$$= 1.80 \times 10^{-2} \text{ kg/mol, or } 18.0 \text{ g/mol}$$

As expected the molar mass of the unknown gas is much larger than that of H_2 because at the same temperature the unknown gas molecules were moving much slower.

The molar mass of the unknown gas is 18.0 g/mole.

Method 2

We can also use Equation 8.17 to solve for the molar mass of the unknown gas.

$$\frac{\mu_A}{\mu_B} = \sqrt{\frac{\mathcal{M}_B}{\mathcal{M}_A}} \tag{8.17}$$

If we make the faster moving gas A, it will be easier to solve for $\mathcal{M}_B$.

We do not know the actual root-mean-square speed of either H_2 or the unknown gas. We do know that the unknown gas moves only one-third as fast as the H_2 gas or that H_2 is moving three times faster on the average than the unknown gas.

In other words

$$\frac{\mu_{H_2}}{\mu_{\text{unknown}}} = 3$$

Now we can use the Equation 8.17.

The molar mass of H_2 is 2.0 g/mol. Since this equation has two molar masses, the units will cancel and it does not matter whether we use g/mol or kg/mol as long as both molar masses have the same units.

$$\frac{\mu_{H_2}}{\mu_{unknown}} = \sqrt{\frac{\mathcal{M}_{unknown}}{\mathcal{M}_{H_2}}}$$

$$3 = \sqrt{\frac{\mathcal{M}_{unknown}}{2 \text{ g/mol}}}$$

Squaring both sides:

$$9 = \frac{\mathcal{M}_{unknown}}{2 \text{ g/mol}}$$

$$\mathcal{M}_{unknown} = (9)(2 \text{ g/mol}) = 18.0 \text{ g/mol}$$

The molar mass of the unknown gas is 18.0 g/mol.

115. This problem can be solved by application of Graham's law.

The molar mass of $^{13}CO_2$ is 45.0 g/mol.

The molar mass of $^{12}CO_2$ is 44.0 g/mol.

The lighter of these gases ($^{12}CO_2$) will diffuse slightly faster.

To find the how much faster use Graham's law:

$$\frac{\text{rate } ^{12}CO_2}{\text{rate } ^{13}CO_2} = \sqrt{\frac{\mathcal{M}_{^{13}CO_2}}{\mathcal{M}_{^{12}CO_2}}}$$

$$\frac{\text{rate } ^{12}CO_2}{\text{rate } ^{13}CO_2} = \sqrt{\frac{45 \text{ g/mol}}{44 \text{ g/mol}}} = \sqrt{1.0227} = 1.01$$

$^{12}CO_2$ effuses 1.01 times as fast as $^{13}CO_2$.

117. In both cases, the gas inside the balloon has smaller particles than air. The gases inside the balloon (H_2 and He) will effuse out of the balloon faster than the air outside can move inside, so the volume of the balloon decreases with time.

According to Graham's law the lighter the gas particles (smaller molar mass or atomic weight) the faster the rate of effusion. Since the molar mass of H_2 is 2.0 g/mol and the atomic weight of He is 4.0 g/mol, hydrogen molecules have a faster effusion rate than helium, so more H_2 molecules effuse out of the H_2 balloon than He atoms effuse out of the helium balloon. After 24 hours the hydrogen balloon has a smaller volume.

119. Gas particles are in constant motion. As H_2S travels there will be many collisions between the gas particles. These collisions can change the direction of travel of H_2S molecules and/or slow them down. The net result is that gas molecules take a longer time to reach a destination point than is expected on the basis of the root-mean-square velocity.

121. At low temperatures real gases do not behave ideally because the particles are both close together and also move past each other at a much lower rate. At low temperatures (near the boiling point of a liquid) the molecules tend to interact with one another due to the forces of attraction between particles (note that these forces of attraction eventually cause a real gas to form a liquid as the temperature decreases).

At high pressure real gases do not behave ideally because the particles are close to each other. The total volume occupied by the particles becomes a significant fraction of the total volume. This results in less space for the particles to move. We need to include a volume correction to account for this reduced space.

122. The constant a is associated with the attractive forces between gas particles.

The attractive forces present in water (a polar molecule) are much stronger than the attractive forces present in N_2 gas (a nonpolar molecule). Water should have a much higher van der Waals a constant.

123. When gas particles are attracted to each other they collide less often with the walls of the container. The observed pressure should be less than the ideal pressure.

124. The van der Waals b constant is related to the volume occupied by the gas particles. As the atomic number of the noble gases increases so does the size of the gas particles and the volume they occupy.

125. The van der Waals a constant is related to the intermolecular forces of attraction associated between particles of a particular gas. In general, increasing size (molar mass or atomic weight) and molecular complexity increases the strength of these intermolecular forces, which results in larger van der Waals a values.

126. The more polar a molecule, the stronger the intermolecular forces of attraction and the larger the van der Waals a value.

127. When intermolecular forces are more important, $PV/RT < 1$. When the volume occupied by the gas is more important, $PV/RT > 1$.

Volume effects are more important for helium gas because the plot of PV/RT is always greater than 1.

129. For H_2, the van der Waals constants are as follows:

$a = 0.244 \text{ L}^2\cdot \text{atm/mol}^2$

$b = 0.0266$ L/mol

Information given:

mass of $H_2 = 40.0$ g

$T = 20°\text{C} = 293$ K

$V = 1.00$ L

The van der Waals equation solved for P is

$$P = \frac{nRT}{V - nb} - \frac{n^2a}{V^2}$$

We need to convert mass of H_2 in grams to moles before using this equation. The molar mass of H_2 is 2.00 g/mol.

$$\text{moles of } H_2 = 40.0 \text{ g } H_2 \times \frac{1 \text{ mol } H_2}{2.00 \text{ g } H_2} = 20.0 \text{ mol } H_2$$

Substituting the known values we get

$$P = \frac{(20.0 \text{ mol})\left(0.0821\frac{\text{L} \cdot \text{atm}}{\text{mol} \cdot \text{K}}\right)(293 \text{ K})}{1.00 \text{ L} - (20.0 \text{ mol})\left(0.0266\frac{\text{L}}{\text{mol}}\right)}$$

$$- \frac{\left((20.0 \text{ mol})^2\, 0.244\frac{\text{L}^2 \cdot \text{atm}}{\text{mol}^2}\right)}{(1.00 \text{ L})^2}$$

$$P = \frac{481.1 \text{ L} \cdot \text{atm}}{1.00 \text{ L} - 0.532 \text{ L}} - \frac{97.6 \text{ L}^2 \cdot \text{atm}}{1.00 \text{ L}^2}$$

$$= \frac{481.1 \text{ atm}}{0.468} - 97.6 \text{ atm}$$

$$= 1028 - 97.6 = 930 \text{ atm}$$

The pressure of H_2 under these conditions is 930 atm.

If H_2 behaved ideally, the calculated pressure would be

$$P = \frac{nRT}{V} = \frac{(20.0 \text{ mol})\left(0.0821\frac{\text{L} \cdot \text{atm}}{\text{mol} \cdot \text{K}}\right)(293 \text{ K})}{1.00 \text{ L}}$$

$$= 481 \text{ atm}$$

This value is only about half of the true value of the pressure. Under these conditions H_2 gas does not behave ideally.

CHAPTER 9

Intermolecular Forces and Liquids: Water, Nature's Universal Solvent

1. Particles in the gaseous state experience far fewer interactions relative to those in the solid and liquid states.
2. Since the solid state has the most interactions between particles and the gaseous state has the least, the process of sublimation will require the most disruption of intermolecular interactions.
3. All molecules experience London forces.
4. No. In order for a molecule to have the ability to form hydrogen bonds the molecule must first have a hydrogen atom. Secondly, that hydrogen atom must be bonded to an N, O, or F atom.
5. First, methane is a nonpolar compound. Secondly, in methane there is no hydrogen atom bonded to N, O, or F.
6. Ion–dipole interactions tend to be stronger than dipole–dipole interactions because one of the interacting species has a full positive or full negative charge. Larger charges generally lead to stronger attractions. Coulomb's law states that the force between two charged particles is directly proportional to the product of the charges and inversely proportional to the square of the distance between the two particles.

$$\text{force} \propto \frac{(\text{charge 1})(\text{charge 2})}{(\text{distance})^2}$$

7. Hydrogen bonds are a type of dipole–dipole interaction. The reason why they are considered a separate type of bond is because of their strength relative to most other dipole–dipole interactions.
8. The polar compound will most likely have the higher boiling point. The nonpolar compound will only have London forces between molecules, while the polar compound will have London forces and dipole–dipole forces. Since the polar compound will have more (and stronger) interactions, overcoming these interactions when the compound boils will require more energy (a higher temperature) for the polar compound compared to the nonpolar compound.
9. As molecular size increases, the number of electrons increases, and the size of the electron cloud generally increases. London forces tend to increase as the size of the electron cloud surrounding a group of atoms increases. The larger an electron cloud is, the more easily it can be distorted and hence experience larger temporary dipoles.
10. The difference in solubility comes from the fact that the ions in $CaSO_4$ are more difficult to separate relative to those in NaCl. $CaSO_4$ is made up Ca^{2+} and SO_4^{2-} and the magnitudes of the charges on these ions are double those found in NaCl. Larger charges lead to stronger interparticle forces—Coulomb's law. Stronger interparticle forces lead to higher boiling points.
11. Ar has a larger electron cloud than He. The larger electron cloud of He is more polarizable and hence Ar has stronger London forces than He. The van der Waals constant a accounts for intermolecular attractions, so we would expect Ar to have a larger van der Waals constant a because it has stronger interparticle (in this case, London) forces.
12. The van der Waals constant a relates to intermolecular attractions. CO_2 and CS_2 are both nonpolar compounds. CS_2 will have larger London dispersion forces than CO_2. The larger London dispersion forces of CS_2 would lead

to a larger value of *a* for CS_2. We expect CS_2 to have a value of *a* greater than 3.59 $L^2 \cdot atm/mol^2$.

13. The simple answer is that NH_3 has the ability to form hydrogen bonds and PH_3 does not. We must be careful not to believe that hydrogen bonds are always stronger than London forces. If we continued down Group 5 and examined the series of hydrides (XH_3), at some point molecular London forces would be greater than the hydrogen bonding forces. Moving down the table we have AsH_3 with a b.p. of −62.5°C and SbH_3 with a b.p. of −18.3°C. Note how much larger the molar mass must be in order to have the London forces outweigh the hydrogen bonding.

14. Water has the ability to make more hydrogen bonds compared to methanol because water has two hydrogens attached to its oxygen while methanol has only one. Also, water has a larger dipole moment than methanol. Both of these factors explain why water has the higher boiling point.

15. This is due to the greater London forces in CH_2Cl_2 relative to CH_2F_2. The larger electron clouds of the Cl atoms are more polarizable and this leads to stronger London forces. Note that CH_2F_2 does not have any H–F bonds and therefore is not able to hydrogen bond.

16. The larger boiling point is due to the much larger electron cloud on HBr. This larger electron cloud in HBr results in stronger London forces and hence a higher boiling point. Based on the given dipole moments, HCl has stronger dipole–dipole forces than H–Br, but HCl has weaker London dispersion forces than H–Br. We must be careful not to believe that dipole–dipole forces are always stronger than London dispersion forces.

17. The London forces will be similar since the molar masses are similar. The difference is that Br_2 is nonpolar and ICl polar. The nonpolar Br_2 has London forces while ICl has dipole–dipole forces along with London forces. The dipole–dipole forces of ICl are stronger than the London forces of Br_2. Stronger forces in ICl yield a higher boiling point.

18. The interactions that need to be overcome in boiling pure NaCl are ion–ion interactions. Ion–ion interactions are much stronger than the ion–dipole and hydrogen-bonding interactions that must be overcome when boiling water in NaCl(*aq*).

19. All of these ionic compounds contain the Br^- ion, so we are really comparing the cations. Smaller ions result in greater attractions and larger magnitude charges also result in greater attractions. Sr^{2+} will exhibit greater attractions than either K^+ or Cs^+. K^+ will exhibit greater attractions than Cs^+ because of the relatively small ionic radius of K^+. The order of increasing interactions is CsBr < KBr < $SrBr_2$.

21. The melting points should decrease with increasing atomic number of the halide. As atomic number increases, the size of the halide ions increases. The force between the ions decreases as the distance between the ions increases. Weaker forces lead to lower melting points.

23. $$K(s) + \frac{1}{2}\,Cl_2(g) \rightarrow KCl(s)$$

We have been asked to calculate the lattice energy of KCl. This means we have been asked to calculate the energy change associated with the reaction of gas phase ions to form one mole of the ionic compound. We have been given the energy changes associate with five reactions. We have all the pieces necessary to carry out the calculation. We were given the energy changes needed convert K(*s*) and $Cl_2(g)$ into $K^+(g)$ and $Cl^-(g)$. We were also given the energy associated with converting K(*s*) and $Cl_2(g)$ into KCl(*s*). Since lattice energy is the energy change associated with converting $K^+(g)$ and $Cl^-(g)$ into KCl(*s*), all we have to do is apply Hess's law to obtain the lattice energy.

Reaction	***Energy Change***
$K(s) \rightarrow K(g)$	89 kJ
$K(g) \rightarrow K^+(g) + 1\,e^-$	425 kJ
$\frac{1}{2}\,Cl_2(g) \rightarrow Cl(g)$	240 kJ/mol × $\frac{1}{2}$ mol = 120 kJ
$Cl(g) + 1\,e^- \rightarrow Cl^-(g)$	−349 kJ
$K^+(g) + Cl^-(g) \rightarrow KCl(s)$	? (lattice energy)
$K(s) + \frac{1}{2}\,Cl_2(g) \rightarrow KCl(s)$	−438 kJ

89 kJ + 425 kJ + 120 kJ + (−349 kJ) + lattice energy = −438 kJ
lattice energy of KCl = −723 kJ

25. a. CCl_4 because of its larger and more polarizable electron cloud
    b. C_3H_8 because of its larger and more polarizable electron cloud

27. CO_2 will have the weakest interactions between its molecules because it is the only nonpolar compound in the group. Molecules of CO_2 have only London forces, while all the others have London forces and dipole–dipole interactions.

29. a. CCl_4 is nonpolar while $CHCl_3$ is polar. We expect $CHCl_3$ to be more soluble than CCl_4 in H_2O, a polar compound. Remember the general rule "likes dissolve likes" or in this case "like dissolves in like."
    b. The Lewis structures for these two molecules are shown below and both are polar molecules. CH_3OH will be more soluble than $C_6H_{13}OH$ in H_2O because a larger portion of this molecule is similar to H_2O. Only a relatively small portion of $C_6H_{13}OH$ is similar to water. Another factor is that $C_6H_{13}OH$ has a

relatively large nonpolar portion, the C_6H_{13}-part. The large nonpolar portion of $C_6H_{13}OH$ would make it more soluble in C_6H_{14} than in CH_3OH.

```
    H                    H   H   H   H   H   H
    |    ..              |   |   |   |   |   |    ..
H — C — O — H        H — C — C — C — C — C — C — O — H
    |    ..              |   |   |   |   |   |    ..
    H                    H   H   H   H   H   H
```

c. The higher lattice energy of MgO will make it less soluble than NaCl in water. The interactions between the doubly charged ions of MgO are stronger than the interactions between the ions in NaCl, and water is not as capable of disrupting the ion–ion interactions of MgO.

31. We would expect CaO to be the least soluble in water because it has the highest lattice energy of the group. It is more difficult for water to disrupt the ion–ion interactions of CaO than those of the other three compounds.

33. As temperature increases, the average kinetic energy of the molecules in a sample of liquid increases. The increased average kinetic energy allows more of the molecules to overcome the attractive forces keeping them in the liquid state, so the number of molecules in the vapor phase and the vapor pressure increase.

34. The liquid boils. The boiling point of a liquid is defined as the temperature at which the vapor pressure of a liquid equals the atmospheric pressure. Note that the normal boiling point of a compound is the temperature at which a liquid boils when the pressure is 1 atm.

35. The compound with the stronger attractive forces will have the lower vapor pressure. Less molecules of this sample will have the kinetic energy necessary to escape the liquid phase. The order of increasing vapor pressure is $CH_3CH_2OH < CH_3OCH_3 < CH_3CH_2CH_3$. The order is explained by the fact that CH_3CH_2OH is a polar molecule that also is capable of hydrogen-bonding, CH_3OCH_3 is a polar compound without hydrogen bonding, and $CH_3CH_2CH_3$ interacts only through London forces.

37. Calculate the mole fraction of water

$$\chi_{H_2O} = \frac{3.5}{3.5 + 1.5}$$
$$= 0.70$$

Since the sum of all mole fractions must be equal to 1

$$\chi_{C_6H_{12}O_6} = 1 - 0.70$$
$$= 0.30$$

We use Raoult's law to calculate the vapor pressure of the solution. We also need to make the assumption that glucose is a nonvolatile solute.

$$P_{\text{Solution}} = \chi_{\text{Solvent}} \times P_{\text{Solvent}}$$
$$P_{\text{Solution}} = 0.70 \times 23.8 \text{ torr}$$
$$= 16.66 \text{ torr}$$
$$= 17 \text{ torr (2 significant figures)}$$

39. a. Solids are most likely to exist at low temperatures and under high pressures. These factors tend to bring particles close together and allow for maximum interactions.
    b. Gases are most likely to exist under low pressures and at high temperatures. These conditions tend to allow for a large separation of particles and decreased interactions.

40. In order for a substance to be considered a supercritical fluid, it must exist at a temperature above the critical temperature at a pressure above the critical pressure.

41. Yes. If the pressure was above the pressure for the triple point, the solid water would first liquefy, then turn into a gas. Sublimation is the process in which a substance goes directly from the solid state to the gaseous state.

42. No. A substance must have a triple point in order for it to be able to sublime.

43. Refer to the phase diagram of water in the text (Figure 9.12). At 100°C and 5.0 atm, water exists as a liquid. If the temperature was held constant and pressure was reduced to 0.5 atm, the sample would be converted from a liquid into a gas.

45. At temperatures below −56.6°C, the triple point of CO_2, there is no liquid phase of CO_2. So at any temperature below −56.6°C, there is a solid/gas phase boundary line. By lowering the pressure below that indicated on the boundary line, solid CO_2 can be converted into gaseous CO_2.

47. Consult the phase diagram in the text (Figure 9.12), or try to answer the questions without its aid.
    a. At twice the atmospheric pressure, but only barely above the normal boiling point of water, we expect water to be in the liquid state.
    b. Since we are at half the atmospheric pressure, but are only 20° lower than the normal boiling point, we expect water to be in the gaseous phase.
    c. At 0.010°C and 0.0060 atm, the sample would be at the triple point of water. At 0°C and 0.0060 atm water would be in the solid phase.

49. Look at the phase diagram for water (Figure 9.12 in the text) and focus on the solid/liquid phase boundary line.

By examining the boundary line we see that the melting point for water decreases with increasing pressure.

51. Note that the sketch below is not drawn to scale.

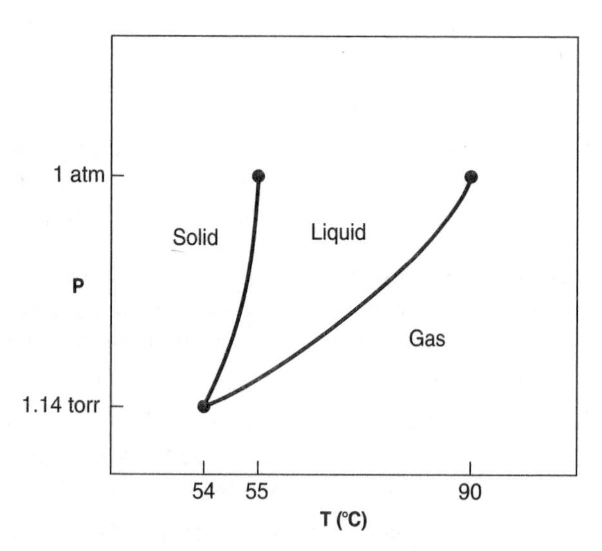

53. The slope of the line for compound Y along the solid/liquid phase boundary is similar to that of water. Therefore, we expect compound Y to expand as it freezes, just like water.

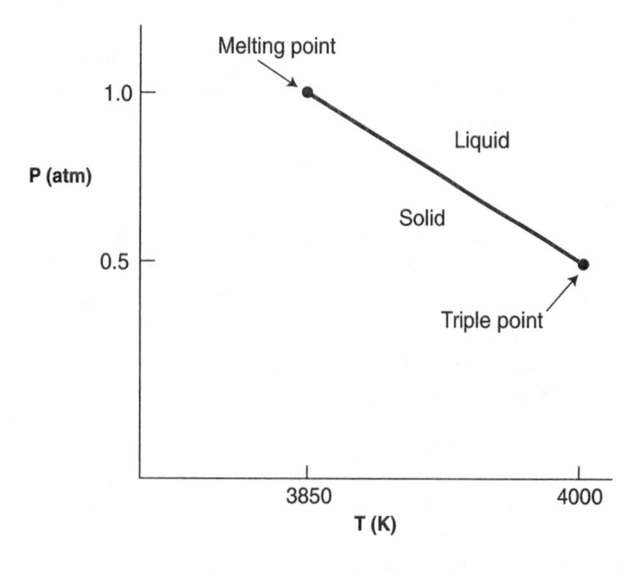

55. Water has a very high surface tension due to the strengths of the hydrogen bonds that water can form. Although both H_2O and CH_3OH can form hydrogen bonds, H_2O can form more of a network of interactions because of the two lone pairs on the oxygen and the two hydrogens coming off the oxygen. This allows H_2O to make more hydrogen bonds per molecule than CH_3OH. With fewer attraction forces in CH_3OH, we expect it to have a lower surface tension than H_2O. The surface tension of CH_3OH must not be strong enough to support a needle, while that of H_2O is strong enough.

56. In a narrow glass tube, there is less mass of water per cm climbed than in a wider tube. The adhesive and cohesive forces in a tube pull the water up, while the force of gravity pulls it down. So, the narrower tube should be able to support a greater height of water, but not a greater mass of water since the narrower tube contains less H_2O than a wider tube filled to a comparable height.

57. The type and strength of intermolecular forces will differ for different liquids. The densities of different liquids also differ. Since the liquids will rise until the adhesive and cohesive forces balance the force of gravity, we expect these heights to be different for different substances.

58. Ice floats on water because it is less dense than water. Ice is less dense than water because the rigid framework of hydrogen bonds that form simply take up more space than when they were in the liquid state.

59. Below 0°C, liquid water may freeze. Since ice is less dense than water, it will expand to a larger volume and possibly break the pipes while trying to expand.

60. As temperature increases, we know that the average kinetic energy of the particles in a sample increases too. More particles will now have the energy necessary to counteract the adhesive and cohesive forces that control capillary action. We expect a hot liquid to rise to a lower height than a cold liquid.

61. It is unlikely, due to the lower surface tension of the hot water. The cold needle could not cool the water sample to the point where the surface tension was large enough to support the needle, especially as water has a very high specific heat.

62. Yes, nonpolar liquids have surface tensions because they have London forces. The surface tensions of these liquids tend to be small in comparison to those of polar compounds. This is due to the fact that London forces tend to be weaker than dipole–dipole forces. Mercury is an example of a nonpolar species with a high surface tension. Note that it has a high molar mass.

63. Mercury is nonpolar and will not have strong adhesive interactions with the glass, SiO_2. In addition, the sample of mercury will maximize its cohesive forces by assuming a shape with minimum surface area for its volume (a sphere) and this results in a convex shape.

64. Assuming that the capillary tube is glass, we know that it is polar and the mercury is nonpolar. Mercury will not have strong adhesive forces with the glass surface and hence will not be drawn into and up the walls of the

capillary tube. The mercury will also resist going in to the tube because it will try to maximize its cohesive forces. Lastly, mercury also has a very high surface tension that also helps prevent it from flowing into the capillary tube.

65. The water will rise higher than the ethanol. The water is the more polar compound of the two and capable of forming a network of hydrogen bonds. The water will have stronger adhesive and cohesive forces compared to ethanol and this will allow the water to climb higher.

CHAPTER 10 The Solid State: A Molecular View of Gems and Minerals

1. In order for an X-ray pattern to contain distinct peaks, the sample studied must have a lattice, or repeating array of atoms, ions, or molecules. The repeating structure allows information to be gained about the repeating unit (unit cell) in the structure. The number of particles, types of particles, sizes of the particles, and spatial distribution within the repeating unit cell can all be determined by X-ray diffraction of a crystalline substance.

2. The atoms, molecules, and/or ions present in a solution are in constant random motion. X-ray diffraction requires that the sample have a lattice, or repeating array of atoms, ions, or molecules. If the particles are not even locked into a position, let alone a repeating array, they will not be able to act like a diffraction grating.

3. X-ray diffraction, like all diffraction, requires a grating. In X-ray diffraction, repeating arrays of atoms, molecules, or ions make up the grating. You may remember from physics class that diffraction tends to work best when the spaces in the grating and the wavelength of the light are close in size. Microwaves have much longer wavelengths than the distances between atoms, ions, and molecules in the solid state so they cannot be used like X-rays for studying crystalline solids.

4. A crystallographer may change the wavelength of the X-ray source if it appears that the crystal to be examined has larger or smaller spacings between its layers than is appropriate for the X-ray source in use. Diffraction tends to work best when the spaces in the grating and the wavelength of the light are close in size, so changing X-ray sources can help the crystallographer obtain better data. Based on the Bragg equation, the value of θ will be larger for $\lambda = 154$ pm than for $\lambda = 71$ pm.

5. $$n\lambda = 2d \sin\theta$$

   For given values of n and λ, $2\, d \sin\theta$ must be constant. As d increases, $\sin\theta$ must decrease. As $\sin\theta$ decreases, so must the value of θ. Therefore we expect NaCl to have greater values of 2θ relative to KCl.

7. $$n\lambda = 2d \sin\theta$$

   We were given λ and two values of 2θ, then asked to calculate d. Obtaining values for θ is easy, but we have to do a little work to obtain the values of n.

   If $2\theta = 13.98^\circ$, $\theta = 6.99^\circ$
   If $2\theta = 21.25^\circ$, $\theta = 10.63^\circ$

   Comparing the two values of θ, we see that $10.63/6.99 \approx 1.5$. Since n must be an integer, the values of n could be $2/3$, or $4/6$, or $6/9$, etc. We'll start with the lowest possible values of n and see if we obtain similar values of d.

   $$d = n\lambda/(2 \sin\theta)$$

   If $\theta = 6.99^\circ$

   $$d = (2)(71.7 \text{ pm})/(2 \sin 6.99) = 589 \text{ pm}$$

   If $\theta = 10.63^\circ$

   $$d = (3)(71.7 \text{ pm})/(2 \sin 10.625) = 583 \text{ pm}$$

   The two values of d that we obtain are fairly similar and we find the average value of the spacing between layers in PbS to be 586 pm.

9.
$$n\lambda = 2d \sin\theta$$

We'll use the Bragg equation, given above, to solve for the smallest possible value of θ. The smallest value of θ will be obtained when $n = 1$. Substituting into the equation we obtain

$$n\lambda/(2d) = \sin\theta$$
$$(1)(154 \text{ pm})/(2)(1855 \text{ pm}) = \sin\theta$$
$$0.0415 = \sin\theta$$
$$\theta = 2.38 \text{ degrees}$$

By convention, the angle of diffraction is given as $2\theta = 4.76$.

11. Species at the corners, faces, and edges of a unit cell are not completely within a single unit cell. Instead they are split between the cells that meet at the point on the unit cell where they are located. Particles on faces are split between two cells and hence we count $1/2$ of the particle per unit cell. Particles on the edges of cubic structures are split between four cells and hence we count $1/4$ of the particle per unit cell. Particles at the corners of cubic structures are split between eight cells and hence we count $1/8$ of the particle per unit cell. Particles contained within the body of the unit cell count as a whole.

12. A body-centered cubic structure has the more efficient packing. See Problems 33 and 34 for a supporting calculation.

13. A face-centered cubic unit cell will have the higher density since it is a more efficient packing scheme than body-centered cubic. See Problems 33 and 34 for a supporting calculation.

14. No. If we remove all of the Na^+ ions and replace them with Ca^{2+} ions, we see that this cannot work. When we do this and count the ions to obtain the formula of the unit cell, we obtain $Ca_4Cl_4^{4+}$. We know that this is not the formula unit of $CaCl_2$, which has a ratio of Ca^{2+} to Cl^- of 1:2. $CaCl_2$ must adopt a different crystal structure, where the ratio of cations to anions is 1:2. Note that this does not mean that every ionic compound with $+1$ cation and a -1 anion adopts the structure for NaCl.

15. In unit cells, smaller ions tend to fit into the holes made by the larger ions. Using the radius of the larger spheres, the size of various types of holes in various packing schemes can be obtained. The cation-to-anion ionic radius ratio can be used to predict the types of holes into which the smaller ions can fit. See Table 10.2 in the text.

16. In cubic close packing (ccp) and hexagonal close packing (hcp) single layers of spheres are arranged as shown here:

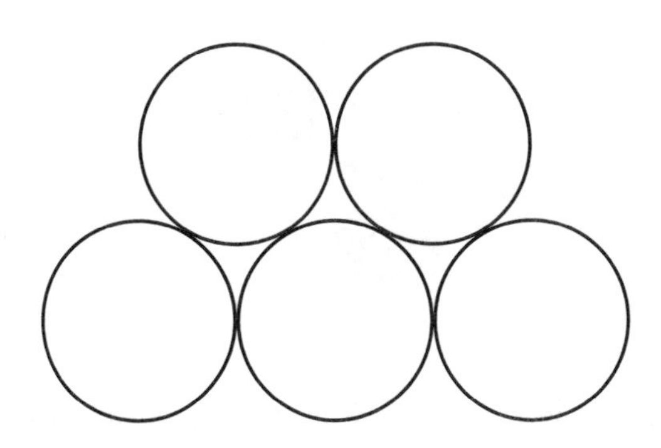

The difference in the two types of packing arises in how layers are stacked upon one another. In cubic close packing, layers are arranged in an ABCABCAB-CABC... pattern. This means that every ion in an "A" layer has another identical ion directly above it and below it in all of the "A" layers. So, every fourth layer is a repeat (ABC*A*, A*BCAB*, ABC*ABC*). (See Figure 10.6 in the text.) In hexagonal close packing, the layers are arranged in an ABABABAB... pattern. This means that every ion in an "A" layer has another identical ion directly above it and below it in all of the "A" layers. Every third layer is a repeat (AB***A***, A***BAB***).

17. Since cations tend to be the smaller ions and fit into the holes, we expect the cation-to-anion ratio of a triangular hole to be larger ($1/3$) than that of a tetrahedral hole ($1/4$).

18. A CsCl unit cell is made of a simple cubic arrangement of chloride ions, and this explains why CsCl is often assigned a simple cubic structure. It is sometimes assigned a body-centered structure since the cesium ion fits in the position where the body-centered atom would be in the body-centered unit cell.

19. In the pattern on the left, there are eight light-colored squares and one dark square per unit cell. The unit cell is as follows:

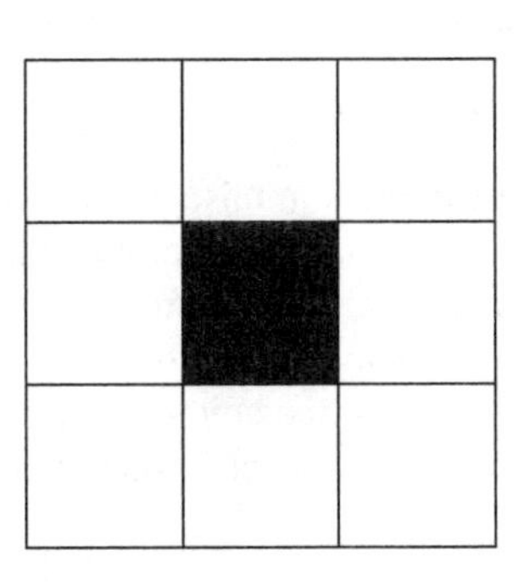

A

In the pattern on the right, there are two light and two dark squares per unit cell. The unit cell is as follows:

B

21. In the given unit cell the number, type, and locations of the species are

$$\text{Corner A} = 4 \rightarrow 4(1/8) = 1/2 \text{ A}$$
$$\text{Total} \quad 1/2 \text{ A}$$

$$\text{Corner B} = 4 \rightarrow 4(1/8) = 1/2 \text{ B}$$
$$\text{Total} \quad 1/2 \text{ B}$$

$$\text{Edge X} = 4 \rightarrow 4(1/4) = 1 \text{ X}$$
$$\text{Total} \quad 1 \text{ X}$$

The unit cell contains half of an A cation, half of a B cation, and one B anion.

23. In the given unit cell the number, type, and locations of the species are

$$\text{Body A} = 4 \rightarrow 4(1) = 4 \text{ A}$$
$$\text{Body B} = 4 \rightarrow 4(1) = 4 \text{ B}$$
$$\text{Corner X} = 4 \rightarrow 4(1/8) = 1/2 \text{ X}$$

The formula of this unit cell is $A_4B_4X_{1/2}$ and the formula unit of this ionic compound would be A_8B_8X.

25. In the given unit cell the number, type, and locations of the species are

$$\text{Corner Mg} = 6 \rightarrow 6(1/12) = 1/2 \text{ Mg}$$
$$\text{Body B} = 1 \rightarrow 1(1) = 1 \text{ B}$$

Note how the corners in this cell are different from the corners in cubic cells. Only one-twelfth of each corner sphere is in this unit cell, compared to one-eighth for the corner spheres in cubic cells.

The formula of this unit cell is $Mg_{1/2}B$ and the formula unit of this ionic compound would be MgB_2.

27. To determine the type of unit cell for copper, all we have to do is calculate the density for each of the three possible cubic structures (simple, body-centered, or face-centered) and match that with the experimental data. Simple cubic cells have one atom per unit cell, body-centered cubic cells have two, and face-centered cubic cells have four. First we will obtain the mass of one copper atom, then the mass of one unit cell.

Mass of one copper atom:

$$1 \text{ Cu atom} \times \frac{\text{mole Cu}}{6.022 \times 10^{23} \text{ Cu atoms}} \times \frac{63.55 \text{ g Cu}}{\text{mole Cu}}$$
$$= 1.055 \times 10^{-22} \text{ g}$$

Mass of simple cubic unit cell of Cu

$$= 1.055 \times 10^{-22} \text{g (1 atom per unit cell)}$$

Mass of body-centered unit cell of Cu

$$= 2.111 \times 10^{-22} \text{g (2 atoms per unit cell)}$$

Mass of face-centered unit cell of Cu

$$= 4.221 \times 10^{-22} \text{g (4 atoms per unit cell)}$$

Next we need the volume of the three unit cells. Using the radius of the copper atom (1.278×10^{-8}am),

Simple cubic cell volume:

$$l = 2r$$

$$\text{Volume} = l^3 = (2r)^3 = 8r^3$$

A

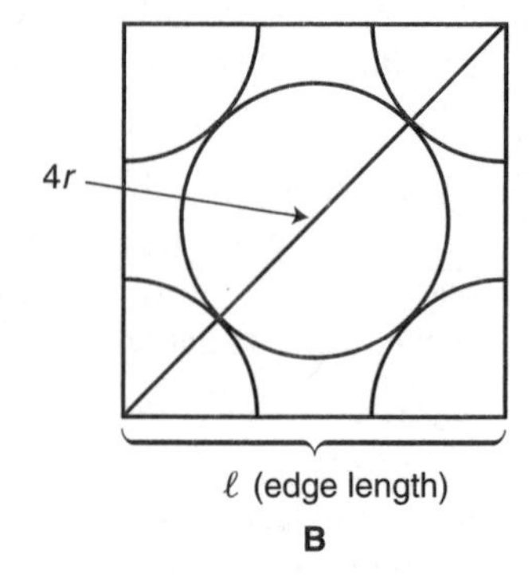

B

Face-centered cubic unit cell volume:

$$l^2 + l^2 = 4r^2$$
$$2l^2 = 16r^2$$
$$l^2 = 8r^2$$
$$l = \sqrt{8}r$$

$$\text{Volume} = l^3 = (\sqrt{8}r)^3 = 8\sqrt{8}r^3$$

Body-centered unit cell volume:

In a bcc the cross body diagonal has a length of $4r$. A cross face diagonal has a length equal to $\sqrt{2}l$, where l is the edge length.

$$(\text{Cross body diagonal})^2 = (\text{edge length})^2 + (\text{cross face diagonal})^2$$

$$(4r)^2 = l^2 + (\sqrt{2}l)^2$$

$$16r^2 = l^2 + 2l^2$$

$$16r^2 = 3l^2$$

$$\frac{16}{3}r^2 = l^2$$

$$\frac{4}{\sqrt{3}}r = l$$

$$\text{Volume} = l^3 = \left(\frac{4}{\sqrt{3}}r\right)^3 = \frac{64}{3\sqrt{3}}r^3$$

Volume of a simple cubic unit cell $= 8\,r^3$

$= 8(1.278 \times 10^{-8}\text{ cm})^3$

$= 1.670 \times 10^{-23}\text{ cm}^3$

Volume of a bcc unit cell $= (64\,r^3)/(3\sqrt{3})$

$= 64(1.278 \times 10^{-8}\text{ cm})^3/(3\sqrt{3})$

$= 2.571 \times 10^{-23}\text{ cm}^3$

Volume of a fcc unit cell $= 8\sqrt{8}\,r^3$

$= 8\sqrt{8}\,(1.278 \times 10^{-8}\text{ cm})^3$

$= 4.722 \times 10^{-23}\text{ cm}^3$

Dividing the mass of the cubic cell by the volume gives:

Density of a simple cubic Cu unit cell $= 6.317\text{ g/cm}^3$

Density of a bcc Cu unit cell $= 8.211\text{ g/cm}^3$

Density of a fcc Cu unit cell $= 8.939\text{ g/cm}^3$

Within error limits it seems that Cu must crystallize in an fcc unit cell.

29. ReO_3 has 8 Re atoms at the corners $(8 \times 1/8)$ for a total of 1 Re atom in the unit cell. There are 12 edge O atoms in a ReO_3 unit cell for a total of 3 O atoms $(12 \times 1/4)$. Each unit cell of ReO_3 contains one formula unit of ReO_3. Given that the atoms touch along the edges, we can obtain the edge length.

$$\text{Edge length of } ReO_3 = 2\,(\text{radius}_{Re}) + 2\,(\text{radius}_{O})$$

$$= 2(137\text{ pm}) + 2(74\text{ pm})$$

$$= 422\text{ pm}$$

Given the edge length and the formula, we can calculate the density.

Mass of one unit cell

$$1 \text{ Formula unit of } ReO_3 \times \frac{234.21\text{ g } ReO_3}{\text{mole } ReO_3} \times \frac{1\text{ mole } ReO_3}{6.022 \times 10^{23}\text{ formula units of } ReO_3}$$

$$= 3.8892 \times 10^{-22}\text{g } ReO_3$$

$$\text{Density} = \frac{\text{g}}{\text{cm}^3}$$

$$= \frac{3.8892 \times 10^{-22}\text{ g}}{(422\text{ pm})^3} \times \left(\frac{1 \times 10^{10}\text{ pm}}{\text{cm}}\right)^3$$

$$= 5.18\text{ g/cm}^3$$

31. Eu crystallizes in a bcc unit cell. Basic trigonometry yields

$$\text{edge length} = 4r/\sqrt{3}$$

$$240.6\text{ pm} = 4r/\sqrt{3}$$

$$\text{radius of Eu} = 104.2\text{ pm}$$

33. To calculate the packing efficiency we need to obtain the volume of the spheres and the total volume of a simple cubic unit cell. In a simple cubic unit cell there is a total of one atom per unit cell. The volume of this atom would be the volume of a sphere $= (4/3)\pi r^3$. Since the atoms touch along the edge, the volume of the whole unit cell is $8r^3$.

$$\text{packing efficiency} = \frac{(4/3)\pi r^3}{8r^3} \times 100$$

$$= 52.36\%$$

35. The formula will be BaF_2. There will be a total of one F^- ion and 1/2 of a Ba^{2+} ion per unit cell. A simple cubic arrangement of F^- contains 8 corner F^-, which makes a total of one F^- per unit cell. Each single unit cell contains one cubic hole. If half of the cubic holes are occupied, there is 1/2 a Ba^{2+} ion in each unit cell.

37. There will be a total of four O^{2-} ions in the unit cell. The cubic closest-packed arrangement of oxide ions has eight tetrahedral holes and four octahedral holes. Since half of the octahedral holes contain Al^{3+} and one fourth of the tetrahedral holes contain Mg^{2+}, there are two Al^{3+} ions and one Mg^{2+} ion in the unit cell. The formula will be $MgAl_2O_4$. This mineral is called aluminate.

39. To determine the density and radius of C_{60} we were given an edge length of 1410 pm and told that C_{60}

crystallizes in an fcc fashion. Basic trigonometry yields the relationship between edge length and radius:

$$\text{edge length} = r\sqrt{8}$$

$$1410 \text{ pm} = r\sqrt{8}$$

$$\text{radius of } C_{60} = 499 \text{ pm}$$

$$\text{density} = \frac{\text{mass of unit cell}}{\text{volume of unit cell}}$$

$$= \frac{4(\text{mass of } C_{60} \text{ molecule})}{(\text{edge length})^3}$$

Mass of unit cell:

$$4\, C_{60} \text{ molecules} \times \frac{1 \text{mole } C_{60}}{6.022 \times 10^{23}\, C_{60} \text{ molecules}}$$

$$\times \frac{720.666 \text{ g } C_{60}}{\text{mole } C_{60}} = 4.787 \times 10^{-21} \text{ g } C_{60}$$

Volume of a fcc unit cell:

$$8\sqrt{8}\, r^3 = 8\sqrt{8}\,(4.99 \times 10^{-8} \text{ cm})^3$$

$$= 2.81 \times 10^{-21} \text{ cm}^3$$

Density of an fcc C_{60} unit cell:

$$= (4.787 \times 10^{-21} \text{ g})/(2.81 \times 10^{-21} \text{cm}^3) = 1.70 \text{ g/cm}^3$$

41. We can use the density of Pd and the fact that it crystallizes in a ccp fashion to calculate the atomic radius of Pd.
 Mass of unit cell

$$4 \text{ Pd atoms} \times \frac{\text{mole of Pd atom}}{6.022 \times 10^{23} \text{ Pd atoms}} \times \frac{106.42 \text{ g Pd}}{\text{mole Pd}}$$

$$= 7.069 \times 10^{-22} \text{ g Pd}$$

Use the density of Pd to calculate the volume of the unit cell:

$$7.069 \times 10^{-22} \text{ g Pd} \times \frac{1 \text{ cm}^3}{11.99 \text{ g}} = 5.896 \times 10^{-23} \text{ cm}^3$$

$$\text{volume of an fcc unit cell} = 8\sqrt{8}\, r^3$$

$$5.896 \times 10^{-23} \text{ cm}^3 = 8\sqrt{8}\, r^3$$

$$\text{Pd radius} = 1.376 \times 10^{-8} \text{ cm}$$

$$= 137.6 \text{ pm}$$

$$\text{radius ratio} = \frac{r_{\text{smaller}}}{r_{\text{larger}}}$$

$$= \frac{74.1 \text{ pm}}{137.6 \text{ pm}}$$

$$= 0.538$$

With a radius ratio of 0.538, we expect D atoms to fit into the octahedral holes, not the tetrahedral.

43. In Problem 39 we calculated the radius of C_{60} and found it to be 499 pm. Using the radius ratio we can determine into which type of hole the K^+ ions are able to fit.

$$\text{radius ratio} = \frac{r_{\text{cation}}}{r_{\text{anion}}}$$

$$= \frac{138 \text{ pm}}{499 \text{ pm}}$$

$$= 0.277$$

With a radius ratio of 0.277, we expect the K^+ ions to occupy tetrahedral holes. In a ccp arrangement of C_{60}, there will be four tetrahedral holes and 3/4 of them will be occupied by K^+ ions.

45. The angular molecular geometry of each S in S_8 prevents the eight sulfurs from being able to flatten out in an octagon shape. If we attempt to make a typical octagon with the S_8, it will be puckered. A flat, symmetrical octagon with sides of equal length should have interior angles of 135°, but the S—S—S bond angles in S_8 are expected to be smaller than 109.5° because of lone-pair repulsion. To accommodate optimal S—S—S bond angles, the octagon must be puckered. Try building an S_8 model. If a model set was not a required course material, ask your professor or teaching assistant to build an S_8 molecule.

46. We expect any elemental form of phosphorus to have either three or five bonds to other phosphorus atoms. A phosphorus atom with three bonds to other phosphorus atoms would also have a lone pair. The lone pair would yield a trigonal pyramid arrangement of P atoms. The puckered nature of the rhombohedral form of elemental phosphorus is due to a lone pair on each P atom. In graphite, the sp^2 hybridized C atoms all lie within the same plane.

47. Allotropes are different forms of the same element that have different structures and properties. Polymorphs are different chemical compounds with the same formula. Since these are both pure forms of the same element, they are allotropes and not polymorphs.

48. These two different forms of iron would be considered allotropes since they are different crystalline forms of iron. The major difference between these two forms would be the packing efficiency.

49. In this simple cubic form of phosphorus, one unit cell contains one P atom and the edge length is given to us as 238 pm or 2.38×10^{-8} cm.

Mass of one unit cell:

$$1\ \text{P atom} \times \frac{\text{mole of P atoms}}{6.022 \times 10^{23}\ \text{P atoms}} \times \frac{30.97\ \text{g P}}{\text{mole P}}$$

$$= 5.143 \times 10^{-23}\ \text{g P}$$

$$\text{density} = \frac{\text{mass}}{\text{volume}}$$

$$= \frac{5.143 \times 10^{-23}\ \text{g P}}{(2.38 \times 10^{-8}\ \text{cm})^3}$$

$$= 3.81\ \text{g/cm}^3$$

51. To determine the density of ionic ice, we will use the ionic radius of O^{2-} and the fact that the oxide ions are arranged in a bcc fashion. Therefore there are two oxide ions and four hydrogen ions in one unit cell of ionic ice. Basic trigonometry yields the relationship between the O^{2-} radius and volume of a bcc unit cell:

$$\text{volume of a bcc unit cell} = (64\,r^3)/(3\sqrt{3})$$

$$= 64(1.26 \times 10^{-8}\ \text{cm})^3/(3\sqrt{3})$$

$$= 2.46 \times 10^{-23}\ \text{cm}^3$$

$$\text{density} = \frac{\text{mass of unit cell}}{\text{volume of unit cell}}$$

$$= \frac{(\text{mass of 2 } H_2O \text{ molecules})}{(2.46 \times 10^{-23}\ \text{cm}^3)}$$

Mass of H_2O per unit cell:

$$2\ H_2O\ \text{molecules} \times \frac{1\ \text{mole } H_2O}{6.022 \times 10^{23}\ H_2O\ \text{molecules}} \times \frac{18.015\ \text{g } H_2O}{\text{mole } H_2O} = 5.983 \times 10^{-23}\ \text{g } H_2O$$

density of a bcc unit cell of ionic ice unit cell

$$= \frac{5.983 \times 10^{-23}\ \text{g } H_2O}{2.46 \times 10^{-23}\ \text{cm}^3}$$

$$= 2.44\ \text{g/cm}^3$$

Lewis structure for ionic ice: $[H]^+ [:\ddot{\underset{..}{O}}:]^{2-} [H]^+$

53. To replace all of the Al^{3+} in $Al_2(OH)_4(Si_2O_5)$ with Mg^{2+}, we need to determine the total charge on all of the Al^{3+}. The two Al^{3+} ions have a total charge of +6. Three Mg^{2+} ions will be needed to make up this charge and the formula of antigorite is $Mg_3(OH)_4(Si_2O_5)$.

55. We know the metal copper is a good conductor of electricity. Both the electron-sea model and band theory can be used to account for these properties. In the electron-sea model, copper nuclei are assumed to be dispersed in a sea of electrons that are free to move throughout the bulk sample of copper. Band theory is based on molecular orbital (MO) theory and says that a large number of atoms in a bulk sample will form two bands of MOs. One band will be of bonding MOs and the other of antibonding MOs. In copper, the bonding MOs will be partially filled and the antibonding MOs will be empty. In materials that are conductors, the two bands are very close in energy and electrons are able to move from one band to another. This allows the electrons freedom of movement throughout the bulk sample.

56. These observations tell us that the ionic interactions in NaCl are stronger than the metallic bonds in sodium metal. Stronger intermolecular forces yield higher melting points.

57. The figure shows is the MO diagrams for an H_2 molecule and for a molecule with the formula H_n, where n is a very large number. For the H_2 molecule, two atomic orbitals combine to make two MOs. For the H_n molecule, n atomic orbitals combine to make n MOs. In H_2, the bonding MOs are filled; we expect the same for H_n.

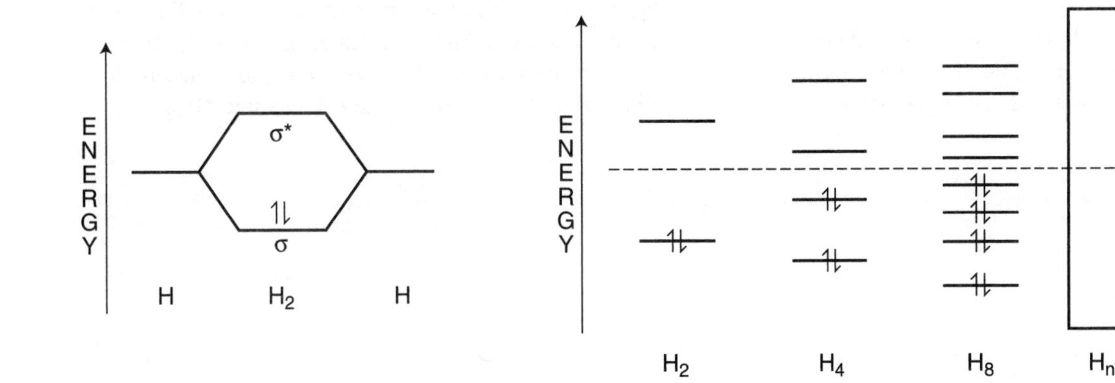

59. Systems with partially filled d orbitals often have color. They appear to have color because the energies that they absorb correspond to wavelengths in the visible spectrum. The wavelengths that are not absorbed are reflected back and that is what we see.

60. The orbitals of d_{z^2} and $d_{x^2-y^2}$ in an octahedral field are pointed directly toward the atoms or ions defining the octahedron. The repulsions between these corner atoms or ions and the lobes of the d_{z^2} and $d_{x^2-y^2}$ orbitals of the transition metal raise the energy of these two orbitals relative to the other three d orbitals (d_{xy}, d_{xz}, and d_{yz}).

61. The orbitals of $d_{x^2-y^2}$ in a square planar field have their lobes pointed directly toward the atoms or ions that form the square plane. The repulsions between these corner atoms or ions and the lobes of the $d_{x^2-y^2}$ orbitals of the transition metal raise the energy of this orbital relative to the other d orbitals.

62. The value of Δ_o helps determine if a system is high spin or low spin. Systems with large values of Δ_o usually have their highest-energy d electrons in the lower-energy d orbitals. This arrangement promotes the pairing up of electrons in the lower-energy d orbitals. Systems with small values of Δ_o have their highest-energy d electrons spread out more in the d orbitals. This often results in many unpaired electrons in a system. For example, Mn^{2+} has five unpaired electrons with small values of Δ and one unpaired electron with large values of Δ. Mn^{2+} contains five 3d electrons.

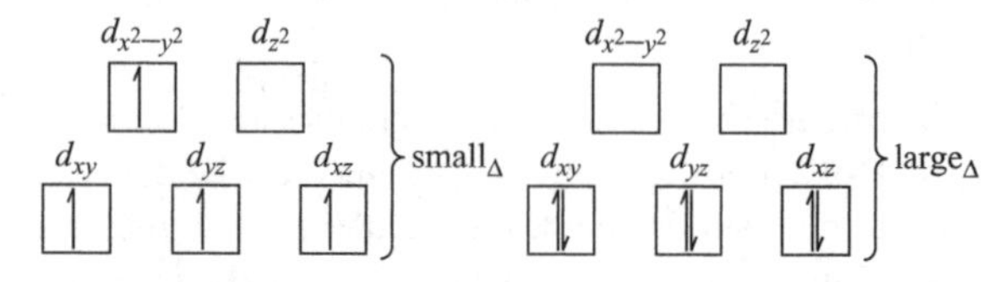

63. Zn^{2+} ions have the configuration [Ar] $4s^0\,3d^{10}$. Since their 3d subshell is completely filled, this does not allow for transitions that correspond to wavelengths within the visible range.

64. Ti^{4+} is isoelectronic with Ar. The transitions of electrons in Te^{4+} will not be seen within the visible spectrum. This is because of the noble gas configuration of Te^{4+}.

65. In an octahedral field:

    Fe^{2+} (d^6) can have 4 unpaired electrons (high spin) or 0 unpaired electrons (low spin).
    Fe^{3+} (d^5) can have 5 unpaired electrons (high spin) or 1 unpaired electron (low spin).
    Mn^{4+} (d^3) will have 3 unpaired electrons regardless of the size of Δ_o.
    Cr^{3+} (d^4) can have 4 unpaired electrons (high spin) or 2 unpaired electrons (low spin).

67. The relationship between crystal field splitting energy and wavelength is: $\Delta_o = hc/\lambda$.

    This means that the smaller wavelength will correspond to a larger splitting energy and vice versa. The compound that absorbs the 450 nm light will have a larger splitting energy than the compound that absorbs 580 nm light.

69. To calculate the crystal field splitting energy, we use the formula $\Delta_o = hc/\lambda$.

    Remember that the wavelength substituted must be in meters!

    For $\lambda = 450$ nm, $\Delta_o = 4.41 \times 10^{-19}$ J
    For $\lambda = 580$ nm, $\Delta_o = 3.42 \times 10^{-19}$ J

    The compound that absorbs the 450 nm light will have a larger splitting energy than the compound that absorbs 580 nm light.

71. In an octahedral field:

    Fe^{3+}(d^5) will have 5 unpaired electrons in a high-spin configuration.
    Rh^{+}(d^8) will have 2 unpaired electrons in a high-spin configuration.
    V^{3+}(d^2) will have 2 unpaired electrons in a high-spin configuration.
    Mn^{3+}(d^4) will have 2 unpaired electrons in a low-spin configuration.

73. The complexes in this problem are Co^{2+}(d^7) and Co^{3+} (d^6). The number of unpaired electrons is the clue to figuring out the relative magnitudes of Δ_o. The Co^{2+} complex must be high spin (small Δ_o) in order to have 3 unpaired electrons. The Co^{3+} complex must be low spin (large Δ_o) in order to have 0 unpaired electrons.

CHAPTER 11 Thermochemistry and the Quest for Energy

1. Energy is the capacity to do work. For example, consider a water being held behind a dam. The water has potential energy (stored energy). If the water is allowed to flow through conduits near the top of the dam, flowing water can be delivered to a turbine. This water can then turn rotors in the turbine to create electricity. The potential energy is converted to work, creating electricity.

2. Kinetic energy is energy associated with moving objects. Potential energy is stored energy (a function of position).

3. A state function depends on the initial and final conditions, but does not depend on the path taken to get from the initial to the final set of conditions.

4. Potential energy is a state function. It is related to position not how the object of interest got to that position. Kinetic energy is not a state function. The total kinetic energy will depend on the pathway taken.

5. Potential energy is stored energy.
   a. A battery has stored potential energy (in the form of chemical energy). When the CD player is turned on, the potential energy is converted to direct current that is used to play the CD.
   b. The potential energy here is stored in the chemical bonds present in gasoline. As the gasoline molecules react, old bonds are broken and new ones are formed. The result is that chemical energy is transformed into heat and kinetic energy. The kinetic energy is used to move your car.
   c. The potential energy in a wave depends on the height of the wave. The higher the wave the greater the potential energy present.

7. No. The general formula of an alkane is C_nH_{2n+2}. Because various alkanes have different empirical formulas they cannot have the same percent composition. Any substances that have the same empirical formula (simplest, smallest whole-number ratio) have the same percent composition.

8. No, the general formula for an alkane is C_nH_{2n+2}. By substituting for n in this general formula, we can see that alkanes cannot have the same empirical formula. A few examples of alkanes and their empirical formulas are as follows:

Name	***Molecular Formula***	***Empirical Formula***
Methane	CH_4	CH_4
Ethane	C_2H_6	CH_3
Propane	C_3H_8	C_3H_8
Butane	C_4H_{10}	C_2H_5
Octane	C_8H_{18}	C_4H_9

All these alkanes have different empirical formulas.

9. A methylene group, H—C—H, consists of a carbon atom that is bonded to 2 hydrogen atoms and 2 other atoms.

```
   H
   |
 —C—     A methylene group
   |
   H
```

A methyl group consists of a carbon atom that is bonded to 3 hydrogen atoms, and only 1 other atom.

```
    H
    |
 H—C—     A methyl group
    |
    H
```

10. No, the carbon chains in butane are not linear. The bond angle connecting each central carbon to 2 other carbon atoms is about 109°. It is called a linear alkane to distinguish it from branched alkanes. In a linear alkane each carbon atom is bonded to 2 other carbon atoms.

11. Alkanes are nonpolar substances. The only interactions between molecules are London (dispersion) forces. These forces increase with molecular size (normally increasing molar mass means larger molecule size) because the larger the molecule, the more polarizable is its charge cloud of bonding electrons. As molecules increase in size more energy is required to overcome these attractive forces. The result is that the boiling points of alkanes tend to increase with molar mass.

12. Alkanes are nonpolar species. London (dispersion) forces are the only intermolecular forces present.

13. Any substance that has the general formula C_nH_{2n+2} is an alkane.
    a. CH_4 is an alkane.
    b. C_2H_2 is not an alkane (the alkane that contains 2 carbon atoms has the molecular formula C_2H_6.
    c. $C_{16}H_{32}$ is not an alkane because it does not fit the formula C_nH_{2n+2}. However, in Chapter 12 you will learn that this formula matches that of a cycloalkane (a series of carbon atoms connected in a ring) and also an alkene. The linear alkane that contains 16 carbon atoms has the molecular formula $C_{16}H_{34}$.
    d. C_4H_{10} is an alkane. It is the formula predicted for the alkane that contains 4 carbon atoms.

15. The C—C—C bond angle in propane is about 109°.

17. The system includes the chemical and physical changes being studied. The surroundings are everything else except the system.

18. An exothermic process gives off heat (from the system) to the surroundings. An endothermic process absorbs heat from the surroundings.

19. The internal energy is the sum of all kinetic and potential energies of the components of the system.

20. i. Doing work on the system. The internal energy can be calculated using the formula $\Delta E = q + w$. If the surroundings do work on the system, then work by definition has a positive value. This makes ΔE larger. Anytime a gas is compressed, the surroundings perform work on the system, for example if the volume of an ideal gas is decreased from 5.0 L to 2.5 L at constant pressure.
    ii. Adding heat to the system from the surroundings; that is, the system absorbs heat from the surroundings. In this process q is positive (endothermic). This also increases the value of ΔE. Heating a beaker of water is an endothermic process.

21. In the combustion of methane, the system is the reaction of methane with oxygen (combustion). All combustion processes release heat to the surroundings so this process is exothermic and q is negative.

    In the freezing of water to make ice the system is the water. To freeze water, heat must be removed from the system; that is, heat must flow from the system (hotter) to the surroundings (cooler). This is also an exothermic process and q is negative.

    When you touch a hot stove heat must flow from the hot stove to the colder person. If the stove is considered to be the system, then heat flows from the system (hot stove) to the surroundings (the person). This process would be exothermic. If the person is considered to be the system, then the person is absorbing heat from the surroundings. This process is endothermic, so q is positive.

22. In driving an automobile, the automobile is the system. Heat is transferred from the system to the surroundings due to the combustion of gasoline and also friction created between the road surface and the tires. This process is exothermic, so q is negative.

    When applying ice to a sprained ankle, if the ankle is considered to be the system, then heat flows from the ankle (system) to the ice (surroundings), so the ankle (the system) loses heat. This process is exothermic so q is negative.

    Note: If one considered the ice to be the system (this is a possible interpretation) and the ankle to be the surroundings, then all heat processes would be reversed. The process would be endothermic and q would be positive.

    When cooking a hot dog, the hot dog is the system. It absorbs heat from the surroundings. This process is endothermic and q is positive.

23. We can make the following generalizations about exothermic and endothermic phase changes (changes between solids, liquids, and gases). If a solid changes to liquid or a liquid changes to a gas (vapor), the substance absorbs heat. These processes are endothermic.

    If a gas (vapor) changes to a liquid or a liquid changes to a solid, the substance loses heat. These processes are exothermic.
    a. Burning a match is an exothermic process because heat is released from the system (the burning match) to the surroundings (the air nearby).
    b. Molten is a term used to describe a metal in the liquid state. The process here is transformation of a liquid to a solid. For this process to occur, the metal (the

system) must lose heat (to the surroundings). This is an exothermic process.

c. Your skin feels cold when rubbing alcohol evaporates because the alcohol is absorbing heat from the skin to vaporize; that is, heat is flowing from your skin to the alcohol. This leaves less heat in your skin, making it feel colder. If the system is your skin, the process is exothermic. If the system is the alcohol, the process is endothermic.

25. Internal energy is determined from the equation $\Delta E = q + w$. At the boiling point, heat is absorbed from the surroundings. In this case q (heat) is a positive quantity and the internal energy increases.

27. The formula for work is $w = P\Delta V$. The pressure needs to be in atmospheres and ΔV needs to be in liters. This gives us work units of L · atm.

$$P = 1.00\,\text{atm}$$
$$\Delta V = 750.\,\text{mL} - 250.\,\text{mL}$$
$$= 500\,\text{mL}$$
$$= 0.500\,\text{L}$$
$$w = (1.00\,\text{atm})(0.500\,\text{L})$$
$$= 0.500\,\text{L}\cdot\text{atm}$$

The conversion factor needed to convert this quantity to joules is

$$101.3\,\text{J} = 1.00\,\text{L}\cdot\text{atm}$$
$$0.500\,\text{L}\cdot\text{atm} \times \frac{101.3\,\text{J}}{1.00\,\text{L}\cdot\text{atm}} = 50.6\,\text{J}$$

Note: Since work is done by the system on the surroundings, the actual sign on the work is negative. That is

$$w = -0.500\,\text{L}\cdot\text{atm}$$
$$= -50.6\,\text{J}$$

29. $\Delta E = q + w$

a. $q = 100\,\text{J}; w = -50\,\text{J}$

$$\Delta E = 100\,\text{J} + (-50\,\text{J}) = 50\,\text{J}$$

b. $q = 6.2\,\text{kJ}; w = 0.7\,\text{J}$

We need to convert either 6.2 kJ into joules or 0.7 J into kilojoules. In this problem, we convert kilojoules to joules because the work value (0.7 J) would be a very small number if converted to kilojoules.

$$q = 6.2\,\text{kJ}$$
$$= 6200\,\text{J}$$
$$\Delta E = 6200\,\text{J} + 0.7\,\text{J}$$
$$= 6200.7\,\text{J}$$
$$= 6201\,\text{J or } 6.2\,\text{kJ}$$

Rounded to two significant figures the answer is 6.2 kJ.

c. $q = -615\,\text{J}; \quad w = -325\,\text{J}$

$$\Delta E = (-615\,\text{J}) + (-325\,\text{J}) = -940\,\text{J}$$
$$= -940\,\text{J}$$

31. If a gas releases 210 kJ of heat energy on combustion then the process is exothermic.

$$q = -210\,\text{kJ}$$

If the gas does 65.5 kJ of work on the surroundings then the work is given a negative sign.

$$w = -65.5\,\text{kJ}$$
$$\Delta E = (-210\,\text{kJ}) + (-65.5\,\text{kJ}) = 276\,\text{kJ}$$
$$= 276\,\text{kJ}$$

33. From the gas laws, we determined that the volume of a gas (at constant pressure) is directly proportional to the number of moles of gas present. If the number of moles of gas increases, then the volume of the container must increase (pressure remains constant).

We can determine the work done at constant pressure by looking at the coefficients of the gaseous reactants and products.

If there are more moles of gaseous products than reactants, the volume increases and the system does work on the surroundings.

If there are more moles of gaseous reactants than products, the volume decreases and the surroundings perform the work on the system.

If there is an equal number of moles of gas on both sides of the reaction arrow, then there is always the same number of moles of gas in the system. The volume does not change and no work is performed.

a. $CH_4(g) + 2\,O_2(g) \rightarrow CO_2(g) + 2\,H_2O(g)$

There are 3 moles of gaseous products (2 mol H_2O and 1 mol CO_2).

There are 3 moles of gaseous reactants (1 mol CH_4 and 2 mol O_2).

Since there is an equal number of moles of gas on both sides of the arrow, the volume will not change. In this case no work is performed and $w = 0$.

b. $C_3H_8(g) + 5\,O_2(g) \rightarrow 3\,CO_2(g) + 4\,H_2O(g)$

There are 7 moles of gas on the product side (3 moles of CO_2 and 4 moles of H_2O).

There are 6 moles of gas on the reactant side (1 mole of C_3H_8 and 5 moles of O_2).

There are more moles of gas on the product side. The volume expands to accommodate more gas particles, and the system does work on the surroundings.

c. $$N_2(g) + 2\,O_2 \rightarrow 2\,NO_2(g)$$

There are 2 moles of gaseous products (2 moles of NO_2).
There are 3 moles of gaseous reactants (1 mole of N_2 and 2 moles of O_2).
The volume decreases because there are fewer gas particles present after reaction, and the surroundings do work on the system.

35. Enthalpy change is defined as the heat change at constant pressure.

36. The enthalpy change, ΔH, is defined as the heat change at constant pressure. The internal energy change, ΔE, measures the changes in the kinetic and potential energies of the components of a system. However, this definition is not typically how we determine ΔE.

 The formula for internal energy is $\Delta E = q + w$, while the formula for enthalpy, ΔH, is given by $\Delta H = \Delta E + P\Delta V$.

 Heat is defined as an energy transfer due to differences in temperature between the system and surroundings. Work (usually pressure-volume work in this chapter) is associated with a change in volume. The magnitude of the change in internal energy is a combination of the heat transfer (at constant pressure) and work associated with a given process. If there is no work performed—that is, the volume does not change (as in a bomb calorimeter)—then ΔE is defined as the heat change under conditions of constant volume.

 Typically, the numerical values of ΔH and ΔE are quite close for constant pressure processes.

37. In an exothermic process, the system gives off heat to the surroundings.

 If the system loses heat, then the energy content of the system must be lower after the change. Since the final energy content is lower then the initial one, the change in ΔH is a negative quantity.

$$\Delta H = \Delta q = q_{\text{final}} - q_{\text{init}}$$

38. When a process is reversed, the sign of ΔH changes, but not the numerical value of ΔH.

39. The surroundings are warming up. This means that the system is giving off heat to the surroundings. This process is exothermic so the sign of ΔH is negative.

41. It requires energy to break the bond between the two oxygen atoms. Any process that requires energy from an outside source (the surroundings) is endothermic, so this process is endothermic and the sign of ΔH is positive.

43. For hydrogen to go from a gas to a solid, heat energy must be lost (as it is for any substance changing from a gas to either a liquid or a solid). This process must be exothermic and the sign of ΔH is negative.

45. Specific heat is the heat required to increase the temperature of *1 g of a substance* by 1°C (or 1 K). Heat capacity is the heat required to raise the temperature of *a sample* by 1°C (or 1 K).

 The difference is that specific heat is an amount of heat per gram and is therefore an intensive property of the substance, whereas the heat capacity depends on the sample size.

46. The heat capacity doubles if the mass of the material being studied is doubled, because heat capacity is dependent on the mass of the sample being studied. The specific heat, being an intensive property of the material, is unchanged if the mass is doubled.

47. No. It typically requires much more energy to vaporize a substance than it does to melt it (i.e., the heat of vaporization is usually much larger than the heat of fusion for a particular substance).

48. The change in temperature can be calculated using the formula

$$q = (\text{heat capacity})(\Delta T)$$

Solving for temperature we find that

$$\Delta T = q/\text{heat capacity}$$

If q is the same for two processes, then the metal with the smaller heat capacity results in the the greater temperature change.

 The metal with the smaller heat capacity has the greater change in temperature and has a higher final temperature.

49. Water has a higher molar heat capacity than air. Rearranging Equation 11.11 we get $\Delta T = \frac{q}{\text{heat capacity}}$. The larger the heat capacity, the smaller the temperature change.

 Car manufacturers would choose water rather than air as a coolant because water should keep the engine cooler.

50. The larger the thermal conductivity, the more quickly heat passes through a substance. The thermal conductivity of molten sodium is about 233 times greater than that of water ($1.42\ \text{J/cm}\cdot\text{s}\cdot\text{K}/6.1 \times 10^{-3}\ \text{J/cm}\cdot\text{s}\cdot\text{K}$). That is, the sodium will absorb heat 233 times faster than will water. This can more than compensate for the fact that it requires less heat energy to heat 1 mole of sodium than 1 mole of water. That is, 1 mole of water absorbs 2.7 times as much heat energy as does 1 mole of sodium ($75.31\ \text{J/mol}\cdot\text{K}/28.28\ \text{J/mol}\cdot\text{K}$).

 However, the greater thermal conductivity of sodium allows the sodium to absorb heat much more quickly than water, a substance with a lower thermal conductivity. The greater thermal conductivity of sodium is an advantage in nuclear reactors, where rapid cooling is critical for safety. Using a large quantity of sodium could compensate for its lower heat capacity.

51. The formula to use is

$$q = nc_p\Delta T$$

where,

n is the number of moles
c_p is the molar heat capacity
ΔT is the change in temperature

$$\Delta T = T_{\text{final}} - T_{\text{init}}$$
$$= 100°\text{C} - 30°\text{C}$$
$$= 70°\text{C}$$

The molar heat capacity of H_2O (c_p) is 75.3 J/mol°C

The mass of water is 100. g. Before substituting into the equation, we need to convert grams of water to moles of water (molar mass H_2O = 18.0 g/mol).

$$\text{moles } H_2O = 100.\text{ g } H_2O \times \frac{1 \text{ mol } H_2O}{18.0 \text{ g } H_2O}$$
$$= 5.56 \text{ mol } H_2O$$

We can now determine the quantity of heat needed for this process.

$$q = (5.56 \text{ mol})(75.3 \text{ J/mol} \cdot °\text{C})(70°\text{C})$$
$$= 2.93 \times 10^4 \text{ J, or } 29.3 \text{ kJ}$$

53.

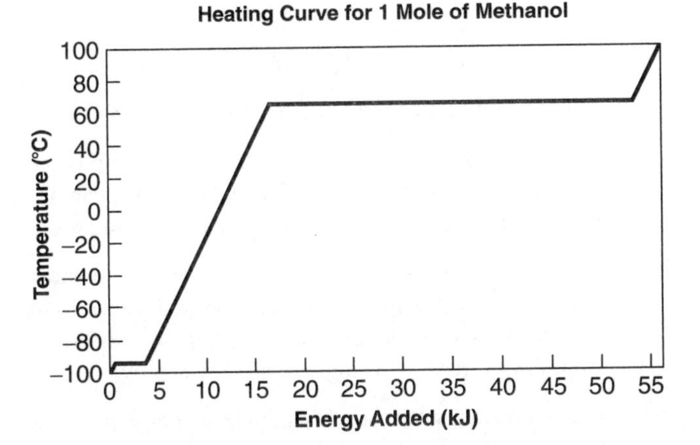

The flat portions on the curve are transition states. The shorter of the two flat portions corresponds to solid methanol melting. The longer portion corresponds to methanol vaporizing.

The three diagonal portions correspond to heating the solid, liquid, and vapor phases.

To determine the energy associated with each of the five steps, we need to use the formulas:

$$q = nc_p\Delta T \quad \text{and} \quad q = n\Delta H_{\text{phase change}}$$

For heating the solid, liquid, and vapor use $q = nc_p\Delta T$.

Since we are not given the molar heat capacity for each phase of methanol, we will assume that the heat capacity given refers to its most common state, the liquid state.

To determine the heat energy required to warm the solid and gaseous forms of methanol, it would require the molar heat capacity of methanol for these phases to be known. Without this information, we cannot determine the heat needed for these processes and we cannot accurately draw the portion of the above curve that describes these processes. However, we can estimate the values for the portion of the figure that shows how much heat is needed to warm the solid and the vapor by assuming that the molar heat capacities of all three states of matter are similar.

To heat the solid from −100°C to −94°C (the melting point):

$$q = (1 \text{ mol})(81.1 \text{ J/mol} \cdot °\text{C})(6°\text{C})$$
$$= 0.487 \text{ kJ}$$

To melt the solid methanol:

$$q = (1 \text{ mol})(3.18 \text{ kJ/mol})$$
$$= 3.18 \text{ kJ}$$

To warm liquid methanol from −94°C to 65°C (the boiling point):

$$q = (1 \text{ mol})(81.1 \text{ J/mol} \cdot °\text{C})(159°\text{C}) = 12.9 \text{ kJ}$$

To vaporize liquid methanol:

$$q = (1 \text{ mol})(37.0 \text{ kJ/mol})$$
$$= 37.0 \text{ kJ}$$

To warm the vapor from 65°C to 100°C:

$$q = (1 \text{ mol})(81.1 \text{ J/mol} \cdot °\text{C})(35°\text{C})$$
$$= 2.84 \text{ kJ}$$

55. The amount of heat generated at a phase change is given by the equation

$$q = n\Delta H_{\text{phase change}}$$

In this problem $q = 2000.$ kJ.

The phase change is the vaporization of water.

$$\Delta H_{\text{vap}} \text{ for water} = 40.67 \text{ kJ/mol (given in text)}$$

Solving the above equation for moles of water we get

$$n = \frac{q}{\Delta H_{vap}} = \frac{2000.\ kJ}{40.67\ kJ/mol} = 49.18\ mole$$

Lastly, we need to convert this quantity from moles to grams (molar mass of $H_2O = 18.0$ g/mol).

$$\text{mass of } H_2O = 49.18\ mol \times \frac{18.0\ g\ H_2O}{1\ mol\ H_2O} = 885\ g\ H_2O$$

The athlete would have to evaporate 885 g of water to dissipate 2000. kJ of heat.

57. In this question the heat lost by the skillet is absorbed by the cold water. This leads to a sign problem; that is, the heat lost term is negative but the heat gained must be a positive quantity. In formula form $-q_{heat\ lost} = q_{heat\ gained}$. In other words, to equate these two quantities they must have the same sign. The only way this can be accompalished is to change the sign on the heat lost term to a positive quantity. That is $|-q_{heat\ lost}| = q_{heat\ gained}$

Adding 10.0 mL (10.0 g) of water cools the skillet from a higher temperature to a lower temperature. We do not know initial temperature of the skillet. The object here is to solve for the change in temperature (ΔT). We can do this by equating the heat lost by the skillet and the heat gained by the water, then solving for ΔT.

Heat lost by the skillet:

(moles of iron)(molar heat capacity of skillet)(ΔT)

We know that the skillet has a mass of 1.20 kg (1.20×10^3 g) and the molar heat capacity of iron is 25.19 J/mol · °C.

First we need to convert the grams of iron to moles of iron.

$$\text{moles of Fe} = 1.20 \times 10^3\ g\ Fe \times \frac{1\ mol\ Fe}{55.85\ g\ Fe} = 21.5\ mol\ Fe$$

The heat lost by the skillet is

$$q_{lost} = (21.5\ mol\ Fe)(25.19\ J/mol \cdot {}^\circ C)(\Delta T) = (542\ J/{}^\circ C)(\Delta T)$$

Heat gained by the water:

The water must be warmed from 25°C to 100°C and then it is vaporized. We assume the final temperature of the steam is 100°C, since the problem did not say the steam was warmed above 100°C.

The heat needed to warm the water from 25°C to 100°C is

(moles of water)(molar heat capacity of water)(ΔT)

The heat required to vaporize the water is (moles of water)(ΔH_{vap})

$$c_p = 75.3\ J/mol \cdot {}^\circ C$$

$$\Delta H_{vap} = 40.67\ kJ/mol$$

$$\text{moles of water} = 10.0\ g\ H_2O \times \frac{1\ mol\ H_2O}{18.0\ g\ H_2O} = 0.556\ mol\ H_2O$$

Now we can determine the heat gained by the water.

$$\begin{aligned}\text{heat gained} &= \text{heat required to warm water from 25°C to 100°C + heat required to vaporize 10 g of water} \\ &= (0.556\ mol)(75.3\ J/mol \cdot {}^\circ C) \\ &\quad \times (100^\circ C - 25^\circ C) + (0.556\ mol) \\ &\quad \times (40.67\ kJ/mol) \\ &= 3140\ J + 22.6\ kJ \text{ (Note units are not the same.)}\end{aligned}$$

The total heat gained by the water is 3140 J + 22,600 J = 25740 J

Now we can solve for ΔT using $q_{lost} = q_{gained}$

$$(542\ J/{}^\circ C)(\Delta T) = 25740\ J$$

$$\Delta T = \frac{25740\ J}{(542\ J/{}^\circ C)} = 47.5^\circ C$$

The temperature change of the skillet is −47.5°C.

59. The enthalpy change depends on the number and types of bonds broken and formed. The only way we can determine these quantities is from the balanced equation.

60. The structures must be known because we need to know which atoms are bonded to one another and the strength of multiple bonds (double and triple bonds) is different from the strength of a single bond between two atoms.

61. Fuel value measures the amount of heat energy released per gram of combusted fuel.

It requires energy to convert liquid butane to gaseous butane. This energy would then be subtracted from the energy released when the butane is burned. The result is that less energy is produced per gram of liquid butane than per gram of gaseous butane and gaseous butane has a higher fuel value.

62. The question being asked here is to explain why the combustion of C(*s*) is an exothermic process. On an atomic scale to form CO_2, we need to vaporize carbon atoms and also break apart oxygen-oxygen double bonds so that there are oxygen atoms. The gaseous carbon and oxygen atoms then combine to form carbon dioxide (CO_2) molecules. The overall process is

$$C(s) + O_2 \rightarrow CO_2(g)$$

For this process, $\Delta H =$ energies of bonds broken (O_2) + energy required to vaporize solid carbon − energies of bonds formed (CO_2). It requires energy to vaporize both carbon atoms and to break apart an oxygen molecule. Energy is released when the carbon and oxygen atoms combine to form carbon dioxide. In this process more energy is released in forming the CO_2 molecules than is absorbed in vaporizing carbon and breaking apart an O_2 molecule.

The energy needed to break 2 moles of C=O doubles is an exothermic quantity. The forming of these bonds from carbon and oxygen atoms will release the same quantity of energy (an endothermic quantity) as needed to break them.

63. The enthalpy change for any reaction can be approximated using the equation

$$\Delta H = \Delta H_{\text{bonds broken}} - \Delta H_{\text{bonds formed}}$$

It requires energy to break chemical bonds; that is, breaking chemical bonds is an endothermic process.

When new bonds form, energy is released because the newly formed substance is lower in energy than the free atoms would be. This is an exothermic process.

To determine the energy associated with bond breaking and bond formation, one needs to know the structures of all the substances involved in the process. Drawing Lewis structures is very helpful in determining the type of bonds broken and formed. (*Note:* We do not need to include lone pairs, only the bonds in the structures.)

All bond energies are given for gaseous species.

If a species is not listed as a gas in the equation, one must determine the amount of energy required to vaporize this substance before using the bond energy formula. This amount of heat energy must be included in your calculation.

a. $$N_2(g) + 3\,H_2(g) \rightarrow 2\,NH_3(g)$$

N_2 contains 1 N≡N triple bond.
H_2 contains 1 H−H single bond.
NH_3 contains 3 N−H single bonds. (The Lewis structure of NH_3 includes a lone pair of electrons that is not shown.)

```
   H
   |
H—N—H
```

In this process we need to break 1 N≡N triple bond and 3 H−H single bonds.

This process forms 6 N−H single bonds. (Each mole of NH_3 forms 3 N−H single bonds.)

The bond energies for each bond can be found in Table 11.2.

N≡N bond energy is 946 kJ/mol
H−H bond energy is 436 kJ/mol
N−H bond energy is 388 kJ/mol

For this reaction we can calculate the approximate value of ΔH.

$$\begin{aligned}\Delta H &= (1\text{ mol})(946\text{ kJ/mol}) + (3\text{ mol})(436\text{ kJ/mol}) \\ &\quad - (6\text{ mol})(388\text{ kJ/mol}) \\ &= 946\text{ kJ} + 1308\text{ kJ} - 2328\text{ kJ} \\ &= -74\text{ kJ}\end{aligned}$$

b. $$N_2(g) + 2\,H_2(g) \rightarrow H_2NNH_2(g)$$

N_2 contains 1 N≡N bond.
H_2 contains 1 H−H bond.
The Lewis structure of H_2NNH_2 is

```
   H  H
   |  |
H—N—N—H
```

Note: Lone pairs on N atoms are not shown.

In this process we need to break 1 N≡N triple bond and 2 H−H single bonds.

This process forms 4 N−H single bonds and 1 N−N single bond.

The bond energies for each bond can be found in Table 11.2.

N≡N (946 kJ/mol)
H−H (436 kJ/mol)
N−H (388 kJ/mol)
N−N (163 kJ/mol)

For this reaction we can calculate the approximate value of ΔH.

$$\begin{aligned}\Delta H &= (1\text{ mol})(946\text{ kJ/mol}) + (2\text{ mol})(436\text{ kJ/mol}) \\ &\quad -[(4\text{ mol})(388\text{ kJ/mol}) + (1\text{ mol})(163\text{ kJ/mol})] \\ &= 946\text{ kJ} + 872\text{ kJ} - [1552\text{ kJ} + 163\text{ kJ}] \\ &= 103\text{ kJ}\end{aligned}$$

c. $$2\,N_2(g) + O_2(g) \rightarrow 2\,N_2O(g)$$

N_2 contains 1 N≡N triple bond.
O_2 contains 1 O=O double bond.
N_2O is a species that has several resonance forms. The Lewis structure for three resonance forms of

N_2O are shown with formal charges listed below each atom.

$N-N\equiv O$	$N=N=O$	$N\equiv N-O$
-2 $+1$ $+1$	-1 $+1$ 0	0 $+1$ -1

Note: Lone pairs on N and O atoms are not shown.

The general rules for determining how important an individual resonance structure is to the overall bonding are:

1. The most electronegative atom should have a negative formal charge.
2. Smaller formal charges are better than higher ones. The best structure has a formal charge of zero on all atoms. Formal charges of $+2$ or -2 are less desirable than formal charges of $+1$ or -1.
3. Positive and negative charges should be on adjacent atoms. Separation of charges is not desirable.

Of these three structures, the third one, $N\equiv N-O$, should contribute most to the overall structure.

The reason is nitrogen is less electronegative than oxygen; the best structure should have the nitrogen atom with a positive formal charge and the oxygen atom with a negative formal charge. This structure is the only that gives the oxygen atom a negative formal charge.

We will assume that this structure dominates the overall structure. However, it is very likely that a true answer would need to include a weighted value for each of the structures. We do not have this information and it is not likely to be known.

Now we are ready to determine the bonds broken and formed.

Bonds broken: 2 $N\equiv N$ bonds and 1 $O=O$ bond
Bonds formed: 2 $N\equiv N$ bonds and 2 $N-O$ bonds

The bond energies for each bond can be found in Table 11.2.

$N\equiv N$ (946 kJ/mol)
$O=O$ (497 kJ/mol)
$N-O$ (157 kJ/mol)

For this reaction, we can calculate the approximate value of ΔH.

$$\begin{aligned}\Delta H &= (2\text{ mol})(946\text{ kJ/mol}) + (1\text{ mol})(497\text{ kJ/mol})\\ &\quad -[(2\text{ mol})(946\text{ kJ/mol}) + (2\text{ mol})(157\text{ kJ/mol})]\\ &= 1892\text{ kJ} + 497\text{ kJ} - [1892\text{ kJ} + 314\text{ kJ}]\\ &= 2389\text{ kJ} - 2206\text{ kJ}\\ &= 183\text{ kJ}\end{aligned}$$

65. $2\,CO(g) + O_2(g) \rightarrow 2\,CO_2(g)$ $\Delta H_{comb} = -566\text{ kJ}$

CO_2 contains 2 $C=O$ double bonds: $O=C=O$
CO contains 1 $C\equiv O$ triple bond.
O_2 contains 1 $O=O$ double bond.

Bonds broken: 2 $C\equiv O$ bonds and 1 $O=O$ bond
Bonds formed: 4 $C=O$ bonds (note the coefficient 2 in front of CO_2 in the balanced chemical equation)

Bond energies given in problem: $O=O$ (497 kJ/mol); $C=O$ (799 kJ/mol)

We were also given that $\Delta H_{comb} = -566\text{ kJ}$

We could look up the bond energy for $C\equiv O$ in Table 11.2, but the problem asks us to calculate this value from the information given.

We can use the equation

$$\Delta H_{comb} = \Delta H_{\text{bonds broken}} - \Delta H_{\text{bonds formed}}$$

Bonds broken: 2 $C\equiv O$ bonds and 1 $O=O$ bond
Bonds formed: 4 $C=O$ bonds

Substituting the values given in the problem

$$-566\text{ kJ} = (2\text{ mol})\Delta H_{C\equiv O} + (1\text{ mol})(497\text{ kJ/mol})$$
$$-(4\text{ mol})(799\text{ kJ/mol})$$
$$-566\text{ kJ} = (2\text{ mol})\Delta H_{C\equiv O} + 497\text{ kJ} - 3196\text{ kJ}$$

Solving for $\Delta H_{C\equiv O}$

$$(2\text{ mol})\Delta H_{C\equiv O} = -566\text{ kJ} - 497\text{ kJ} + 3196\text{ kJ}$$
$$(2\text{ mol})\Delta H_{C\equiv O} = 2133\text{ kJ}$$
$$\Delta H_{C\equiv O} = \frac{2133\text{ kJ}}{2\text{ mol}} = 1166.5\text{ kJ/mol}$$

Note: This is reasonably close to the value listed in Table 11.2 (1076 kJ/mol). The table lists average bond energies. It is likely that bond energies vary from compound to compound.

67. The balanced reaction for the process that produces carbon monoxide (CO) instead of carbon dioxide is:

$$CH_4(g) + 3/2\,O_2(g) \rightarrow CO(g) + 2\,H_2O(g)$$

The Lewis structures for CH_4, O_2, CO, and H_2O (lone pairs on central atoms not shown) are:

$$\begin{array}{c}H\\|\\H-C-H\\|\\H\end{array}\quad O=O\quad C\equiv O\quad H-O-H$$

The balanced equation for the reaction that produces CO_2 and H_2O is:

$$CH_4(g) + 2\,O_2(g) \rightarrow CO_2(g) + 2\,H_2O(g)$$

The Lewis structure of CO_2 is: $O=C=O$

The question asks us to find the difference in energy released in these two reactions (how much less energy is given off in the incomplete combustion).

Reaction 1 $CH_4(g) + 3/2\,O_2(g) \rightarrow CO(g) + 2\,H_2O(g)$

Bonds broken: 4 C—H single bonds and 1.5 O=O double bonds

Bonds formed: 4 O—H single bonds and 1 C≡O triple bond

Reaction 2 $CH_4(g) + 2\,O_2(g) \rightarrow CO_2(g) + 2\,H_2O(g)$

Bonds broken: 4 C—H single bonds and 1 O=O double bond

Bonds formed: 2 C=O double bonds and 4 H—O single bonds

The bond energies are

C—H single bond = 413 kJ/mol
O=O double bond = 497 kJ/mol
O—H single bond = 463 kJ/mol
C=O double bond = 799 kJ/mol
C ≡ O triple bond = 1076 kJ/mol

$$\Delta H = \Delta H_{\text{bonds broken}} - \Delta H_{\text{bonds formed}}$$

ΔH for the process: $CH_4(g) + 3/2\ O_2(g) \rightarrow CO(g) + 2\ H_2O(g)$

$$\Delta H = [(4\,\text{mol})(413\,\text{kJ/mol}) + (1.5\,\text{mol})(497\,\text{kJ/mol})]$$

$$[(4\,\text{mol})(463\,\text{kJ/mol}) + (1\,\text{mol})(1076\,\text{kJ/mol})]$$

$$\Delta H = [1652 + 745.5] - [1852 + (1076)]$$

$$= 2397.5 - 2928\,\text{kJ}$$

$$= -530.5\,\text{kJ}$$

ΔH for the process: $CH_4(g) + 2\ O_2(g) \rightarrow CO_2(g) + 2\ H_2O(g)$

$$\Delta H = [(4\,\text{mol})(413\,\text{kJ/mol}) + (2\,\text{mol})(497\,\text{kJ/mol})]$$

$$[(4\,\text{mol})(463\,\text{kJ/mol}) + (2\,\text{mol})(799\,\text{kJ/mol})]$$

$$\Delta H = [1652 + 994] - [1852 + (1598)]$$

$$= 2646 - 3450\,\text{kJ}$$

$$= -804\,\text{kJ}$$

The second process is more exothermic by 804 − 530.5 kJ = 273.5 kJ.

273.5 kJ less energy is obtained from the process that produces CO.

69. $4\ NH_3(g) + 7\ O_2(g) \rightarrow 4\ NO_2(g) + 6\ H_2O(g)$

We need to determine the Lewis structure for each substance in the reaction and then determine the bonds broken and the bonds formed. Note that lone pairs are not shown.

NH_3	O_2	H_2O
H–N–H (with H above N)	O=O	H—O—H

The Lewis structure for NO_2 contains an unpaired electron on the nitrogen atom because this species has an odd number of valence electrons. The Lewis structure for this substance is

O—N=O

Note that the lone pairs on oxygen atoms are not shown and the N atom also has an unpaired single electron on it. The lone pair is on nitrogen rather than oxygen because this Lewis structure results in each atom having zero formal charge (the most favorable structure).

Bonds broken in NH_3: 12 N—H bonds (each NH_3 molecule contains 3 N—H bonds)

Bonds broken in O_2: 7 O=O bonds

Bonds formed in NO_2: 4 N—O bonds
4 N=O bonds
(Each NO_2 molecule contains an N—O bond and an N=O bond.)

Bonds formed in H_2O: 12 O—H bonds
(Each water molecule contains 2 O—H bonds.)

Bond energies: N—H (388 kJ/mol)
O=O (497 kJ/mol)
O—H (463 kJ/mol)
N=O (630 kJ/mol)
N—O (157 kJ/mol)

For the process

$$4\ NH_3(g) + 7\ O_2(g) \rightarrow 4\ NO_2(g) + 6\ H_2O(g)$$

$$\Delta H = (12\ \text{mol})(388\ \text{kJ/mol}) + (7\ \text{mol})(497\ \text{kJ/mol})$$

$$- [(4\ \text{mol})(630\ \text{kJ/mol}) + (4\ \text{mol})(157\,\text{kJ/mol})$$

$$+ (12\ \text{mol})(463\ \text{kJ/mol})]$$

$$= [4656 + 3479] - [2520 + (628) + (5556)]$$

$$= 8135 - 8704\ \text{kJ}$$

$$= -569\ \text{kJ}$$

71. A bomb calorimeter device consists of a bomb (a heavy metal container inside which a substance is combusted), water (the bomb sits inside this water), and an insulating material (this prevents any heat from being lost to the rest of the surroundings and does not absorb any heat). The heat capacity of the calorimeter includes the

bomb and the surrounding water. One simply needs to measure the temperature change of the entire system to determine how much heat is released. The heat capacity of the calorimeter assumes that it contains the same amount of water for each combustion reaction perfomed.

The alternative is to determine the mass of your bomb and the molar heat capacity heat of the bomb and the same quantities for the surrounding water. From this information and the temperature change you could determine the amount of heat absorbed by the bomb and the water. There is no logical reason to determine these quantities when determining the heat capacity of the calorimeter and its temperature change will give the same results.

To determine the heat capacity of the calorimeter, one needs to know only the heat of combustion of a substance accurately. This method is much easier to determine.

72. In theory, an endothermic reaction could be used. In this process the temperature of the calorimeter would decrease because the system would be absorbing heat energy from the surroundings (the calorimeter).

In reality, it is much more difficult to determine this information. We could not use a combustion process because all combustion processes are exothermic. Finding a method to start an endothermic reaction would not be easy. Combustion reactions simply require a spark. This is provided by the wires connected to the bomb.

Lastly, most combustion processes give off a considerable quantity of heat, and this results in an easily measurable temperature change. Most endothermic processes absorb significantly smaller quantities, making it difficult to get an accurate temperature change.

73. The heat capacity of a calorimeter measures the amount of heat absorbed to raise the entire calorimeter by 1°C. If the amount of water in a calorimeter is not the same for each measurement, the heat capacity of the calorimeter will change. The same would be true if the liquid used in the calorimeter was not the same for each measurement.

The basic idea is to get a system that is always identical. By doing this, one can simply measure the temperature change of the calorimeter and use the heat capacity to determine how much heat is released during the combustion.

The amount of heat released in a bomb calorimetry experiment is given by the expression

$$q = (\text{heat capacity of the bomb calorimeter})(\Delta T)$$

However, the amount of heat released during a specific combustion (let's say 1 g of a specific substance) will not be changed. However, if a different liquid or amount of water is used from experiment to experiment, then the temperature change will vary with each one. If ΔT changes, then the heat capacity of the calorimeter must also change if q is to be the same.

74. For a calorimeter $q = C\Delta T$.

If only a small quantity of a substance is being combusted, we would expect the heat (q) generated to be a small quantity. To get a measurable change in temperature, a small value of C is preferable, since it would result in a larger value of ΔT.

$$\Delta T = q/C$$

The smaller the value of C, the larger the value of ΔT.

75. The following information is given:

The heat of combustion of benzoic acid is 26.38 kJ/g (see p. 536 of the textbook).

ΔT (the temperature change of the calorimeter) is 16.397°C.

The mass of benzoic acid combusted is 5.000 g.

From the information given, we can calculate the total heat generated when 5.000 g of benzoic acid is combusted. The total heat generated corresponds to q in the formula $q = C\Delta T$

The heat generated in the combustion of 5.000 g of benzoic acid is

$$5.000 \text{ g benzoic acid} \times \frac{26.38 \text{ kJ}}{\text{g benzoic acid}} = 131.9 \text{ kJ}$$

We can now determine the heat capacity of the calorimeter because we know both q and ΔT.

$$C = \frac{q}{\Delta T}$$

$$= \frac{131.9 \text{ kJ}}{16.397°\text{C}}$$

$$= 8.044 \text{ kJ/°C}$$

77. In this problem, we are given the heat capacity of the calorimeter and the change in temperature of the calorimeter. From this information, we can determine how much heat is given off when 1.200 g of cinnamaldehyde is combusted.

Information given:

1.200 g of cinnamaldehyde is combusted.
C (heat capacity of the calorimeter) = 3.640 kJ/°C
ΔT (temperature change of calorimeter) = 12.79°C

We are asked to find the molar heat of combustion of cinnamaldehyde.

The formula to use is $q = C\Delta T$

$$q = (3.640 \text{ kJ/°C})(12.79°\text{C})$$

$$= 46.56 \text{ kJ}$$

So 46.56 kJ of heat energy is given off when 1.200 g of cinnamaldehyde is combusted. We need to determine how much heat energy is released when 1 mole of cinnamaldehyde is combusted.

Let's find the how many mole of cinnamaldehyde there are in 1.2 g. The molar mass of cinnamaldehyde (C_9H_8O) is 132 g/mol.

$$\text{moles of cinnamaldehyde} = 1.2 \text{ g cinnamaldehyde} \times \frac{1 \text{ mol cinnamaldehyde}}{132 \text{ g cinnamaldehyde}} = 0.00909 \text{ mol cinnamaldehyde}$$

Lastly divide the heat generated (46.56 kJ) by the moles of cinnamaldehyde combusted.

$$\Delta H_{\text{comb}} \text{ (cinnamaldehyde)} = \frac{46.56 \text{ kJ}}{0.00909 \text{ mol}} = 5122 \text{ kJ/mol}$$

The molar heat of combustion of cinnamaldehyde is 5122 kJ/mol.

79. Information given:

Molar heat of combustion of dimethylphthalate ($C_{10}H_{10}O_4$) = 4685 kJ/mol
C (heat capacity of calorimeter) = 7.854 kJ/°C
1.00 g of dimethylphthalate ($C_{10}H_{10}O_4$) is combusted.

We are asked to find ΔT (change in temperature of the calorimeter).

The formula to use is $q = C\Delta T$.

The question asks us to find ΔT. We know the heat capacity of the calorimeter (C). If we also know q, then we can solve for ΔT.

We are given the molar heat of combustion and are told that we combust 1.00 g of dimethylphthalate. From this information we can determine the heat generated in this process.

To find q for this process, divide the molar heat of combustion by the molar mass of dimethylphthalate ($C_{10}H_{10}O_4$, molar mass 194.0 g/mol) to get the quantity of heat released when 1.00 g of dimethylphthalate is combusted.

$$q = \frac{4685 \text{ kJ/mol}}{194.0 \text{ g/mol}} = 24.15 \text{ kJ/g}$$

We can now solve for ΔT.

$$\Delta T = \frac{q}{C} = \frac{24.15 \text{ kJ}}{7.854 \text{ kJ/°C}} = 3.075°\text{C}$$

The change in temperature for the calorimeter is 3.075°C when 1.00 g of dimethylphthalate is combusted. The initial temperature of the calorimeter was 20.215°C, so the final temperature of the calorimeter is 23.290°C.

81. No. Only one form of each element can be in the standard state, for which the heat of formation is zero. For oxygen, the standard state is O_2.

82. It is very difficult to measure the absolute heat value of any substance. It is relatively easy to measure the change in heat of a system. In order to have a yardstick to measure these changes we define for each element a condition known as the standard state (the most stable form of an element under ordinary conditions), and arbitrarily set the heat of formation of this standard state to be zero. By using Hess's law (when one adds reactions, the heats associated with reactions can also be summed) we can measure the change in heat in any process, relative to the standard state.

83. A formation reaction is one in which a substance is formed from elements in their standard state (the most stable form of an element).

For reactions *a* and *d* ΔH_{rxn} corresponds to the heat of formation.

Reactions shown in *b* and *c* are not standard formation reactions because they both contain compounds as a reactant. In addition, in reaction *c* more than one substance is produced.

For the reaction in part *a*, the elements are in their standard states $C(s)$ and $O_2(g)$.

For the reaction in part *d*, the elements are in their standard states $H_2(g)$ and $C(s)$.

Note: There are several forms of carbon solid. The standard state is the solid form of carbon called graphite.

85. One can find heats of formation in the table in the appendix.

For the substances in the reaction:

$$4\,H_2(g) + CO_2(g) \rightarrow CH_4(g) + 2\,H_2O(g)$$

ΔH_f°(in kJ/mol) 0.0 −393.5 −74.8 −241.8

The formula to use is

$$\begin{aligned}\Delta H^\circ_{\text{rxn}} &= \sum n\Delta H^\circ_{\text{f, products}} - \sum m\Delta H^\circ_{\text{f, reactants}}\\ &= [\Delta H^\circ_{\text{f}CH_4(g)} + 2\Delta H^\circ_{\text{f}H_2O(g)}] - [4\Delta H^\circ_{\text{f}H_2(g)} + \Delta H^\circ_{\text{f}CO_2(g)}]\\ &= [(-74.8) + 2(-241.8)] - [4(0) + (-393.5)]\\ &= [(-74.8) + (-483.6)] - [-393.5]\\ &= -558.4 + 393.5\\ &= -164.9 \text{ kJ}\end{aligned}$$

The enthalpy change for this reaction, ΔH_{rxn}, is −164.9 kJ.

87. One needs to first determine the formula for ammonium nitrate and then write a balanced equation.
    Ammonium nitrate: NH_4NO_3
    The balanced equation is

$$NH_4NO_3(s) \rightarrow N_2O(g) + 2\,H_2O(g)$$

One can find heats of formation in the table in the appendix.
    For the substances in the reaction:

$$NH_4NO_3(s) \rightarrow N_2O(g) + 2\,H_2O(g)$$

ΔH_f° (in kJ/mol) –365.6 +82.1 −241.8
    The formula to use is

$$\begin{aligned}
\Delta H^\circ_{rxn} &= \sum n\Delta H^\circ_{f,\text{ products}} - \sum m\Delta H^\circ_{f,\text{ reactants}} \\
&= [\Delta H^\circ_{fN_2O(g)} + 2\Delta H^\circ_{fH_2O(g)}] - [\Delta H^\circ_{fNH_4NO_3(g)}] \\
&= [(+82.1) + 2(-241.8)] - [-365.6] \\
&= (+82.1) + (-483.6) - [-365.6] \\
&= -(401.5) - (-365.6) \\
&= -401.5 + 365.6 \\
&= -35.9\text{ kJ}
\end{aligned}$$

The enthalpy change for this reaction, ΔH_{rxn}, is −35.9 kJ.

89. One can find heats of formation in the table in the appendix.
    For the substances in the reaction:

$$3\,NH_4NO_3(g) + C_{10}H_{22}(l) + 14\,O_2(g) \rightarrow 3\,N_2(g) + 10\,CO_2(g) + 17\,H_2O(g)$$

ΔH_f° (in kJ/mol) are:

$NH_4NO_3(s) = -365.6$; $C_{10}H_{22}(l) = -249.7$; $O_2(g) = 0.0$; $N_2(g) = 0.0$; $CO_2(g) = -393.5$; $H_2O(g) = -241.8$.
The formula to use is

$$\begin{aligned}
\Delta H^\circ_{rx} &= \sum n\Delta H^\circ_{f,\text{ products}} - \sum m\Delta H^\circ_{f,\text{ reactants}} \\
&= [3\Delta H^\circ_{fN_2(g)} + 10\Delta H^\circ_{fCO_2(g)} + 17\Delta H^\circ_{fH_2O(g)}] \\
&\quad - [3\Delta H^\circ_{fNH_4NO_3(s)} + \Delta H^\circ_{fC_{10}H_{22}(l)} + 14\Delta H^\circ_{fO_2(g)}] \\
&= [3(0.0) + 10(-393.5) + 17(-241.8)] \\
&\quad - [3(-365.6) + (-249.7) + 14(0.0)] \\
&= [(-3935) + (-4110.6)] - [(-1096.8) \\
&\quad + (-249.7)] \\
&= (-8045.6) - (-1346.5) \\
&= -8045.6 + 1346.5 \\
&= -6699.1\text{ kJ}
\end{aligned}$$

The enthalpy change for this reaction, ΔH_{rxn}, is −6699.1 kJ.

91. The fuel value is a heat of combustion reported in units of kJ/g rather than kJ/mol.

92. kJ/g

93. To convert from a molar heat of combustion to a fuel value, divide the heat of combustion by the molar mass of the fuel.

94. No. The fuel value of liquid propane is less than the fuel value of gaseous propane because the liquid propane must absorb heat energy to change from the liquid state to the gaseous state. This energy must be subtracted from the heat of combustion of gaseous propane (a result of the law of conservation of energy and Hess's law).

$C_3H_8(l) \rightarrow C_3H_8(g)$ This process is endothermic

$C_3H_8(g) + 5\,O_2(g) \rightarrow 3\,CO_2(g) + 4\,H_2O(g)$ This process is exothermic

Summing the two steps results in the reaction for the heat of combustion of liquid propane.

$$C_3H_8(l) + 5\,O_2(g) \rightarrow 3\,CO_2(g) + 4\,H_2O(g)$$

The heat of combustion for this reaction is equal to the sum of the ΔH values for the two reactions. One is endothermic and the other is exothermic, so that the sum of the enthalpies for the two reactions is less negative than the heat of combustion of gaseous propane.

95. The balanced equation for the process of burning propane is

$$C_3H_8(g) + 5\,O_2(g) \rightarrow 3\,CO_2(g) + 4\,H_2O(g)$$

Note: The text suggests that all combustion reactions should be written to produce water vapor rather than liquid water. The heats for producing water in different states of matter will not be the same.
    Using this reaction and the heats of formations found in the appendix, we can calculate the heat produced when 1 mole of propane is combusted.

$$\begin{aligned}
\Delta H^\circ_{fC_3H_8(g)} &= -103.9\text{ kJ/mol} \\
\Delta H^\circ_{fO_2(g)} &= 0.0\text{ kJ/mol} \\
\Delta H^\circ_{fCO_2(g)} &= -393.5\text{ kJ/mol} \\
\Delta H^\circ_{fH_2O(g)} &= -241.8\text{ kJ/mol}
\end{aligned}$$

$$\begin{aligned}
\Delta H_{rxn} &= [3\Delta H^\circ_{fCO_2(g)} + 4\Delta H^\circ_{fH_2O(g)}] - [\Delta H^\circ_{fC_3H_8(g)} \\
&\quad + 5\Delta H^\circ_{fO_2(g)}] \\
&= [3(-393.5) + 4(-241.8)] - (-103.9) \\
&= -2043.8\text{ kJ}
\end{aligned}$$

The heat of combustion of propane is −2043.8 kJ/mol C_3H_8.

Next we need to find out how much heat is produced when 1 pound of propane is combusted. To do this, convert 1 pound (453.6 g) to moles of propane and multiply by the molar heat of combustion calculated in the first step.

Moles of propane combusted:

$$\text{moles of } C_3H_8 = 453.6\ C_3H_8 \times \frac{1 \text{ mol } C_3H_8}{44.0 \text{ g } C_3H_8}$$

$$= 10.3 \text{ mol } C_3H_8$$

Total heat produced:

$$\text{total heat produced} = 10.3 \text{ mol } C_3H_8 \times \frac{-2043.8 \text{ kJ}}{\text{mol } C_3H_8}$$

$$= -2.11 \times 10^4 \text{ kJ}$$

This is the total amount of heat released when a pound of propane is combusted.

We could have also arrived at this value using the fuel value of 46.3 kJ/g listed for propane.

Total heat produced using the fuel value of propane:

$$\text{total heat produced} = 453.6 \text{ g} \times \frac{46.3 \text{ kJ}}{1 \text{ g } C_3H_8}$$

$$= 2.10 \times 10^4 \text{ kJ}$$

The slight difference between using the heat of combustion and the fuel value is due to rounding off calculations and quantities.

Now that we know the total heat generated, we need to use this value and the formula, $q = nc_P\Delta T$, to determine how many kilograms of water can be heated from 20°C to 45°C.

The known values are:

$q = 2.11 \times 10^4$ kJ
$\Delta T = 45°C - 20°C = 25°C$
c_p (molar heat capacity of water) = 75.3 J/mol

Note: The molar heat capacity for water can be found in Section 11.6. (p. 526)

The energy units in q and c_P do not match. We need to convert one of them to make the units match. Let's convert q from kJ to J (1 kJ = 1000 J).

$$2.11 \times 10^4 \text{ kJ} = 2.11 \times 10^7 \text{ J}$$

$$2.11 \times 10^7 \text{ J} = (n)(75.3 \text{ J/mol} \cdot °\text{C})(25°\text{C})$$

Solving for n, the number of moles:

$$n = \frac{2.11 \times 10^7 \text{ J}}{(75.3 \text{ J/mol} \cdot °\text{C})(25°\text{C})} = 11{,}200 \text{ mol } H_2O$$

Lastly convert moles of H_2O to grams, then kilograms of H_2O.

$$\text{kilograms of } H_2O = 11{,}200 \text{ mol } H_2O \times \frac{18.0 \text{ g } H_2O}{1 \text{ mol } H_2O} \times \frac{1.00 \text{ kg}}{1000 \text{ g}}$$

$$= 202 \text{ kg } H_2O$$

202 kg of water could be heated from 20°C to 45°C by the combustion of 1 pound of propane gas.

97. The fuel value is a heat of combustion expressed in kJ/g instead of kJ/mol. To convert from a heat of combustion to a fuel value, one divides the heat of combustion by the molar mass of the substance combusted.

a. The molar mass of C_5H_{12} is 72.0 g/mol.

The fuel value is

$$\left(3535 \frac{\text{kJ}}{\text{mol } C_5H_{12}} \times \frac{1 \text{ mol } C_5H_{12}}{72.0 \text{ g } C_5H_{12}}\right)$$

$$= 49.1 \text{ kJ/g } C_5H_{12}$$

b. To determine how much heat is released when 1.00 kg is combusted, multiply the fuel value by 1.00×10^3 g.

$$q = \left(49.1 \frac{\text{kJ}}{\text{g } C_5H_{12}}\right)(1.00 \times 10^3 \text{ g } C_5H_{12})$$

$$= 4.91 \times 10^4 \text{ kJ}$$

An alternate method is to determine the number of moles of C_5H_{12} in 1.00×10^3 g of C_5H_{12} and then multiply by the molar heat of combustion.

$$\text{moles of } C_5H_{12} = 1.00 \times 10^3 \text{ g } C_5H_{12} \times \frac{1 \text{ mol } C_5H_{12}}{72.0 \text{ g } C_5H_{12}}$$

$$= 13.89 \text{ mol } C_5H_{12}$$

$$q = 13.89 \text{ mol } C_5H_{12}\left(3535 \frac{\text{kJ}}{\text{mol } C_5H_{12}}\right)$$

$$= 4.91 \times 10^4 \text{ kJ}$$

c. In this part of the problem, we need to determine the heat required to warm 1.00 kg of water from 20°C to 90°C.

To determine this quantity, use the formula $q = nc_p\Delta T$

Once we know how much heat is required, we can determine how many grams of C_5H_{12} must be combusted to produce this amount of heat.

Let's calculate the heat required to warm 1.00 kg of water:

First find the number of moles in 1.00 kg of water.

$$\text{moles } H_2O = 1000 \text{ g } H_2O \times \frac{1 \text{ mol } H_2O}{18.0 \text{ g } H_2O}$$
$$= 55.6 \text{ mol } H_2O$$

Now we can determine the heat required for this process.

$$q = (55.6 \text{ mol } H_2O)\left(75.3 \frac{\text{J}}{\text{mole} \cdot {}^\circ\text{C}}\right)(70^\circ\text{C})$$
$$= 2.93 \times 10^5 \text{ J or } 293 \text{ kJ}$$

Now we need to determine how many grams (or moles) of C_5H_{12} need to be combusted to produce 293 kJ of heat. We can use the fuel value or the heat of combustion to calculate this quantity.

Using fuel values, we need to divide the heat required by the fuel value.

$$\text{grams of } C_5H_{12} = \left(\frac{293 \text{ kJ}}{49.1 \text{ kJ/g } C_5H_{12}}\right)$$
$$= 5.97 \text{ g } C_5H_{12}$$

Using the heat of combustion (3535 kJ/mol), one determines how many moles of C_5H_{12} are needed to release 293 kJ, and then converts this to grams by multiplying by the molar mass of C_5H_{12}.

To determine the moles of C_5H_{12} needed to produce 293 kJ of heat, divide the 293 kJ by the heat of combustion of C_5H_{12}.

$$\text{moles of } C_5H_{12} = \left(\frac{293 \text{ kJ}}{3535 \text{ kJ/1 mol } C_5H_{12}}\right)$$
$$= 8.29 \times 10^{-2} \text{ mol } C_5H_{12}$$

Convert this quantity to grams:

$$\text{g } C_5H_{12} = 8.29 \times 10^{-2} \text{ mol } C_5H_{12} \times \frac{72.0 \text{ g } C_5H_{12}}{1 \text{ mol } C_5H_{12}}$$
$$= 5.97 \text{ g } C_5H_{12}$$

To warm 1.00 kg of water from 20°C to 90°C requires the combustion of 5.97 g of C_5H_{12}.

99. We need to find the "energy density." The energy density is similar to the fuel value except it is expressed in units of kJ/L instead of kJ/g. First, we need to write balanced chemical equations for the combustion of 1 mole of methane (CH_4) and hexane (C_6H_{14}).

$$CH_4(g) + 2\,O_2(g) \rightarrow CO_2(g) + 2\,H_2O(g)$$

$$C_6H_{14}(l) + 9.5\,O_2(g) \rightarrow 6\,CO_2(g) + 7\,H_2O(g)$$

Remember, when balancing thermochemical equations, fractional coefficients are permitted in the balanced equation.

Next, use the heats of formation in the appendix to calculate the heat of combustion for methane. The heat of combustion for hexane is given in problem 98.

Note: We could also use the fuel value given for methane in the text.

Heats of formation ΔH_f° (in kJ/mol)

$CH_4\,(g) = 74.8$ kJ/mol
$H_2O\,(g) = -241.8$ kJ/mol
$CO_2\,(g) = -393.5$ kJ/mol

Heat of combustion of methane, CH_4, is:

$$\Delta H_{\text{comb}} = [\Delta H^\circ_{fCO_2(g)} + 2\Delta H^\circ_{fH_2O(g)}] - [\Delta H^\circ_{fCH_4(g)} + 2\Delta H^\circ_{fO_2(g)}]$$
$$= [(-393.5) + 2(-241.8)] - [(-74.8) + 2(0)]$$
$$= -802.3 \text{ kJ}$$

$$\text{The fuel value is } \frac{802.3 \text{ kJ/mol}}{16.0 \text{ g } CH_4\text{/mol}} = 50.1 \text{ kJ/g}$$

Heat of combustion of hexane is given in Problem 98. It is 4163 kJ/mol.

$$\text{The fuel value is } \frac{4163 \text{ kJ/mol}}{86.0 \text{ g } C_6H_{14}\text{/mol}} = 48.4 \text{ kJ/g}$$

Lastly convert fuel values from kJ/g to kJ/mL by using the density of each substance. That is, multiply the fuel value by the density.

For methane, CH_4 (density = 0.665 g/L)

$$\frac{50.1 \text{ kJ}}{\text{g } CH_4} \times \frac{0.66 \text{ g } CH_4}{1 \text{ L}} = 33 \text{ kJ/L}$$

For hexane, C_6H_{14} (density = 0.66 g/mL)

$$\frac{48.4 \text{ kJ}}{\text{g } C_6H_{14}} \times \frac{0.66 \text{ g } C_6H_{14}}{1 \text{ mL}} \times \frac{1000 \text{ mL}}{1 \text{ L}} = 3.2 \times 10^4 \text{ kJ/L}$$

The energy density of hexane (3.2×10^4 kJ/L) is much greater than the energy density of gaseous methane (33 kJ/L).

101. The law of conservation of energy states that energy can neither be created nor destroyed in an ordinary chemical process. Hess's law states that the energy of coupled reactions is simply the sum of the individual energy terms for each of these reactions.

Since the energy terms are additive according to Hess's law, it must be consistent with the law of conservation of energy.

102. The reaction that corresponds to the heat of formation of carbon monoxide is

$$C(s) + \tfrac{1}{2}\,O_2(g) \rightarrow CO(g)$$

It is difficult to directly measure the heat of formation of carbon monoxide because it is difficult to react carbon and oxygen to form pure carbon monoxide. If there is an excess of oxygen, the likely product is carbon dioxide. On the other hand, burning carbon in a limited supply of oxygen is more likely to produce carbon dioxide and soot (unburned carbon) than pure carbon monoxide.

103. The following equation, used to find the heat of reaction, is derived from Hess's law.

$$\Delta H^o_{rxn} = \sum n\Delta H^o_{f,\,products} - \sum m\Delta H^o_{f,\,reactants}$$

If one writes a formation reaction for each reactant and product and then adds them up to give an overall reaction, the sum of the ΔH values for the individual reactions equals ΔH^o_{rxn}.

The result is identical to using the above formula.

For example, let's look at the following reaction:

$$C_3H_8(g) + 5\,O_2(g) \rightarrow 3\,CO_2(g) + 4\,H_2O(g)$$

Now, let's write formation reactions for C_3H_8, CO_2, and H_2O.

A formation reaction is one in which a substance is formed from the elements that make up the substance when each element is in its standard state. The three elements in these compounds are carbon, hydrogen, and oxygen. The standard states are C(s) or C(graphite), $H_2(g)$, and $O_2(g)$.

The formation reactions are as follows:

$$3\,C(s) + 4\,H_2(g) \rightarrow C_3H_8(g) \quad \Delta H = \Delta H^o_f$$

$$C(s) + O_2(g) \rightarrow CO_2(g) \quad \Delta H = \Delta H^o_f$$

$$H_2(g) + \tfrac{1}{2}\,O_2(g) \rightarrow H_2O(g) \quad \Delta H = \Delta H^o_f$$

Next, we manipulate these reactions such that when added together we get the reaction

$$C_3H_8(g) + 5\,O_2(g) \rightarrow 3\,CO_2(g) + 4\,H_2O(g)$$

We need to reverse the formation reaction for $C_3H_8(g)$ because propane needs to be a reactant.

$$C_3H_8(g) \rightarrow 3\,C(s) + 4\,H_2(g) \quad \Delta H = -\Delta H^o_f$$

Since we reversed the reaction, the sign of ΔH must be reversed.

We need to multiply the formation reaction for $CO_2(g)$ by 3 because we need 3 moles of carbon dioxide.

$$3\,C(s) + 3\,O_2(g) \rightarrow 3\,CO_2(g) \quad \Delta H = 3\Delta H^o_f$$

Multiplying the reaction through by 3 requires us to multiply the enthalpy change by 3.

We need to multiply the formation reaction for $H_2O(g)$ by 4 because we need 4 moles of water vapor.

$$4\,H_2(g) + 2\,O_2(g) \rightarrow 4\,H_2O(g) \quad \Delta H = 4\Delta H^o_f$$

Multiplying the reaction through by 4 requires us to multiply the enthalpy change by 4.

Adding these three reactions together we get:

$$C_3H_8(g) \rightarrow 3\,C(s) + 4\,H_2(g) \quad \Delta H = -\Delta H^o_f$$

$$3\,C(s) + 3\,O_2(g) \rightarrow 3\,CO_2(g) \quad \Delta H = 3\Delta H^o_f$$

$$4\,H_2(g) + 2\,O_2(g) \rightarrow 4\,H_2O(g) \quad \Delta H = 4\Delta H^o_f$$

$$C_3H_8(g) + 3\,O_2(g) \rightarrow 3\,C(s) + 4\,H_2(g)$$

Summing the enthalpy terms we get

$$\Delta H^o_{rxn} = 3\Delta H^o_{fCO_2(g)} + 4\Delta H^o_{fH_2O(g)} - \Delta H^o_{fC_3H_8(g)}$$

Note: $O_2(g)$ has a ΔH^o_f of 0, so it is not included in the above expression.

This is the same result one gets if you use the formula

$$\Delta H^o_{rxn} = \sum n\Delta H^o_{f,products} - \sum m\Delta H^o_{f,reactants}$$

104. The lattice energy is simply the energy change associated with one of the steps involved in forming an ionic crystal from the elements that make it up. The overall process in forming an ionic crystal can be considered to occur in 5 steps. The total energy change is the heat of formation. We can apply Hess's law to determine the energy change associated with the lattice energy as long as we know the energy change associated with the other steps and the heat of formation. The series of reactions that comprise this process is called the Born–Haber cycle.

For example, let's look at the series of reactions that occur when crystals of NaCl form. The overall process is

$$Na(s) + \tfrac{1}{2}Cl_2(g) \rightarrow NaCl(s) \quad \Delta H = \Delta H^o_f$$

There are a series of steps that occur in order for this substance to form.

Step 1: Solid sodium metal is converted to gaseous sodium.

$$Na(s) \rightarrow Na(g) \quad \Delta H = \Delta H^o_{sublimation}$$

Step 2: The gaseous sodium atom loses an electron; that is, sodium is ionized. Ionization energies are tabulated for gaseous atoms. This is the reason that we needed to convert sodium metal to a gas in Step 1.

$$Na(g) \rightarrow Na^+(g) + e^- \quad \Delta H = \text{ionization energy (IE)}$$

Step 3: A chlorine molecule breaks apart into chlorine atoms. The energy for this process corresponds to the bond energy.

$$\tfrac{1}{2}\,Cl_2(g) \rightarrow Cl(g) \quad \Delta H = \tfrac{1}{2}\text{ bond energy (BE)}$$

Step 4: The chlorine atom gains the electron that the sodium atom lost. The energy for this process corresponds to the electron affinity.

$$Cl(g) + e^- \rightarrow Cl^-(g) \quad \Delta H = \text{electron affinity (EA)}$$

Step 5: The sodium ions and the chloride ions come together to form a solid NaCl unit. The energy associated with this process is called the lattice energy.

$$Na^+(g) + Cl^-(g) \rightarrow NaCl(s) \quad \Delta H = \text{lattice energy}$$

If we add Steps 1–5 together, we should get the heat of formation for NaCl(s).

The only quantity that is not easily measured in this entire process is the lattice energy. We can use Hess's law to solve for the lattice energy.

Adding Steps 1–5 together we get

$$Na(s) \rightarrow Na(g) \quad \Delta H = \Delta H^{o}_{\text{sublimation}}$$
$$Na(g) \rightarrow Na^+(g) + e^- \quad \Delta H = \text{ionization energy (IE)}$$
$$\tfrac{1}{2}Cl_2(g) \rightarrow Cl(g) \quad \Delta H = \tfrac{1}{2}\text{ bond energy (BE)}$$
$$Cl(g) + e^- \rightarrow Cl^-(g) \quad \Delta H = \text{electron affinity (EA)}$$
$$Na^+(g) + Cl^-(g) \rightarrow NaCl(s) \quad \Delta H = \text{lattice energy}$$
$$\overline{Na(s) + \tfrac{1}{2}Cl_2(g) \rightarrow NaCl(s) \quad \Delta H = \Delta H_f^\circ}$$

Using Hess's law we get

$$\Delta H_f^\circ = \Delta H_{\text{subl}} + \text{IE} + \tfrac{1}{2}\text{ BE}_{Cl_2} + \text{EA} + \text{lattice energy}$$

Solving this for lattice energy we get

$$\text{lattice energy} = \Delta H_f^\circ - (\Delta H^{o}_{\text{sub}} + \text{IE} + \tfrac{1}{2}\text{ BE}_{Cl_2} + \text{EA})$$

105. The desired reaction is

$$C(s) + \tfrac{1}{2}O_2(g) \rightarrow CO(g)$$

The reactions to work with are:

1. $CO(g) + \tfrac{1}{2}O_2(g) \rightarrow CO_2(g)$
2. $C(s) + O_2(g) \rightarrow CO_2(g)$

When manipulating equations look for species in the desired reaction that are present in only one of the equations used.

In the preceding reactions (1) and (2) above, $O_2(g)$ is present in both reactions. We do not want to work with O_2 if at all possible. The other two substances in the desired reaction, C(s) and CO(g), are both present in only one of the two reactions.

Let's work with C(s). It is present only in reaction (2), where it is a reactant. It is also a reactant in the desired reaction. Lastly, we have 1 mole of C(s) in reaction (2) and 1 mole of C(s) in the desired reaction. We do not have to do anything to this reaction.

$$C(s) + O_2(g) \rightarrow CO_2(g)$$

Carbon monoxide (CO) is present only in reaction (1). It is present as a reactant in this reaction but as a product in the desired reaction. Lastly, there is 1 mole of CO in both equations. We need to reverse reaction (1).

$$CO_2(g) \rightarrow CO(g) + \tfrac{1}{2}O_2(g)$$

Now we can add them together. They should add up to the desired reaction.

$$C(s) + O_2(g) \rightarrow CO_2(g)$$
$$CO_2(g) \rightarrow CO(g) + \tfrac{1}{2}O_2(g)$$
$$\overline{C(s) + O_2(g) + CO_2(g) \rightarrow CO_2(g) + CO(g) + \tfrac{1}{2}O_2(g)}$$

Canceling the common terms (1 mole CO_2 and $\tfrac{1}{2}$ mole of O_2) yields

$$C(s) + \tfrac{1}{2}O_2(g) \rightarrow CO(g)$$

This is the desired reaction.

To obtain the desired reaction, reverse the first reaction, leave the second reaction as written, and add the two reactions.

107. The desired reaction is

$$S(s) + O_2(g) \rightarrow SO_2(g)$$

The reactions to use are:

1. $2\,SO_2(g) + O_2(g) \rightarrow 2\,SO_3(g) \quad \Delta H = -196\text{ kJ}$
2. $2\,S(s) + 3\,O_2(g) \rightarrow 2\,SO_3(g) \quad \Delta H = -790\text{ kJ}$

Solid S(s) is present in reaction (2) as a reactant. It is also present in the desired reaction as a reactant. However, we only want 1 mole S(s) in the desired reaction. We need to multiply reaction (2) by $\tfrac{1}{2}$ to get only 1 mole of S(s). Lastly, we need to multiply the ΔH value by $\tfrac{1}{2}$.

$$\tfrac{1}{2}\,(2\,S(s) + 3\,O_2(g) \rightarrow 2\,SO_3(g) \quad \Delta H = -790\text{ kJ})$$

or

$$S(s) + \tfrac{3}{2}O_2(g) \rightarrow SO_3(g) \qquad \Delta H = \tfrac{1}{2}(-790) = -395\text{ kJ}$$

$O_2(g)$ is present in both reactions (1) and (2). Skip this substance for the time being.

$SO_2(g)$ is present only in reaction (1), where it is a reactant. We want it as a product, so we need to reverse reaction (1). Additionally we only want 1 mole of SO_2. We need to also multiply reaction (1) through by $\frac{1}{2}$.

$$\tfrac{1}{2}(2\ SO_3(g) \rightarrow 2\ SO_2(g) + O_2(g) \quad \Delta H = 196\ \text{kJ})$$

or

$$SO_3(g) \rightarrow SO_2(g) + \tfrac{1}{2}\ O_2(g) \quad \Delta H = \tfrac{1}{2}(196) = 98\ \text{kJ}$$

Adding the two reactions together we get

$$S(s) + \tfrac{3}{2}\ O_2(g) \rightarrow SO_3(g) \qquad \Delta H = -395\ \text{kJ}$$

$$SO_3(g) \rightarrow SO_2(g) + \tfrac{1}{2}\ O_2(g) \quad \Delta H = 98\ \text{kJ}$$

$$S(s) + \tfrac{3}{2}\ O_2(g) + SO_3(g) \rightarrow SO_3(g) + SO_2(g) + \tfrac{1}{2}\ O_2(g)$$

Canceling out common terms (1 mole SO_3 and $\frac{1}{2}$ mole O_2) leaves

$$S(s) + O_2(g) \rightarrow SO_2(g)$$

This is the desired reaction. We only need to add the ΔH values together to determine ΔH for this reaction.

$$\Delta H = (-395) + (98) = -297\ \text{kJ}$$

The heat of formation of SO_2(g) is –297 kJ/mol.

109. The desired reaction is

$$\alpha\text{-LiAlSi}_2\text{O}_6(s) \rightarrow \beta\text{-LiAlSi}_2\text{O}_6(s)$$

The reactions and their ΔH values are as follows:

$$Li_2O(s) + 2\ Al(s) + 4\ SiO_2(s) + \frac{3}{2}O_2(g) \rightarrow 2\,\alpha\text{-LiAlSi}_2\text{O}_6(s) \qquad \Delta H = -1870.6\ \text{kJ}$$

$$Li_2O(s) + 2\ Al(s) + 4\ SiO_2(s) + \frac{3}{2}\ O_2(g) \rightarrow 2\,\beta\text{-LiAlSi}_2\text{O}_6(s) \qquad \Delta H = -1814.6\ \text{kJ}$$

We need to reverse the first reaction and then add the two reactions together.

$$2\,\alpha\text{-LiAlSi}_2\text{O}_6(s) \rightarrow Li_2O(s) + 2\ Al(s) + 4\ SiO_2(s) + \tfrac{3}{2}\ O_2(g) \qquad \Delta H = +1870.6\ \text{kJ}$$

$$Li_2O(s) + 2\ Al(s) + 4\ SiO_2(s) + \tfrac{3}{2}\ O_2(g) \rightarrow 2\,\beta\text{-LiAlSi}_2\text{O}_6(s) \qquad \Delta H = -1814.6\ \text{kJ}$$

$$2\,\alpha\text{-LiAlSi}_2\text{O}_6(s) \rightarrow 2\,\beta\text{-LiAlSi}_2\text{O}_6(s) \quad \Delta H = +1870.6 - 1814.6$$

$$\Delta H = +56.0\ \text{kJ}$$

However, this is for converting 2 moles of the α form into 2 moles of the β form. We need to divide all quantities by 2 (or multiply by $\frac{1}{2}$).

$$\alpha\text{-LiAlSi}_2\text{O}_6(s) \rightarrow \beta\text{-LiAlSi}_2\text{O}_6(s) \quad \Delta H = +28.0\ \text{kJ}$$

The ΔH_{rx} value for the conversion of α-spodumene into β-spodumene is 28.0 kJ.

111. The desired reaction is

$$2\ NOCl(g) \rightarrow N_2(g) + O_2(g) + Cl_2(g)$$

The reactions to use are:

1. $\frac{1}{2}\ N_2(g) + \frac{1}{2}\ O_2(g) \rightarrow NO(g) \qquad \Delta H = 90.3$ kJ
2. $NO(g) + \frac{1}{2}\ Cl_2(g) \rightarrow NOCl(g) \quad \Delta H = -38.6$ kJ

In the desired reaction, there is 1 mole of $N_2(g)$ present on the product side. In reaction (1) there is $\frac{1}{2}$ mole of $N_2(g)$ on the reactant side. If we reverse reaction (1) and multiply this reaction by 2, there will be 1 mole of $N_2(g)$ on the product side.

$$2\ NO(g) \rightarrow N_2(g) + O_2(g) \qquad \Delta H = 2(-90.3\ \text{kJ}) = -180.6\ \text{kJ}$$

There are 2 moles of NOCl on the reactant side in the desired reaction. Reaction (2) contains 1 mole of NOCl as a product. If we reverse reaction (2) and multiply through by 2, there will be 2 moles of NOCl on the reactant side.

$$2\ NOCl(g) \rightarrow 2\ NO(g) + Cl_2(g)$$

$$\Delta H = 2(38.6\ \text{kJ}) = 77.2\ \text{kJ}$$

We now add the two reactions together:

$$2\ NO(g) \rightarrow N_2(g) + O_2(g) \qquad \Delta H = -180.6\ \text{kJ}$$

$$2\ NOCl(g) \rightarrow 2\ NO(g) + Cl_2(g) \qquad \Delta H = 77.2\ \text{kJ}$$

$$2\ NO(g) + 2\ NOCl(g) \rightarrow N_2(g) + O_2(g) + 2\ NO(g) + Cl_2(g) \quad \Delta H = -180.6 + 77.2 = -103.4\ \text{kJ}$$

The common term is 2 moles of NO(g). Canceling this term out gives

$$2\ NOCl(g) \rightarrow N_2(g) + O_2(g) + Cl_2(g)$$

$$\Delta H = -103.4\ \text{kJ}$$

CHAPTER 12 Energy and Organic Chemistry

1. The physical property used to separate the components of crude oil is the volatility or boiling point.

2. No. The general formula for an alkane is C_nH_{2n+2}. The general formula for a cycloalkane is $C_{2n}H_{2n}$.

   Since a cycloalkane has no terminal carbon atoms it has no CH_3 groups. The maximum number of hydrogen atoms that can bond to a carbon of a cycloalkane is two. An alkane has a CH_3 group bonded at each end of the chain. To convert an alkane into a cycloalkane, one of the hydrogen atoms from each of the two CH_3 groups must be removed. Hence, the general formula of the cycloalkane contains two less hydrogen atoms than an alkane containing the same number of carbon atoms.

3. In alkanes, all carbon atoms use sp^3 hybrid orbitals.

   In an alkane all of the carbon–carbon bonds and any other bonds involving carbon are single ones. This means that carbon must form four single bonds with four other atoms. The only method available for this is the formation of four equivalent hybrid orbitals. The combination of orbitals is the $2s$ orbital and the three $2p$ orbitals of carbon combining to form four equivalent sp^3 hybrid orbitals.

4. Hexanes contain six carbon atoms per molecule. The alkane hexane has the formula of C_6H_{14}. The cycloalkane containing six carbon atoms (cyclohexane) has a formula of C_6H_{12}. Since these two substances have different molecular formulas, they cannot be isomers (structural or any other type).

5. Alkanes and cycloalkanes are not planar because each carbon atom is bonded to four other atoms and has tetrahedral geometry. This results in a bond angle of about 109° for all bonds in both alkanes and cycloalkanes so cycloalkanes cannot be planar.

6. Yes. Any hydrocarbon structure that has only carbon-carbon single bonds is said to be a saturated hydrocarbon. If it contains one or more carbon-carbon double or triple bonds, the substance is considered to be unsaturated.

7. Yes. A structural isomer is any substance that has the same molecular formula, but has a different bonding arrangement of the atoms.

8. No. The chemical and physical properties of structural isomers can be quite different. For example, consider the following structural isomers of the formula C_2H_6O

```
    H   H              H       H
    |   |              |       |
H — C — C — O — H  H — C — O — C — H
    |   |              |       |
    H   H              H       H
```

ethanol — Dimethyl ether

   These two substances belong to different classifications of organic substances (ethanol is an alcohol and dimethyl ether is an ether). Ethanol is much more reactive than dimethyl ether, and the boiling point of ethanol is 78.5°C, whereas the boiling point of dimethyl ether is −24.5°C.

9. Each carbon atom is bonded to two other atoms and is sp^2 hybridized. In this bonding arrangement the bond angle is 120° and the molecule must be planar.

10. Aromatic structures are stable because of the delocalization of electrons in the π bonds found in these substances.

11. Information given:

    The vapor pressure of ethanol = 45 torr at 20°C

The vapor pressure of methanol = 92 torr at 20°C
The mass of ethanol present in the mixture is 75 g.
The mass of methanol in the mixture is 25 g.

From the information given, we conclude that we need to use Raoult's law, Raoult's law relates the mole fraction of each substance in a mixture to the vapor pressure it will exert above the mixture.

The vapor pressure exerted by each component in the mixture is given by the expression:

$$P_a = X_a P_a^o$$

where

the subscript a, denotes a component of the mixture
P_a is the vapor pressure of component a above the mixture
X_a is the mole fraction of component a
P_a^o is the vapor pressure of pure component a

In this problem the pressure associated with the methanol above the solution is given by the expression

$$P_{methanol} = X_{methanol} P^o_{methanol}$$

And the corresponding pressure of ethanol is given by the expression

$$P_{ethanol} = X_{ethanol} P^o_{ethanol}$$

The total vapor pressure above the solution is simply the sum of the individual components.

$$P_{total} = X_a P_a^o + X_b P_b^o$$

We need to determine the mole fraction of both methanol and ethanol in the mixture. To find these quantities we first need to determine the number of moles of each substance.

$$\text{moles of methanol } (CH_3OH) = 25 \text{ g } CH_3OH \times \frac{1 \text{ mol } CH_3OH}{32.0 \text{ g } CH_3OH} = 0.78 \text{ mol } CH_3OH$$

$$\text{moles of ethanol } (CH_3CH_2OH) = 75 \text{ g } CH_3CH_2OH \times \frac{1 \text{ mol } CH_3CH_2OH}{46.0 \text{ g } CH_3CH_2OH} = 1.63 \text{ mol } CH_3CH_2OH$$

The mole fraction (X) can be calculated by the expression:

$$X_a = \frac{\text{moles of } a}{\text{total moles}}$$

$$X_{methanol} = \frac{\text{moles of methanol}}{\text{moles of methanol + moles of ethanol}} = \frac{0.78}{0.78 + 1.63} = 0.32$$

$$X_{ethanol} = \frac{\text{moles of ethanol}}{\text{moles of methanol + moles of ethanol}} = \frac{1.63}{0.78 + 1.63} = 0.68$$

Next, find the vapor pressure above the mixture due to each liquid.

$$P_{methanol} = X_{methanol} P^o_{methanol} = (0.32)(92 \text{ torr}) = 29 \text{ torr}$$

$$P_{ethanol} = X_{ethanol} P^o_{ethanol} = (0.68)(45 \text{ torr}) = 31 \text{ torr}$$

The total vapor pressure above the solution is the sum of the individual vapor pressures.

$$P_{total} = 29 \text{ torr} + 31 \text{ torr} = 60. \text{ torr}$$

The total pressure above this solution is 60. torr (or mm Hg).

13. a. Methane CH_4 1 carbon atom
    b. Pentane C_5H_{12} 5 carbon atoms
    c. Cycloheptane C_7H_{14} 7-carbon ring
    d. Cyclodecane $C_{10}H_{20}$ 10-carbon ring

15. $CH_3CH_2CH_2CH_2CH_3$ *n*-pentane

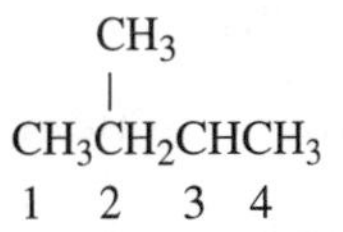

```
      CH3
       |
CH3CH2CHCH3
 1  2  3  4          2-methylbutane
```

Moving the methyl substituent group over to the next carbon atom (the one numbered 3) does not give a different isomer. If one counts from right to left instead of from left to right, we find the methyl substituent is still bonded to carbon-2 (C-2). This is shown in the structure below.

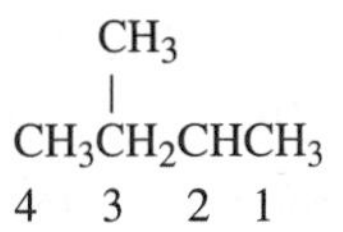

```
      CH3
       |
CH3CH2CHCH3
 4  3  2  1          2-methylbutane
```

There are no other possible isomers for a 4-carbon chain.

Next, let's try a 3-carbon chain.

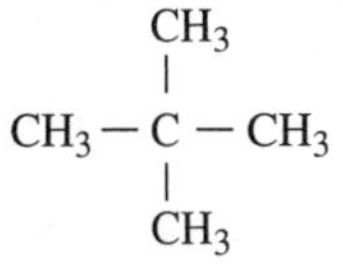

```
       CH3
        |
  CH3 — C — CH3
        |
       CH3           2,2-dimethylpropane
```

Putting an ethyl group on C-2 does not produce a new isomer. As shown in the following structure, the longest chain would contain four carbon atoms and the structure has the name 2-methylbutane (an isomer that has already been shown).

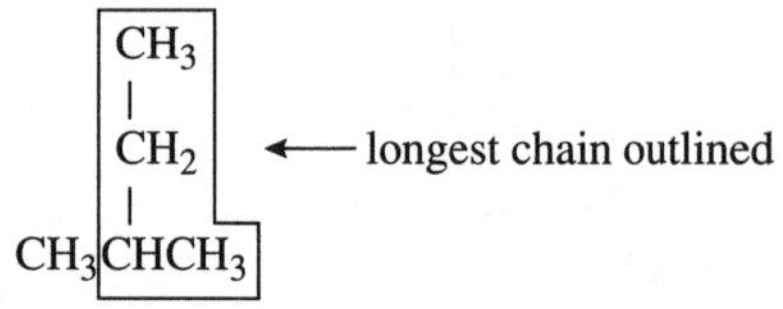

CH_3
|
CH_2 ← longest chain outlined
|
CH_3CHCH_3

There are only three structural isomers with the formula C_5H_{12}.

17. For each line structure, let's write a condensed structural formula. Remember, that each intersection between lines corresponds to a carbon atom and there is a carbon atom at the end of each line.

a. 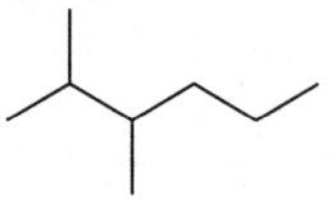 corresponds to

CH_3
|
$CH_3CHCHCH_2CH_2CH_3$
|
CH_3

The name of this substance is 2,3-dimethylhexane. Its molecular formula is C_8H_{18}.

b. 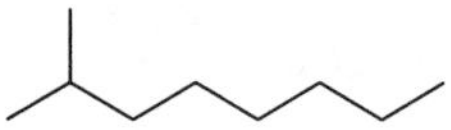 corresponds to

CH_3
|
$CH_3CHCH_2CH_2CH_2CH_2CH_2CH_3$

The name of this substance is 2-methyloctane. Its molecular formula is C_9H_{20}.

c. 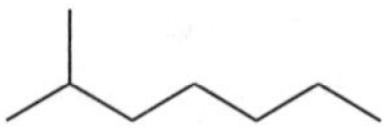 corresponds to

CH_3
|
$CH_3CHCH_2CH_2CH_2CH_2CH_3$

The name of this substance is 2-methylheptane. Its molecular formula is C_8H_{18}.

d. 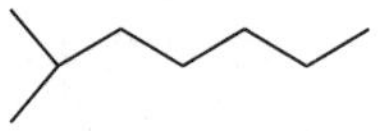 corresponds to

CH_3
|
$CH_3CHCH_2CH_2CH_2CH_2CH_3$

The name of this substance is 2-methylheptane. Its molecular formula is C_8H_{18}.

e. 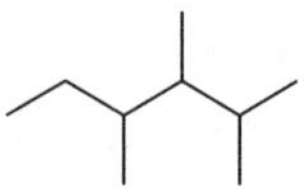 corresponds to

CH_3 CH_3
| |
$CH_3CH_2CHCHCHCH_3$
|
CH_3

The name of this substance is 2,3,4-trimethylhexane. Its molecular formula is C_9H_{20}.

Structures *c* and *d* are the same.

Structures *a* and *c* (or *a* and *d*) are structural isomers. They both have the molecular formula C_8H_{18}. They are isomers of *n*-octane.

Structures *b* and *e* are structural isomers. They both have the molecular formula C_9H_{20}. They are isomers of *n*-nonane.

19. a. 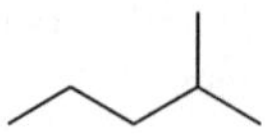 or

CH_3
|
$CH_3CH_2CH_2CHCH_3$

The longest chain contains five carbon atoms.
There is a methyl group bonded to C-2.
The name of this substance is 2-methylpentane.

b. 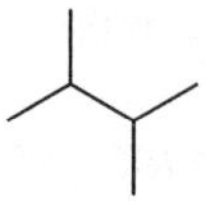 or

CH_3
|
$CH_3CHCHCH_3$
|
CH_3

The longest chain consists of four carbon atoms.
There are methyl groups bonded to C-2 and C-3.
The name of this substance is 2,3-dimethylbutane.

c. 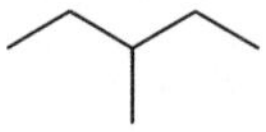 or $CH_3CH_2CHCH_2CH_3$
|
CH_3

The longest chain consists of four carbon atoms.
There are methyl groups bonded to this chain: two methyl groups bonded to C-2 and one methyl group bonded to C-3.
The name of this substance is 2,2,3-trimethylbutane.

d. 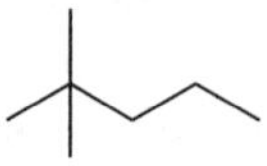 or

CH_3
|
$CH_3CHCH_2CH_2CH_3$
|
CH_3

The longest chain consists of five carbon atoms.
There is one two methyl group bonded to C-2.
The name of this substance is 2,2-dimethylpentane.

Branched structures tend to increase the octane rating significantly as compared to hydrocarbon chains without branches. All structures should significantly increase the octane ratings and thus are more likely to be found in gasoline (see Table 12.3).

21. a. 2-methylpentane

The name pentane indicates a 5-carbon chain. There is a methyl group bonded to C-2. The condensed structural formula is

CH_3
|
$CH_3CHCH_2CH_2CH_3$

The Lewis structure is

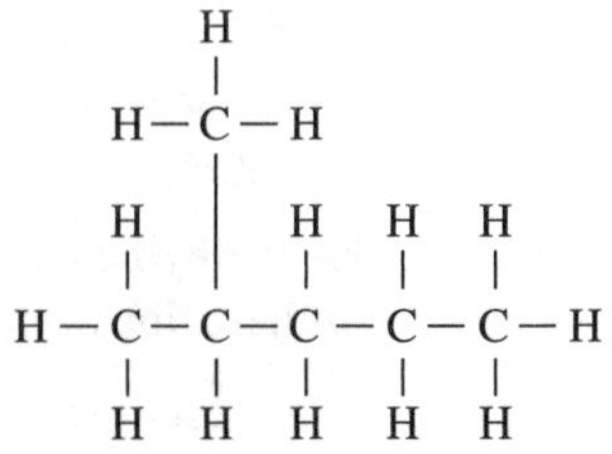

```
            H
            |
         H— C —H
            |
     H      |   H   H   H
     |      |   |   |   |
 H — C  —   C — C — C — C — H
     |      |   |   |   |
     H      H   H   H   H
```

b. 2,3-dimethylhexane

The name hexane indicates a 6-carbon chain. There are methyl groups bonded to C-2 and C-3. The condensed structural formula is

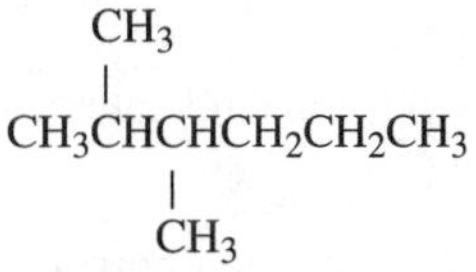

```
    CH3
    |
CH3CHCHCH2CH2CH3
       |
       CH3
```

The Lewis structure is

```
          H
          |
       H— C —H
          |
   H      |   H   H   H   H
   |      |   |   |   |   |
H— C  —   C — C — C — C — C — H
   |      |   |   |   |   |
   H      H   |   H   H   H
              |
           H— C —H
              |
              H
```

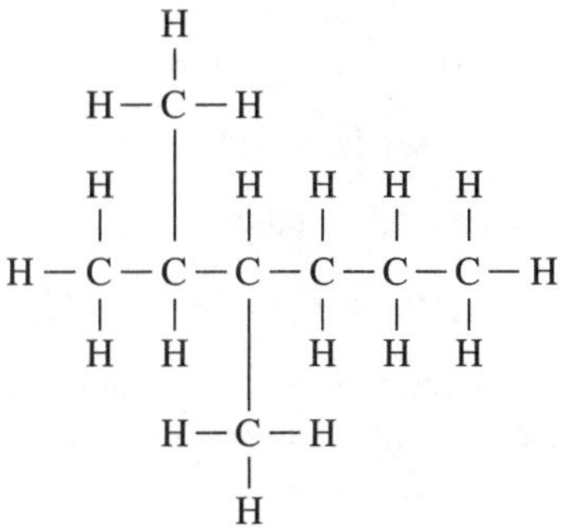

c. 2,3-dimethylbutane

The name butane indicates a 4-carbon chain. There are methyl groups bonded to C-2 and C-3. The condensed structural formula is

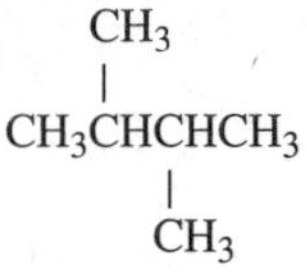

```
    CH3
    |
CH3CHCHCH3
       |
       CH3
```

The Lewis structure is

```
          H
          |
       H— C —H
          |
   H      |   H   H
   |      |   |   |
H— C  —   C — C — C — H
   |      |   |   |
   H      H   |   H
              |
           H— C —H
              |
              H
```

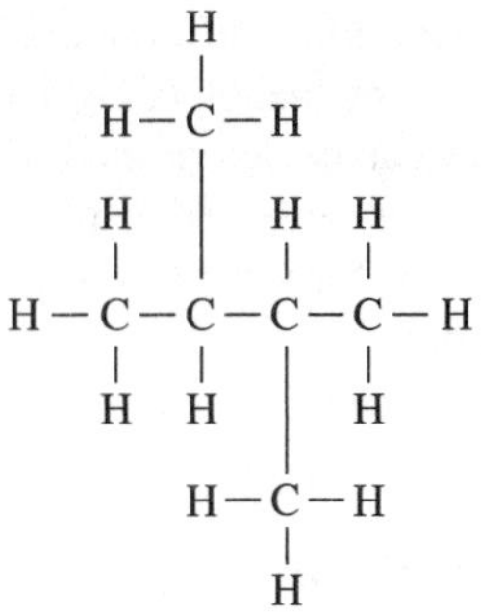

d. 2-methyl-3-ethylpentane

Pentane corresponds to a 5-carbon chain. There is a methyl group bonded to C-2 and an ethyl group bonded to C-3. The condensed structural formula is

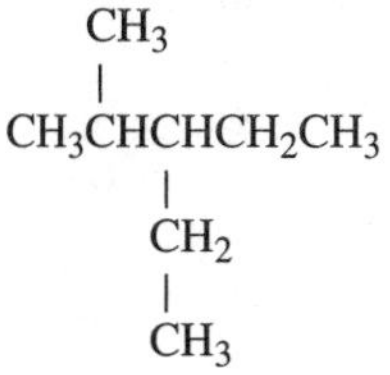

```
    CH3
    |
CH3CHCHCH2CH3
       |
       CH2
       |
       CH3
```

The Lewis structure is

```
          H
          |
       H— C —H
          |
   H      |   H   H   H
   |      |   |   |   |
H— C  —   C — C — C — C — H
   |      |   |   |   |
   H      H   |   H   H
              |
           H— C —H
              |
           H— C —H
              |
              H
```

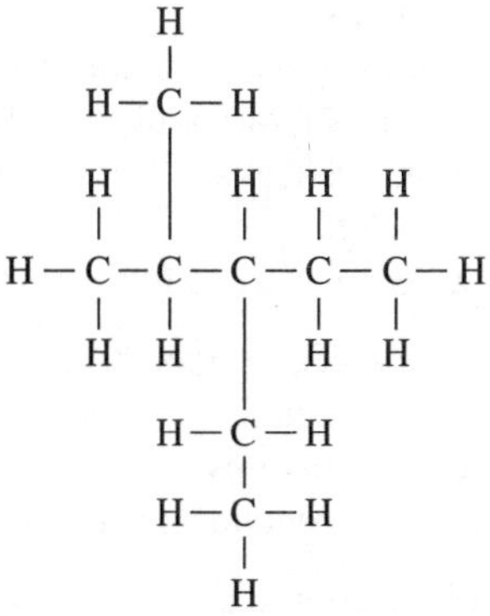

23. The approach here is to draw a condensed structural formula corresponding to the name given and then to rename it correctly.

a. 4-methylpentane

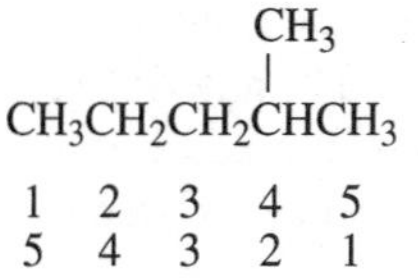

```
             CH3
             |
CH3CH2CH2CHCH3

 1   2   3   4   5
 5   4   3   2   1
```

From the structure shown, we see that the longest chain contains five carbon atoms. Numbering from right to left (instead of from left to right) results in the methyl being bonded to C-2 (instead of C-4). Remember, the chain should be numbered so that the substituent groups are located by the lowest possible number.

The correct name for this substance is 2-methylpentane.

b. 2-ethylbutane

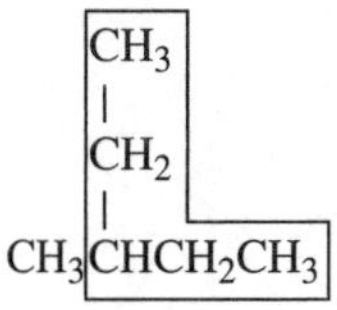

```
   [CH3
   | |
   |CH2
   | |
CH3[CHCH2CH3]
```

The longest chain in this substance contains five carbon atoms (not four carbon atoms). The longest chain is outlined in the structure. The only branch on this chain is a methyl group. It is bonded to the

C-3 regardless of which end is chosen to start the numbering.

The correct name of this substance is 3-methylpentane.

c. 1-methylpropane

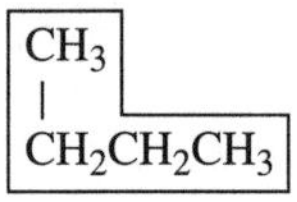

$$\begin{array}{l} CH_3 \\ | \\ CH_2CH_2CH_3 \end{array}$$

The longest chain in this structure contains four carbon atoms.

Note: One can never have an alkyl branch bonded to the first carbon of an alkane. This is because, in all cases, the branch will just be a continuation of the longest chain.

The correct name of this substance is *n*-butane.

25. Structures *c* and *d* contain benzene rings and are considered to be aromatic.

Structure *a* contains only two double bonds. Structure *b* does not contain any double bonds.

27. Each carbon atom in a benzene ring has three bonds to neighboring carbon atoms (one double bond and one single bond), so only one additional group (either an atom, such as H or Cl, or a polyatomic group) can be bonded to each carbon atom.

If the benzene ring is to have two ethyl groups attached, then each one must be bonded to a different carbon atom. For any ring (cyclic) structure, one of the groups bonded to a carbon atom of the ring must be labeled C-1 (we need to start the numbering somewhere).

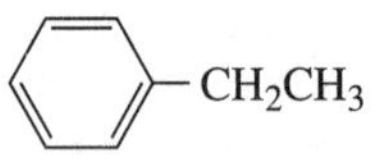

CH_2CH_3 ethylbenzene

The second ethyl group has to be bonded to a different carbon atom.

The only possible arrangements are 1,2-diethylbenzene (*o*-dimethylbenzene), 1,3-diethylbenzene (*m*-diethylbenzene), and 1,4-dimethylbenzene (*p*-dimethylbenzene).

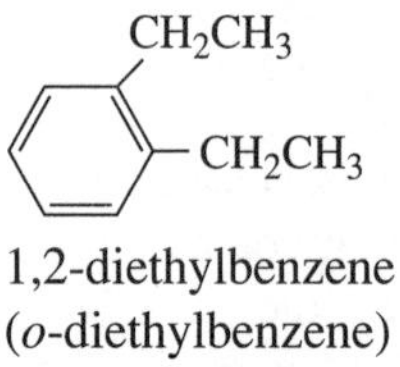

CH_2CH_3 CH_2CH_3

1,2-diethylbenzene
(*o*-diethylbenzene)

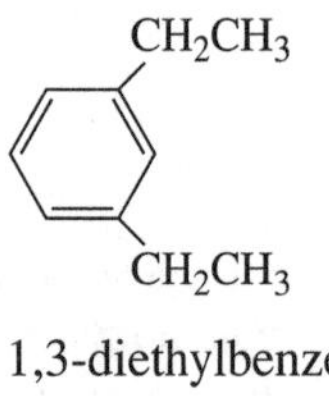

CH_2CH_3 CH_2CH_3

1,3-diethylbenzene
(*m*-diethylbenzene)

CH_2CH_3 CH_2CH_3

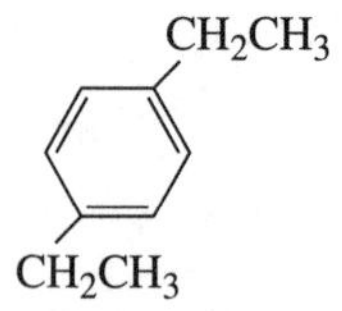

1,4-diethylbenzene
(*p*-diethylbenzene)

Note: 1,5-diethylbenzene is the same as 1,3-diethylbenzene and 1,6-dimethylbenzene is the same as 1,2-dimethylbenzene. The correct name uses the lowest possible numbers to identify the substituents.

The prefix *o* (*ortho*) indicates groups are bonded to C-1 and C-2.

The prefix *m* (*meta*) indicates groups are bonded to C-1 and C-3.

The prefix *p* (*para*) indicates groups are bonded to C-1 and C-4.

29. A combustion reaction corresponds to a reaction with oxygen to produce carbon dioxide and water.

The general formula of an alkane is C_nH_{2n+2}. Undecane has the molecular formula $C_{11}H_{24}$.

The unbalanced equation is

$$C_{11}H_{24} + O_2 \rightarrow CO_2 + H_2O$$

The balanced equation is

$$C_{11}H_{24} + 17\ O_2 \rightarrow 11\ CO_2 + 12\ H_2O$$

31. a. First, determine how much heat energy is required to raise the temperature of water from 20.0°C to 50.0°C, using the formula

$$q(\text{heat}) = (\text{mole of water})(\text{molar heat capacity of water})(\Delta T)$$

The molar heat capacity of water is 75.3 J/(mol · °C).

The moles of water heated are

$$454\ \text{g}\ H_2O \times \frac{1\ \text{mol}\ H_2O}{18.0\ \text{g}\ H_2O} = 25.2\ \text{mol}\ H_2O$$

Now we can calculate the amount of heat needed.

$$q = (25.2\ \text{mol}\ H_2O)\left(75.3\frac{\text{J}}{\text{mol}\cdot{}^\circ\text{C}}\right)(30.0^\circ\text{C})$$

$$= 56{,}972\ \text{J, or } 57.0\ \text{kJ}$$

Next, determine the amount of heat generated when 1 mol of methanol (CH_3OH) is combusted. This quantity can be determined from the balanced equation for the combustion of methanol and by using the heat of formation information found in the text Appendix.

The balanced equation (for 1 mol of CH_3OH combusted) is

$$CH_3OH(l) + \tfrac{3}{2}\ O_2(g) \rightarrow CO_2(g) + 2\ H_2O(g)$$

Note: According to the text, we should assume that water vapor, and not liquid water, is formed in a combustion reaction.

Next, calculate the heat of combustion for this reaction:

$$\Delta H^{\circ}_{comb} = \left[\Delta H^{\circ}_{f,\ CO_2(g)} + 2\Delta H^{\circ}_{f,\ H_2O(g)}\right] - \left[\Delta H^{\circ}_{f,\ CH_3OH(l)} + \tfrac{3}{2}\Delta H^{\circ}_{f,\ O_2(g)}\right]$$

From the Appendix, we find the following heat of formation values (ΔH°_f):

$CO_2(g) = -393.5$ kJ/mol
$CH_3OH(l) = -238.7$ kJ/mol
$H_2O(g) = -241.8$ kJ/mol
$O_2(g) = 0.0$ kJ/mol

Substituting these values, we get

$$\begin{aligned}\Delta H^{\circ}_{comb} &= [(1\text{ mol})(-393.5\text{ kJ/mol}) \\ &\quad + (2\text{ mol})(-241.8\text{ kJ/mol})] \\ &\quad - [(1\text{ mol})(-238.7\text{ kJ/mol}) \\ &\quad + \tfrac{3}{2}\text{ mol }(0.0\text{ kJ/mol})] \\ &= [(-393.5\text{ kJ}) + (-483.6\text{ kJ})] \\ &\quad - [(-238.7\text{ kJ})] = -638.4\text{ kJ}\end{aligned}$$

638.4 kJ of heat energy is released for every mole of methanol combusted.

Now we can find out how many moles of methanol must be combusted to produce 57.0 kJ of heat energy. The following formula allows us to calculate this quantity.

$$\text{moles of methanol combusted} = \frac{\text{total heat energy desired}}{\text{energy per mol}}$$

$$\text{moles of methanol combusted} = \frac{57.0\text{ kJ}}{638.4\text{ kJ/mol}} = 0.0893\text{ mol methanol}$$

We need to combust 0.0893 mol of methanol to produce 57.0 kJ of heat energy.

Lastly, convert moles of methanol combusted to grams of methanol combusted.

$$0.0893\text{ mol } CH_3OH \times \frac{32.0\text{ g } CH_3OH}{1\text{ mol } CH_3OH} = 2.86\text{ g } CH_3OH$$

2.86 g of methanol must be combusted to raise the temperature of 454 g of water from 20.0° to 50.0°.

b. To determine how many grams of carbon dioxide are formed from 2.86 g of methanol, we start with the balanced equation

$$CH_3OH(l) + \frac{3}{2}O_2(g) \rightarrow CO_2(g) + 2\,H_2O(g)$$

We need to convert 2.84 g CH_3OH to mass of CO_2 in grams. First, we convert grams of CH_3OH to moles of CH_3OH, then convert moles of CH_3OH to moles of CO_2. Lastly, we convert moles of CO_2 to grams of CO_2. The mass of CO_2 produced is

$$2.86\text{ g } CH_3OH \times \frac{1\text{ mol } CH_3OH}{32.0\text{ g } CH_3OH} \times \frac{1\text{ mol } CO_2}{1\text{ mol } CH_3OH} \times \frac{44.0\text{ g } CO_2}{1\text{mol } CO_2} = 3.93\text{ g } CO_2$$

3.93 g of CO_2 are produced from 2.86 g of methanol (rounded to 3 significant figures).

33. The structure of ethanol and methanol are similar. They both contain an alcohol functional group (OH) bonded to a carbon atom.

```
    H              H   H
    |              |   |
H — C — OH     H — C — C — OH
    |              |   |
    H              H   H
 Methanol        Ethanol
```

These similar structures mean that similar types of intermolecular forces are present. Both substances are polar and capable of forming hydrogen bonds. Additional intermolecular forces present are dipole–dipole forces and London dispersion forces. Both the hydrogen bonding and dipole-dipole forces should be about the same, because the dipole moments in methanol and ethanol should be similar due to the similar structures. The major difference should be due to the London dispersion forces, since the larger the molecule, the more polarizable the molecule (i.e., the more easily distorted is the electron cloud), and the higher the boiling point. Because size is determined by molar mass typically, the larger the molar mass, the more polarizable the substance, and the higher the boiling point.

The molar mass of ethanol (C_2H_5OH, 46.0 g/mol) is larger than the molar mass of methanol (CH_3OH, 32.0 g/mol), so we would expect ethanol (78.5°C) to have a higher boiling point than methanol (65°C).

34. Ethanol is a polar substance capable of forming hydrogen bonds. Water (H_2O) is also a polar substance that is capable of forming hydrogen bonds. Since ethanol and water contain similar types of intermolecular forces, we expect them to be miscible (mix with each other).

Ethane is a nonpolar substance. The only intermolecular forces present in ethane are London dispersion forces. Because hydrogen bonding is much stronger than London forces, water molecules will be much more strongly attracted to other water molecules than to ethane, so water and ethane do not mix.

The situation is similar with dimethylether

```
    H       H
    |       |
H — C — O — C — H
    |       |
    H       H
```

Dimethylether

Dimethylether is not very polar. It is, however, able to form some hydrogen bonds with water molecules and this allows for limited mixing (about 8 mL of dimethylether can be dissolved per 100 mL of water).

35. Ethanol (CH_3CH_2OH) and dimethylether (CH_3OCH_3) both contain oxygen. Ethane (C_2H_6) contains only the elements carbon and hydrogen. The amount of energy released during combustion depends mainly on the number of carbon atoms available for forming C=O (carbon-oxygen double bonds) bonds in (CO_2) and the number of hydrogen atoms available for forming H_2O molecules.

    The presence of oxygen atoms in these substances adds significantly to their molar mass but adds nothing to the energy produced or the fuel value. The oxygen content of a combustible substance essentially dilutes its energy value. The more "oxygenated" a fuel is, the lower its fuel value.

36. It is likely that the heats of combustion of these two substances are similar but not identical.

    The Lewis structure of diethylether is

```
    H   H       H   H
    |   |       |   |
H — C — C — O — C — C — H
    |   |       |   |
    H   H       H   H
```

The bonding in diethylether consists of:

2 carbon-carbon single bonds
2 carbon-oxygen single bonds
10 carbon-hydrogen single bonds
The Lewis structure of *n*-butanol is

```
    H   H   H   H
    |   |   |   |
H — C — C — C — C — O — H
    |   |   |   |
    H   H   H   H
```

The bonding in *n*-butanol consists of:

3 carbon-carbon single bonds
1 carbon-oxygen single bond
9 carbon-hydrogen single bonds
1 oxygen-hydrogen single bond

The balanced equation for the combustion of both substances is the same, since they both have the molecular formula $C_4H_{10}O$.

$$C_4H_{10}O(l) + 6\,O_2(g) \rightarrow 4\,CO_2(g) + 5\,H_2O(g)$$

The products are the same for both substances. The energy differences will be due to the differences in the bonds present in diethylether and *n*-butanol. Ethanol has an extra carbon-carbon single bond and an oxygen-hydrogen single bond, while diethylether has an extra carbon-oxygen single bond and an extra carbon-hydrogen single bond. These slight differences will result in slightly different heats of combustion for these two substances. To determine the exact difference we would need to look up actual bond energies.

37. A 3-carbon chain cannot have any branches. There are only two possible isomers for the alcohol with the formula C_3H_8O, since the alcohol functional group (OH) can only be bonded to C-1 or C-2.

```
    H   H   H                      H  OH   H
    |   |   |                      |   |   |
H — C — C — C — O — H          H — C — C — C — H
    |   |   |                      |   |   |
    H   H   H                      H   H   H
```

n-propanol (or 1-propanol)   2-propanol (or isopropanol)

Note: The following structure is 1-propanol (or *n*-propanol) and not 3-propanol.

```
        H   H   H
        |   |   |
H — O — C — C — C — H
        |   |   |
        H   H   H
```

39. Let's draw the condensed structural formula for each line structure, then name the structure and determine the molecular formula of each. Substances with the same molecular formula but a different structure (and a different name) will be isomers of each other.

    In this problem we are looking for isomers of *n*-hexanol.

    n-Hexanol consists of a 6-carbon chain with an OH group bonded to one of the end carbon atoms.

    The condensed structural formula of this substance is $CH_3CH_2CH_2CH_2CH_2CH_2OH$. It has the molecular formula $C_6H_{14}O$.

    Any substance that has the molecular formula of $C_6H_{14}O$ (*n*-hexanol) with a different bonding arrangement will be a structural isomer of *n*-hexanol.

    a. OH

The condensed structural formula is

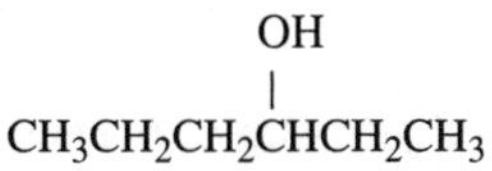

$$\begin{array}{c} \qquad\quad OH \\ \qquad\quad | \\ CH_3CH_2CH_2CHCH_2CH_3 \end{array}$$

The molecular formula of this substance is $C_6H_{14}O$ and it is named 3-hexanol.

This is a structural isomer of *n*-hexanol.

b. O

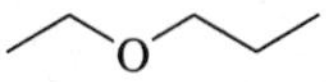

The condensed structural formula is

$$CH_3CH_2-O-CH_2CH_2CH_3$$

The molecular formula of this substance is $C_5H_{12}O$ and it is named ethylpropylether.

This is not a structural isomer of *n*-hexanol.

c. OH

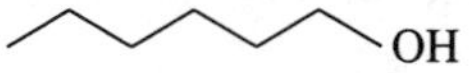

The condensed structural formula is

$$CH_3CH_2CH_2CH_2CH_2CH_2OH$$

The molecular formula of this substance is $C_6H_{14}O$ and it is named *n*-hexanol.

This is an isomer of $C_6H_{14}O$ but since it is *n*-hexanol, it cannot be considered to be an isomer of itself.

d. O

The condensed structural formula is

$$CH_3-O-CH_2CH_2CH_2CH_2CH_3$$

The molecular formula of this substance is $C_6H_{14}O$ and it is named methyl *n*-pentyl ether.

This is a structural isomer of *n*-hexanol.

e. OH

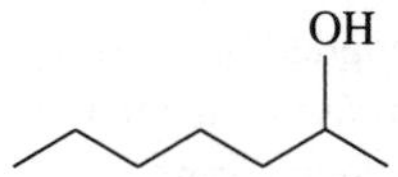

The condensed structural formula is

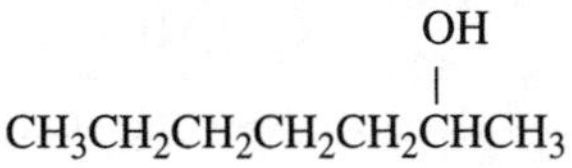

$$\begin{array}{c} \qquad\qquad\quad OH \\ \qquad\qquad\quad | \\ CH_3CH_2CH_2CH_2CH_2CHCH_3 \end{array}$$

The molecular formula of this substance is $C_7H_{16}O$ and it is named 2-heptanol.

It is not a structural isomer of *n*-hexanol.

Structures *a* and *d* are structural isomers of *n*-hexanol.

41. a. 2,3-dimethyl-2-butanol

The longest chain contains four carbon atoms (*but*anol).

The OH group is bonded to C-2 (2-butanol).

There are two methyl groups (dimethyl). One is bonded to C-2 and the other is bonded to C-3 (2,3-dimethyl).

The condensed structural formula is

$$\begin{array}{ccccccc} & & OH & & H & & \\ & & | & & | & & \\ CH_3 & - & C & - & C & - & CH_3 \\ & & | & & | & & \\ & & CH_3 & & CH_3 & & \end{array}$$

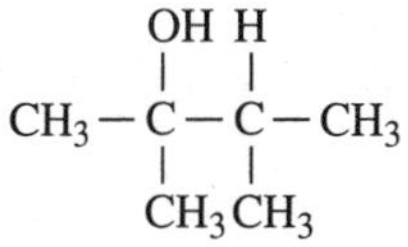

b. 2-ethyl-1-pentanol

The longest chain contains 5 carbon atoms (*pent*anol).

The OH group is bonded to C-1 (1-pentanol).

An ethyl group is bonded to C-2 (the carbon next to the one with the OH).

In naming alcohols (and for any functional group) the longest chain must contain the carbon atom with the functional group. We start counting from the end that places the functional group on the carbon atom with lowest possible number. In this case the OH group is bonded to an end carbon (C-1).

The condensed structural formula of 2-ethyl-1-pentanol is

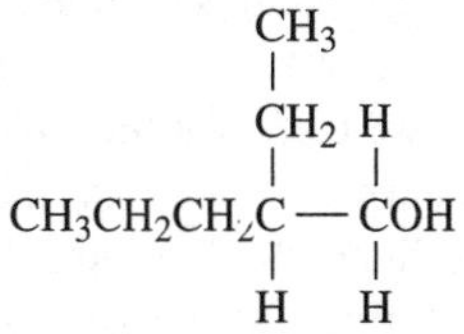

$$\begin{array}{ccccc} & CH_3 & & & \\ & | & & & \\ & CH_2 & & H & \\ & | & & | & \\ CH_3CH_2CH_2 & C & - & COH & \\ & | & & | & \\ & H & & H & \end{array}$$

c. 5,5-diethyl-2-heptanol

The longest chain contains seven carbon atoms (*hept*anol).

The OH group is bonded to C-2 (2-heptanol).

There are two ethyl groups bonded to the chain. They are both bonded to C-5 (5,5-diethyl).

The condensed structural formula of 5,5-diethyl-2-heptanol is

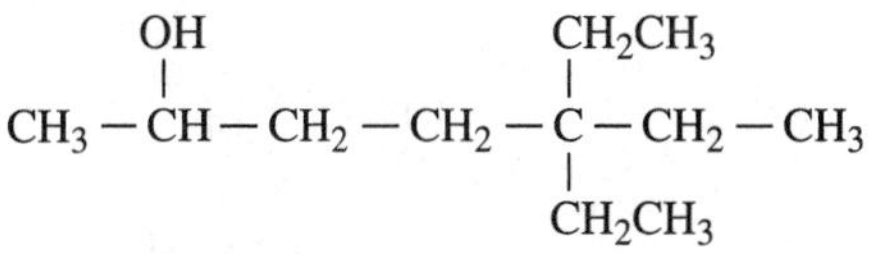

$$\begin{array}{ccccccccccccc} & & OH & & & & & & CH_2CH_3 & & & & \\ & & | & & & & & & | & & & & \\ CH_3 & - & CH & - & CH_2 & - & CH_2 & - & C & - & CH_2 & - & CH_3 \\ & & & & & & & & | & & & & \\ & & & & & & & & CH_2CH_3 & & & & \end{array}$$

43. To name an alcohol, use the following rules:

1. Determine the longest chain to which the OH group is directly bonded (i.e., the OH must be bonded to a carbon atom that is part of the longest chain). Even if there is a longer possible carbon chain, if the OH is not bonded to a carbon of this chain, then it is not designated the longest chain.
2. Replace the *e* of the alkane ending of the longest chain with the ending *ol*.

3. Number the carbon atoms such that the carbon with the OH group receives the lowest possible number.
4. Name and number the substituent groups.

a.

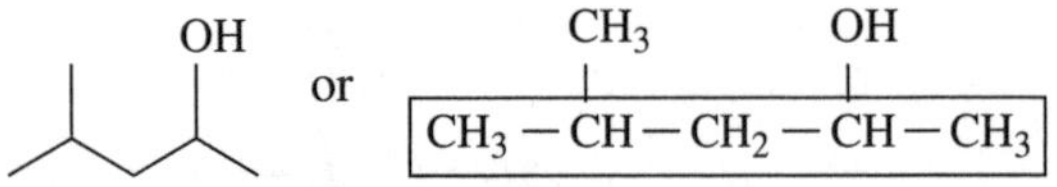

$$CH_3-\underset{}{\overset{CH_3}{CH}}-CH_2-\overset{OH}{CH}-CH_3$$

The longest chain is outlined in the structure. It contains five carbon atoms.

Change the name from pentane to pentanol.

The OH can be bonded to C-4 (counting from left to right) or C-2 (counting from right to left). We number the carbon atoms from right to left (2 is smaller than 4).

This makes the base name 2-pentanol.

The methyl group must be bonded to C-4.

The complete name of this substance is 4-methyl-2-pentanol.

b.

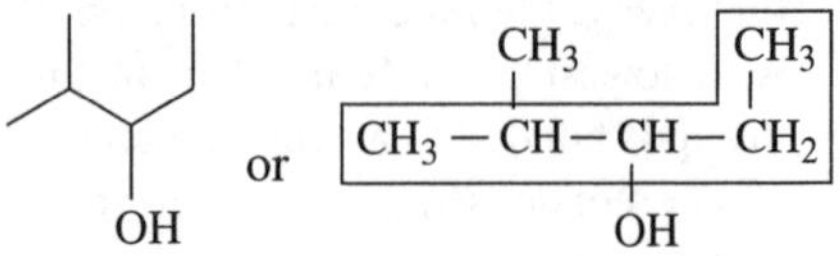

$$CH_3-\overset{CH_3}{CH}-\underset{OH}{CH}-\overset{CH_3}{CH_2}$$

The longest chain is outlined. It contains five carbon atoms.

The base name for this chain is pentanol.

Numbering from left to right, the OH is bonded to C-3 and the methyl group is bonded to C-2.

Numbering from right to left, the OH is still bonded to C-3, but now the methyl group is bonded to C-4.

When numbering from either direction, the OH group is bonded to C-3. However, counting from the left, the methyl group is bonded to C-2. This is lower than the C-4 we start numbering from the right.

The correct name of this substance is 2-methyl-3-pentanol.

c.

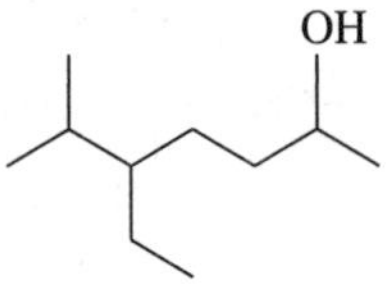

or

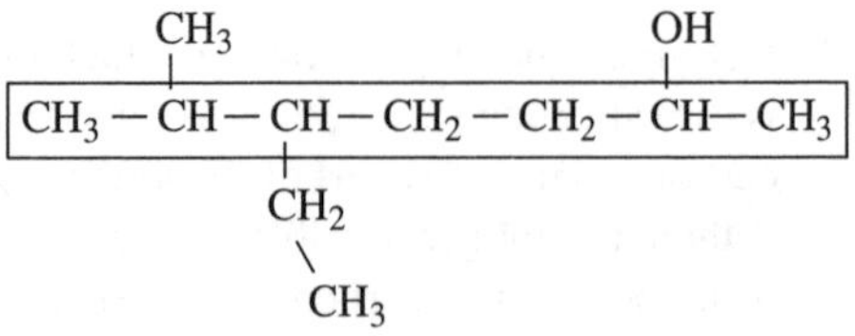

$$CH_3-\overset{CH_3}{CH}-\underset{CH_2-CH_3}{CH}-CH_2-CH_2-\overset{OH}{CH}-CH_3$$

The longest chain is outlined in the structure. It contains seven carbon atoms. The base name is heptanol.

The OH group is bonded to C-2 counting from right to left. (It bonded to C-6 counting from left to right.)

An ethyl group is bonded to C-5 and a methyl group is bonded to C-6.

The correct name of this substance is 5-ethyl-6-methyl-2-heptanol.

d.

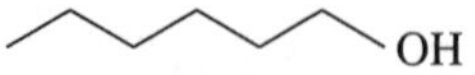

OH or

$CH_3CH_2CH_2CH_2CH_2CH_2OH$

The OH is bonded to an end carbon. There are no branches in this structure. This is a 6-carbon chain with the OH group bonded to C-1.

The correct name of this substance is 1-hexanol.

45. The rules for naming ethers are as follows:

1. Name each of the hydrocarbon chains bonded to the oxygen atom as substituent groups (change *ane* ending to *yl*).
2. Arrange these groups alphabetically placing a space between each.
3. Add the word ether.

a. Ethyl propyl ether

This substance consists of a propyl group ($CH_3CH_2CH_2$—) and an ethyl group (CH_3CH_2—).

The condensed structural formula is $CH_3CH_2CH_2-O-CH_2CH_3$.

The line structure is

O

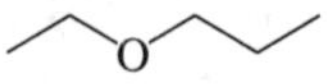

ethyl propyl ether

b. Diisopropyl ether

The prefix *di* tells us that there are two identical groups bonded to the oxygen atom of this ether.

In this case there are two isopropyl groups bonded to the oxygen atom.

An isopropyl group is

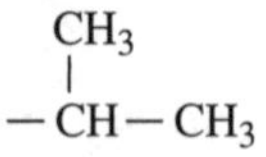

$$-\overset{CH_3}{CH}-CH_3$$

The condensed structural formula is

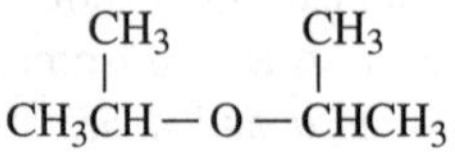

$$CH_3\overset{CH_3}{CH}-O-\overset{CH_3}{CH}CH_3$$

The line structure is

O

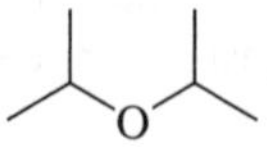

diisopropyl ether

c. Methyl cyclohexyl ether

The two groups bonded to the oxygen atom in this ether are a methyl group and a cyclohexyl group.

The condensed structural formula is

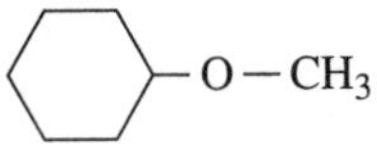

(cyclohexyl)$-O-CH_3$

The line structure is

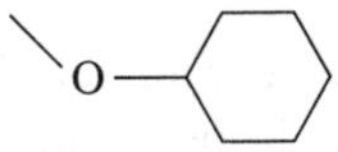

methyl cyclohexyl ether

47. An ether is a compound in which there is an oxygen atom bonded to two different carbon atoms (C−O−C).

a. 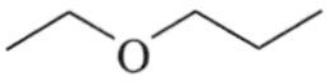 or

ethyl propyl ether

$$CH_3CH_2-O-CH_2CH_2CH_3$$

The alkyl groups bonded to the carbon atoms in this structure are an ethyl group (a 2-carbon chain) bonded to the oxygen atom on the left side and a propyl group (a 3-carbon chain) bonded to the oxygen atom on the right side.

Note: It does not matter which group is bonded on the left side of the oxygen atom and which is bonded on the right.

b. 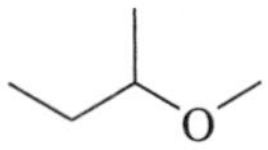 or

sec-butyl methyl ether

$$CH_3CH_2CH(CH_3)-O-CH_3$$

On the left-hand side of the oxygen atom we have a 4-carbon chain that is bonded to the oxygen atom at the second carbon. This group is called *sec*-butyl. On the right-hand side is a methyl group.

c. 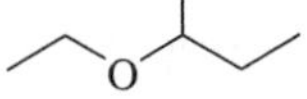 or

sec-butyl ethyl ether

$$CH_3CH_2-O-CH(CH_3)CH_2CH_3$$

CH_3CH_2-
ethyl

$-CH(CH_3)CH_2CH_3$
sec-butyl

49. In a carbonyl functional group, a carbon is double-bonded to an oxygen atom (C=O). Aldehydes and ketones both contain a carbonyl functional group.

In an aldehyde the carbonyl is at the end of the carbon chain (C-1).

In a ketone the carbonyl is on any carbon of the chain, other than an end carbon.

50. In fructose, the carbonyl is not found on a terminal carbon; that is, fructose contains a ketone functional group, which is reflected in the name ketohexose.

In glucose and mannose, the carbonyl is on a terminal carbon atom. These sugars are both aldehydes and are called an aldohexoses.

51. Yes, they can have the same molecular formula. They would be structural isomers because the position of the carbonyl group would be different.

52. A reaction between a carbonyl group on C-1 (of an aldohexose) and the OH bonded to C-3 would result in a 4-membered ring (3 carbon atoms and 1 oxygen atom). This is not a stable structure and could not exist for any length of time before uncoiling back into its linear structure.

Hemiacetal form of glucose

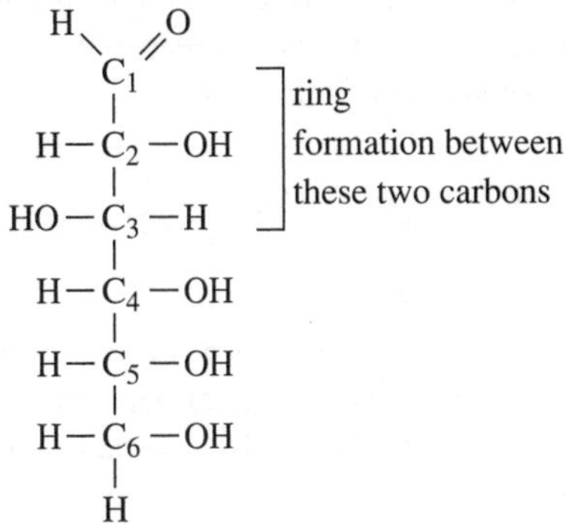

H−C_1=O
H−C_2−OH
HO−C_3−H
H−C_4−OH
H−C_5−OH
H−C_6−OH
H

ring formation between these two carbons

This can be pictured as follows

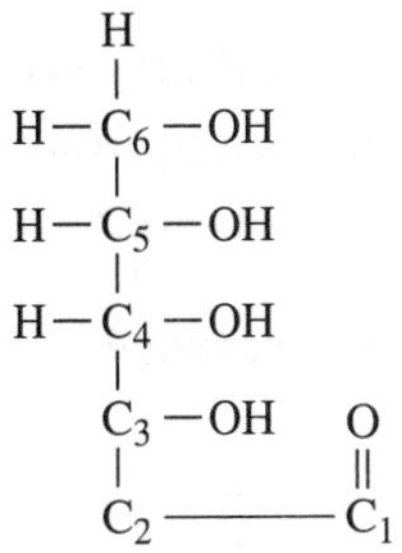

H
H−C_6−OH
H−C_5−OH
H−C_4−OH
C_3−OH   O
C_2 ——— C_1

Note: C's 4,5 and 6 have free rotation.

Closing the ring

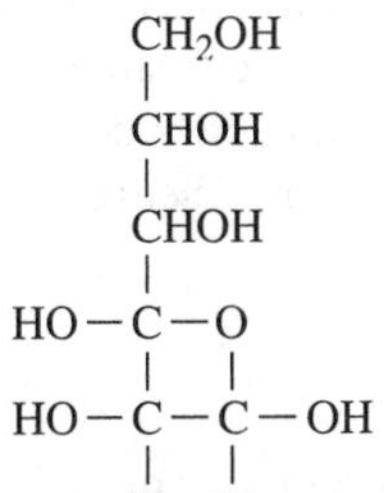

CH_2OH
CHOH
CHOH
HO−C−O
HO−C−C−OH
H H

A bond between C-1 and the OH bonded to C-5 yields a 6-membered ring (5 carbon atoms and 1 oxygen atom). This is a more stable structure than the 4-membered ring, so it is the one formed preferentially.

53. We are looking for substances that have a molecular formula of $C_5H_{10}O$. Any substance that has this molecular formula will be a structural isomer of pentanal ($C_5H_{10}O$).

a. or $CH_3CH(CH_3)CH_2CH(=O)$

$$\begin{array}{c} CH_3 \quad\ \ O \\ | \qquad\ \ \| \\ CH_3CHCH_2CH \end{array}$$

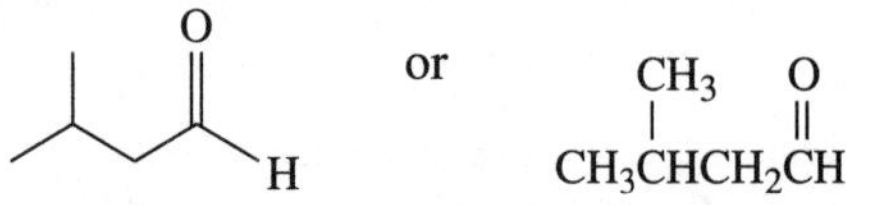

The molecular formula of this substance is $C_5H_{10}O$.

This has the molecular formula $C_5H_{10}O$. It is a structural isomer of pentanal.

The name of this substance is 3-methylbutanal.

b. or

$$\begin{array}{c} \qquad\quad O \\ \qquad\quad \| \\ CH_3CH_2CHCH \\ | \\ CH_3 \end{array}$$

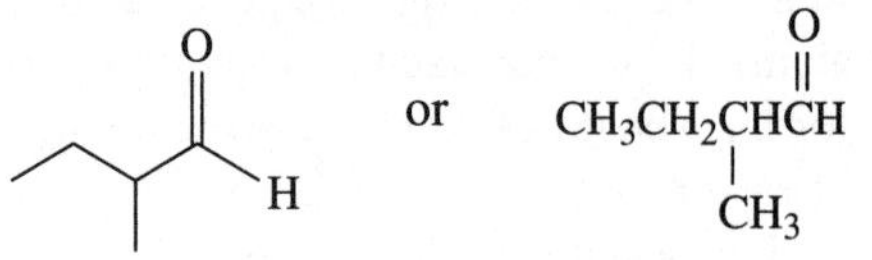

This substance has a molecular formula $C_5H_{10}O$. It is a structural isomer of pentanal.

The name of this substance is 2-methylbutanal.

c. or

$$\begin{array}{c} CH_3\ O \\ |\quad\ \| \\ CH_3-C-C-CH_3 \\ | \\ H \end{array}$$

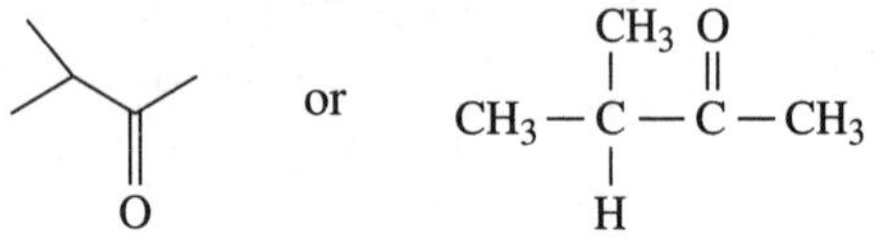

The molecular formula of this substance is $C_5H_{10}O$. It is a structural isomer of pentanal.

The name of this substance is 3-methyl-2-butanone.

d. or

$$\begin{array}{c} \qquad\qquad\quad O \\ \qquad\qquad\quad \| \\ CH_3CH_2CH_2CH_2CH \end{array}$$

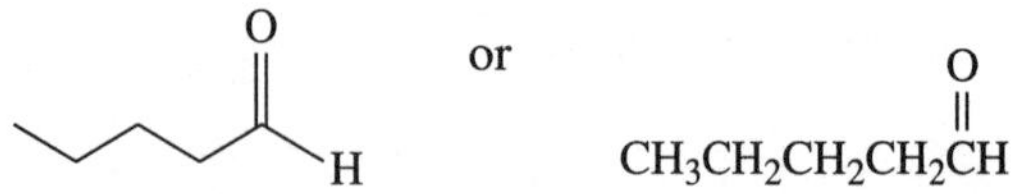

The molecular formula of this compound is $C_5H_{10}O$. It is not a structural isomer of pentanal because it is pentanal itself.

Structures *a*, *b*, and *c* are structural isomers of pentanal.

55. Aldehydes must contain a carbonyl ($\overset{O}{\overset{\|}{C}}$) functional group on the end carbon. This means that an aldehyde cannot be contained in a cyclic structure. Only the last structure (structure *e*) is an aldehyde.

Structure *a* is an ether (C—O—C).

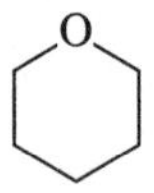

It has the molecular formula $C_5H_{10}O$.
Structure *b* is a ketone (cyclopentanone).

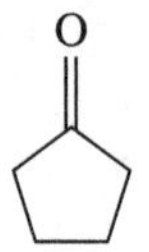

It has the molecular formula C_5H_8O.
Structure *c* is a ketone (2-hexanone).

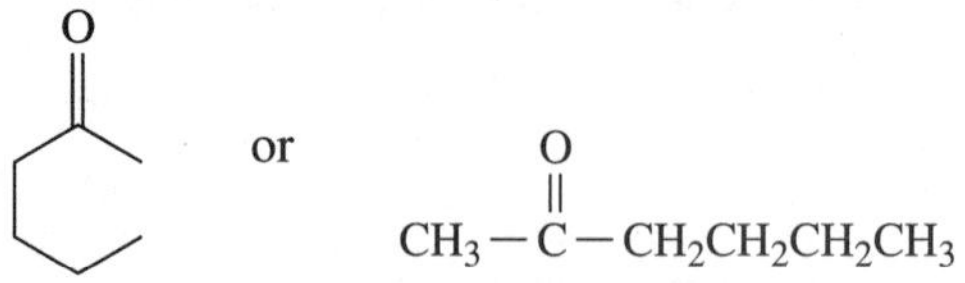

or

$$\begin{array}{c} O \\ \| \\ CH_3-C-CH_2CH_2CH_2CH_3 \end{array}$$

It has the molecular formula $C_6H_{12}O$.
Structure *d* is a ketone (4-methyl-2-pentanone).

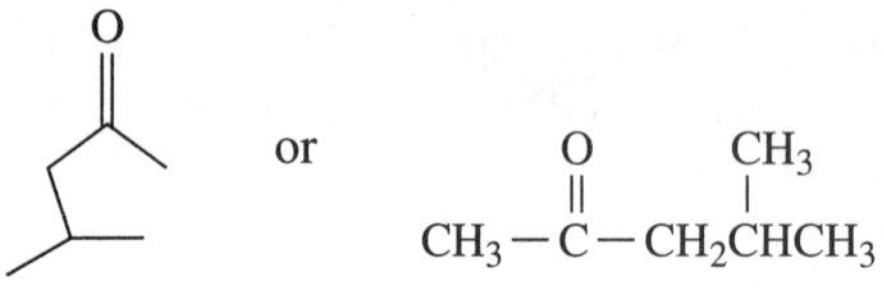

or

$$\begin{array}{c} O \qquad\quad CH_3 \\ \| \qquad\qquad | \\ CH_3-C-CH_2CHCH_3 \end{array}$$

It has the molecular formula $C_6H_{12}O$.
Structure *e* is an aldehyde (3-methylpentanal).

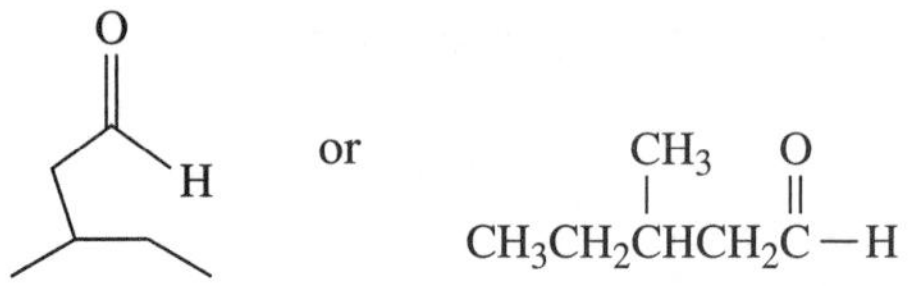

H or

$$\begin{array}{c} CH_3 \qquad O \\ | \qquad\quad \| \\ CH_3CH_2CHCH_2C-H \end{array}$$

It has the molecular formula $C_6H_{12}O$.

Structures *c*, *d*, and *e* are structural isomers of one another.

57. When a 6-carbon aldohexose forms a hemiacetal structure, the OH group bonded to C-5 interacts with the carbonyl carbon (C-1). The OH bonded to C-5 loses a hydrogen atom and the carbonyl oxygen at C-1 gains a hydrogen, so that the carbonyl group at C-1 becomes on OH group, and the OH group at C-5 becomes part of an ether linkage (C—O—C). In the hemiacetal form, C-1 and C-5 are linked by an oxygen atom to form a 6-membered ring consisting of 5 carbon atoms and an oxygen atom.

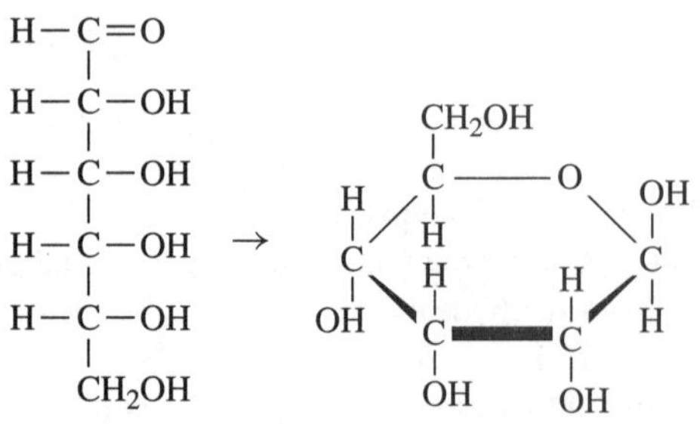

$$\begin{array}{l} H-C{=}O \\ \quad\ | \\ H-C-OH \\ \quad\ | \\ H-C-OH \\ \quad\ | \\ H-C-OH \\ \quad\ | \\ H-C-OH \\ \quad\ | \\ \quad CH_2OH \end{array} \rightarrow$$

Allose

Cyclic hemiacetal

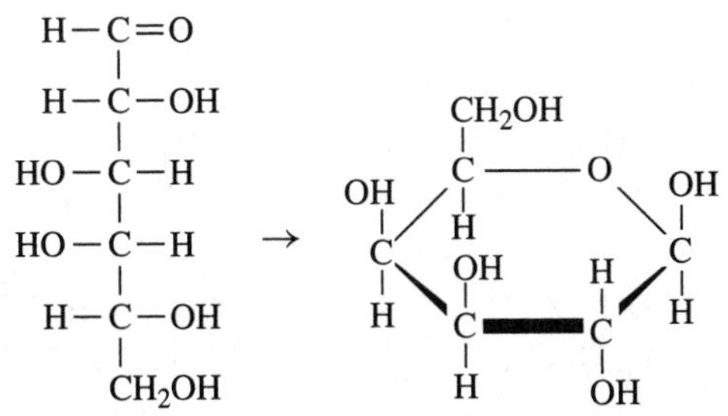

Galactose → Cyclic hemiacetal

59. The α and β forms of a monosaccharide differ only in the orientation of the OH group at C-1. In the alpha (α) form the OH group bonded to C-1 points down and is on the opposite side of the ring from the CH_2OH group bonded to C-5. In the β form, the OH at C-1 points up and is on the same side of the ring as the CH_2OH group.

a.

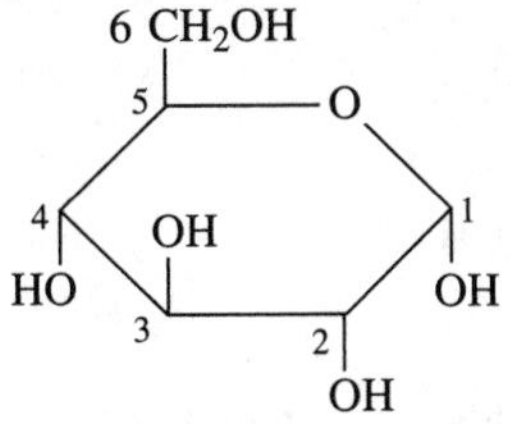

Structure *a* is the α form of a monosaccharide.

b.

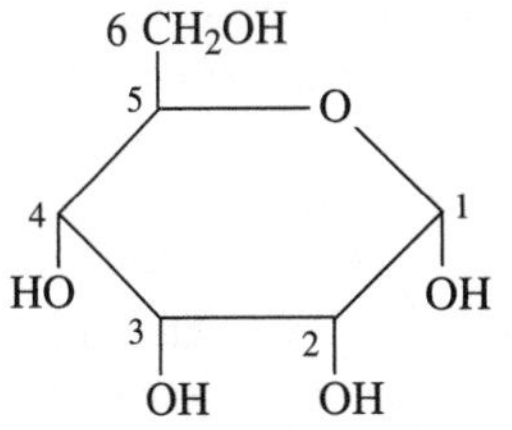

Structure *b* is an α monosaccharide.

c.

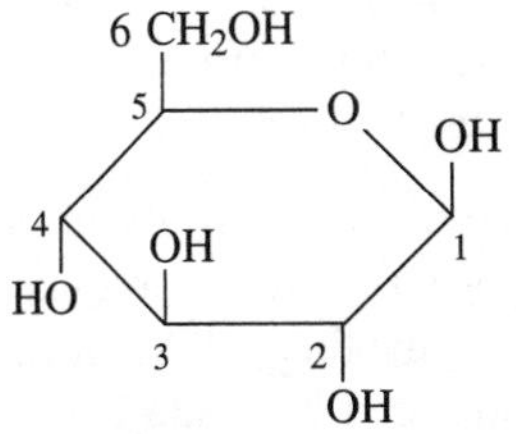

Structure *c* is the β form of a monosaccharide.

The structure *b* is a different monosaccharide from *a* or *c* because the OH group at C-3 points down instead of up. Structures *a* and *c* are α and β forms of the same monosaccharide.

61. Yes, the linear form is capable of forming an α or a β isomer in the cyclic form, and the structure of this linear substance is that of glucose.

63. Starch contains only α-glycosidic linkages, whereas cellulose has only β-glycosidic linkages. The structures are as follows:

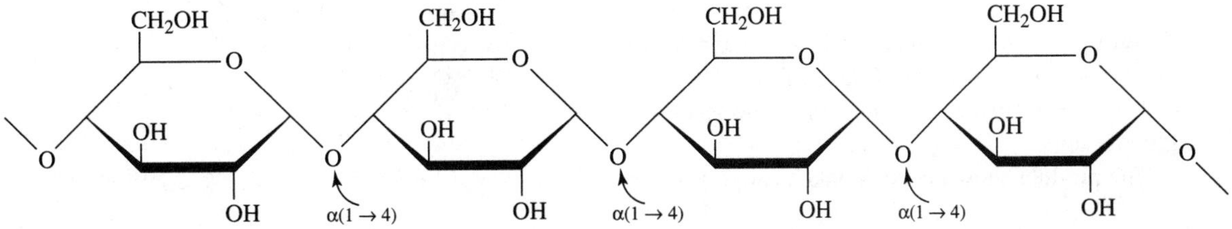

Starch (α-1,4 linkage)

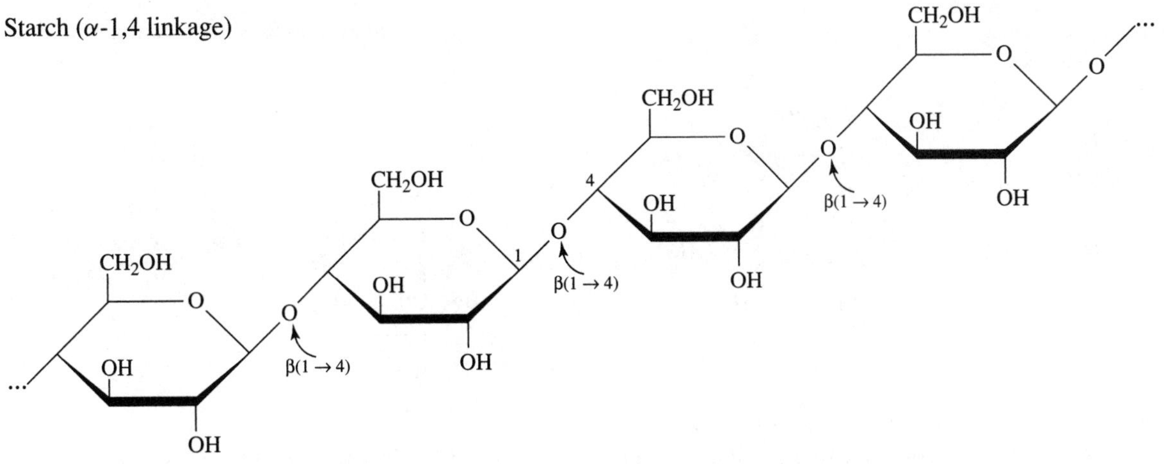

Cellulose (β-1,4 linkage)

64. The fuel value of the constituents of wood (cellulose, hemicellulose, and lignin) is lower than that of alkanes because all the components of wood contain numerous oxygen atoms. Oxygen adds weight to a substance but no fuel value. Therefore the energy produced per gram is lower for these oxygenated substances than for alkanes (which contain no oxygen in their structures).

65. Wood contains a significant amount of water. Some of the heat generated by the combustion process is lost in converting liquid water to steam (water vapor), thus reducing the fuel value of wood.

66. These enzymes can convert cellulose into sugars, which can then be fermented to produce ethanol. The goal is to find an enzyme that will be cheap to make so that sugar and ethanol can be made cheaply. Lowering the cost of ethanol would make it a more competitive source of energy compared to fossil fuels.

67. a. α-glucose + α-glucose $\rightarrow$

The structure of α-glucose is

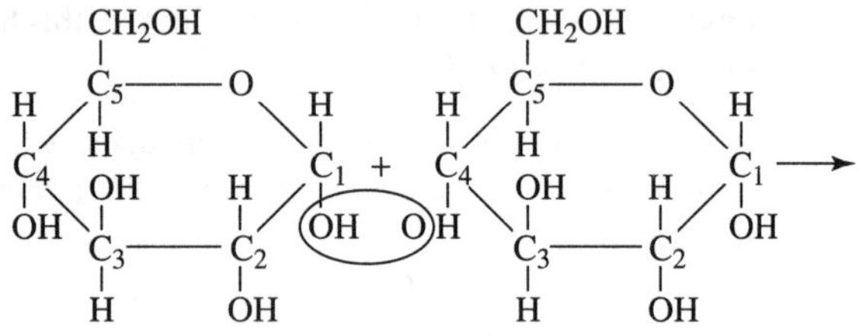

When two α-glucose units condense, water is produced and a new 1,4 C–O–C link is formed between C-1 on one α-glucose unit and C-4 on the second α-glucose unit.

The product shown below contains an α-1,4 link.

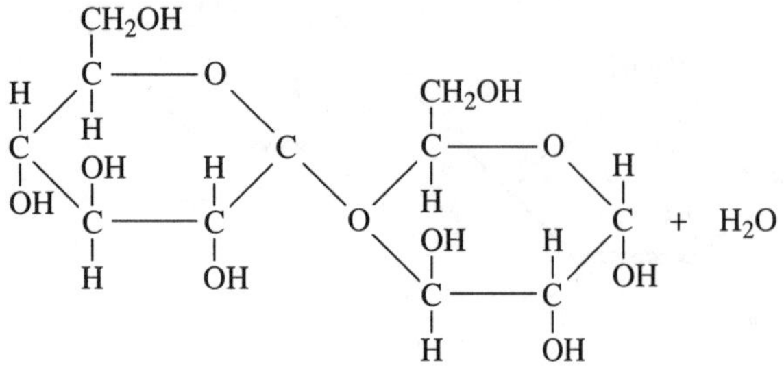

+ H_2O

b. α-glucose + β-glucose (assume a 1,4 link) $\rightarrow$

In this problem, there are four possible substances formed. They are shown below.

1. α-glucose + β-glucose

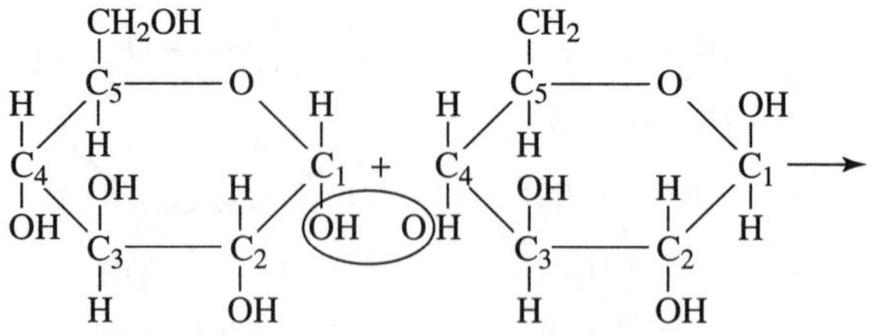

α-glucose β-glucose

The final product shown contains an α-1,4 link.

+ H_2O

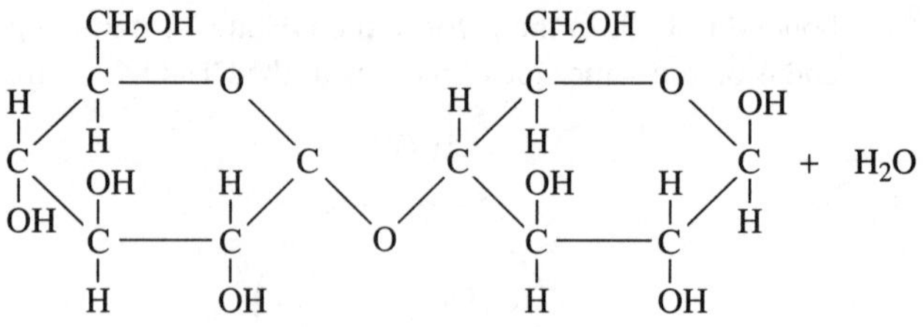

2. β-glucose + α-glucose

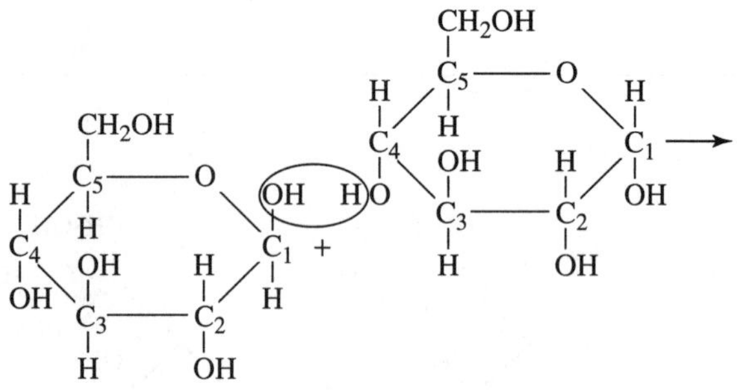

β-glucose α-glucose

The final product shown contains an β-1,4 link.

+ H_2O

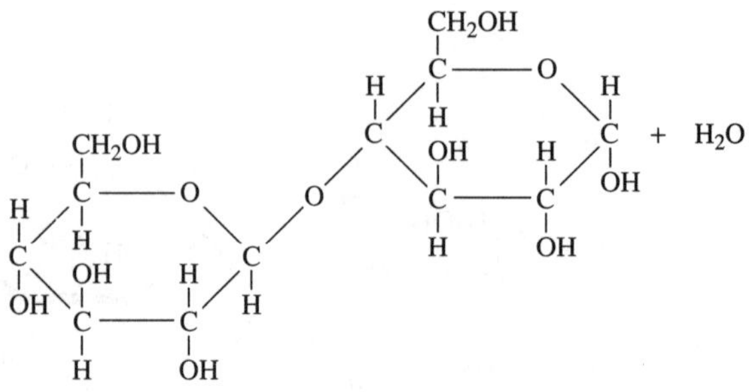

3. α-glucose + α-glucose

This structure is shown in part a of this question.

4. β-glucose + β-glucose

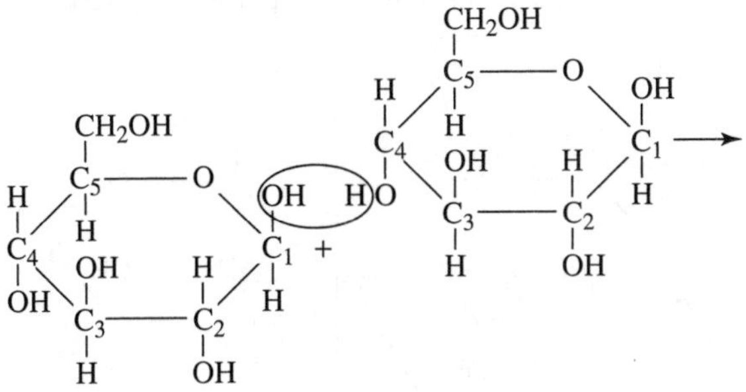

CH_2OH
H C_5 — O OH
CH_2OH H C_4 H C_1 →
C_5 — O OH H O OH H H
H H C_1 + C_3 — C_2
C_4 OH H H H OH
OH C_3 — C_2 H
H OH

β-glucose β-glucose

The final product shown contains an β-1,4 link.

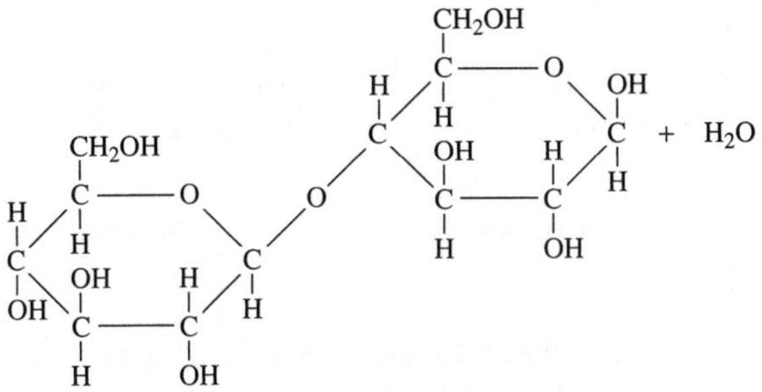

CH_2OH
H C — O OH
CH_2OH H C + H_2O
C — O O OH H H
H H C C — C
C OH H H H OH
OH C — C H
H OH

c. α-glucose + β-mannose (assume a 1,4 link) →

In this problem, there are four possible substances formed. They are shown below.

1. α-glucose + β-mannose

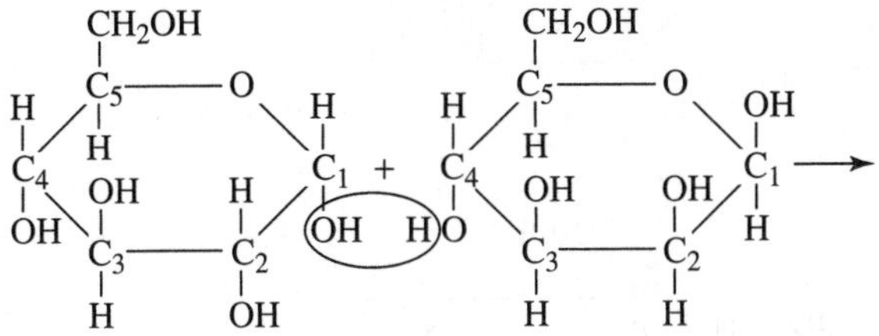

CH_2OH CH_2OH
H C_5 — O H H C_5 — O OH
H H C_1 + H C_1 →
C_4 OH H C_4 OH OH H
OH C_3 — C_2 OH H O C_3 — C_2
H OH H H

α-glucose β-mannose

The product contains an α-1,4 link.

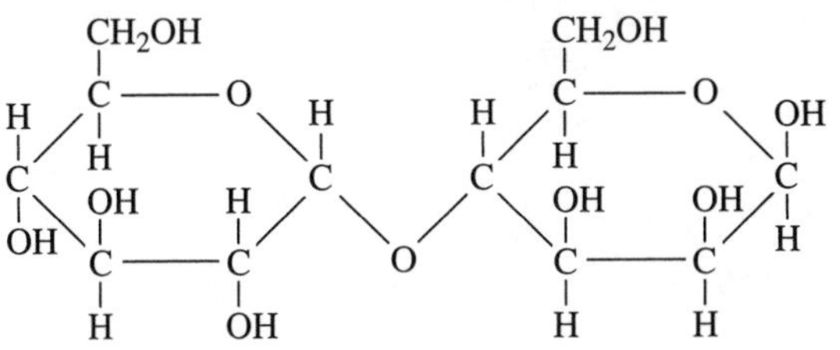

CH_2OH CH_2OH
H C — O H H C — O OH
H C H C
C OH H C OH OH H
OH C — C O C — C
H OH H H

2. β-mannose + α-glucose

CH_2OH
H C_5 — O H
CH_2OH H C_4 C_1 →
H C_5 — O OH H O OH H OH
H H C_1 + C_3 — C_2
C_4 OH OH H H OH
OH C_3 — C_2 H
H H

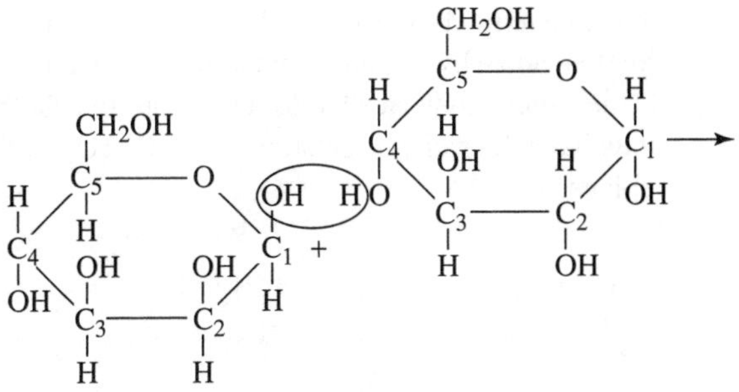

β-mannose α-glucose

The products contains a β-1,4 link.

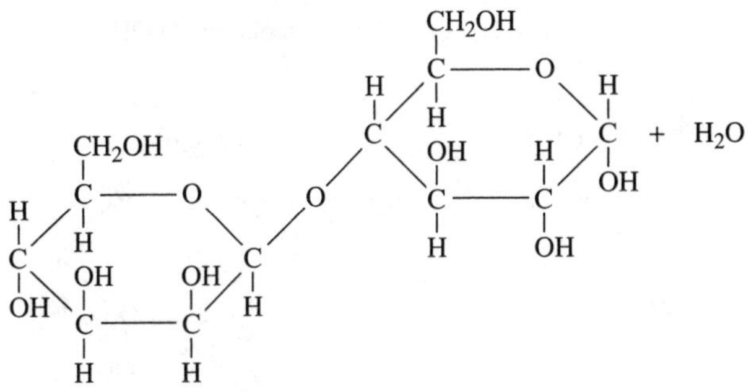

CH_2OH
H C — O H
CH_2OH H C + H_2O
H C — O O C OH H OH
H H C — C
C OH OH C H OH
OH C — C H
H H

3. α-glucose + α-glucose

This structure is shown in part a of this problem.

4. β-mannose + β-mannose

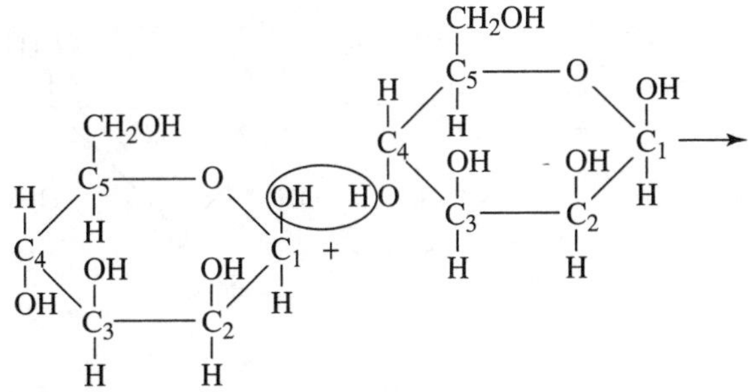

CH_2OH
H C_5 — O OH
CH_2OH H C_4 C_1 →
H C_5 — O OH H O OH OH H
H H C_1 + C_3 — C_2
C_4 OH OH H H H
OH C_3 — C_2 H
H H

β-mannose β-mannose

The product contains an −1,4 link.

CH_2OH
C — O OH
CH_2OH C H C + H_2O
H C — O O OH OH H
H C C — C
C OH OH H H H
OH C — C H
H H

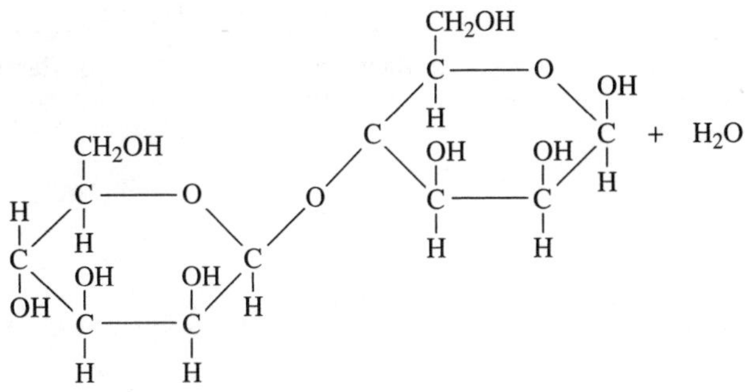

69. To determine the individual monosaccharides, break the disaccharide apart at the bridged oxygen atom. The oxygen atom remains with one of the monosaccharide units. We then add a hydrogen atom to the oxygen and an OH to the other monosaccharide unit. Note that both OH units have the same orientation as the C—O bond they replace.

a.

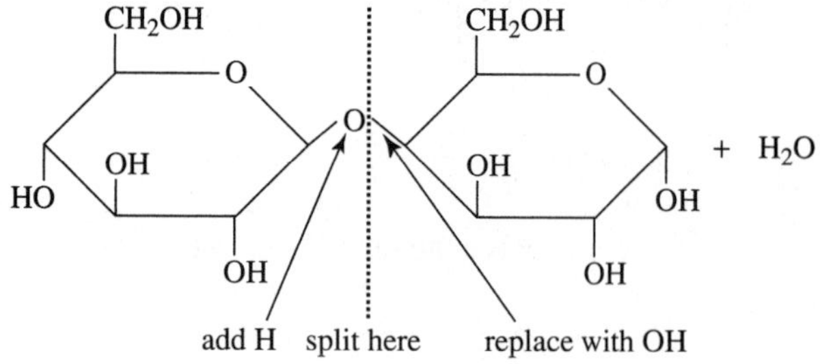

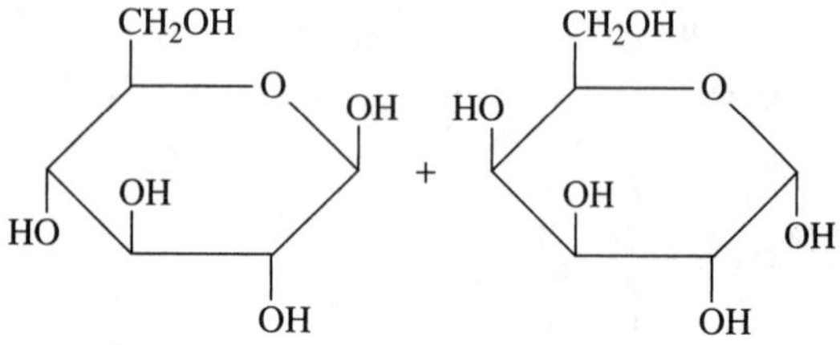

$+ H_2O$

add H split here replace with OH

β-Glucose + α-Galactose

Since the link to the oxygen from both units is up, the original OH groups on the individual monosaccharides must have been pointing up (β).

b.

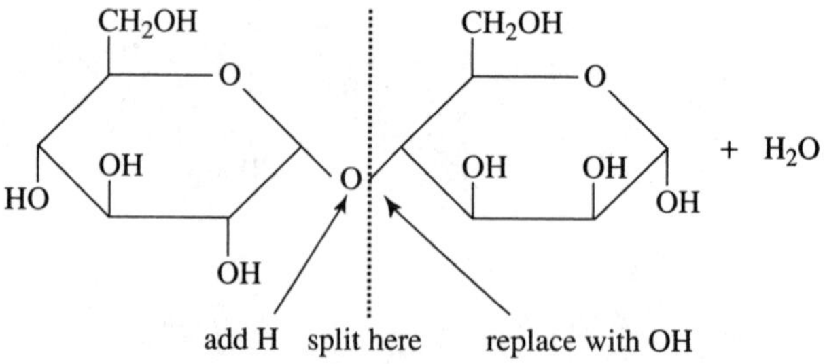

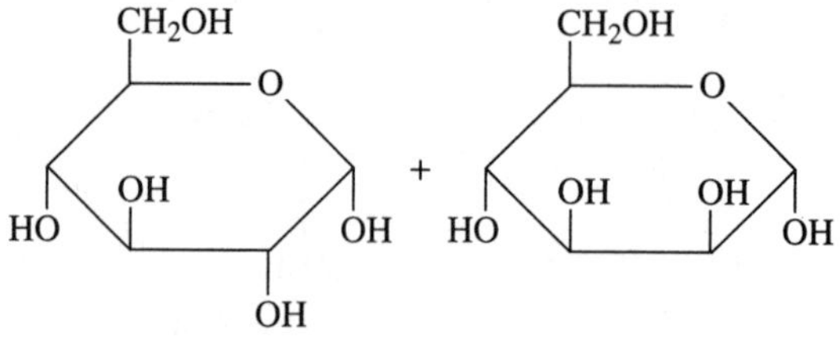

$+ H_2O$

add H split here replace with OH

α-Glucose + α-Mannose

Since the link to the oxygen from both units is down, the original OH groups on the individual monosaccharides must have been pointing down.

c.

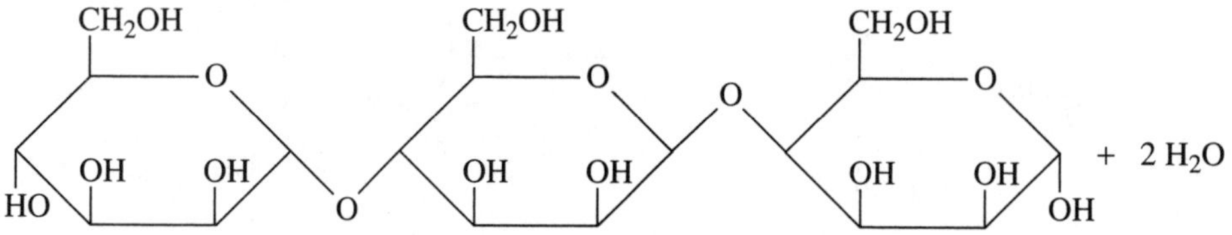

$+ 2\ H_2O$

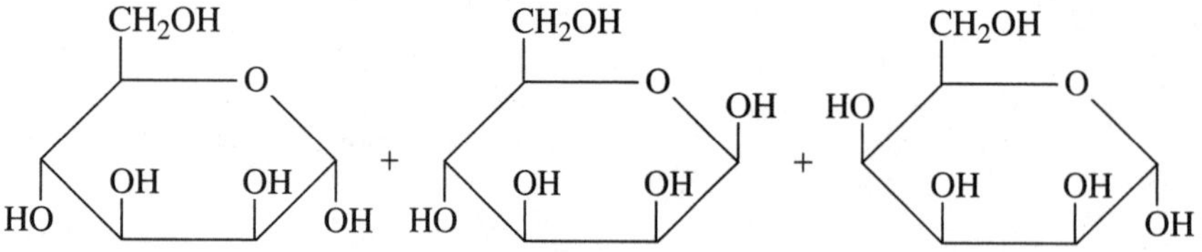

α-Mannose + β-Mannose + α-Talose

There are three monosaccharides in this substance. The linkage between the first two units is down for both. This means for the first monosaccharide the OH bonded to C-1 must have been down and the OH bonded to C-4 of the second monosacchardide is also down.

The linkage between the second monosaccharide is up for both. This means for the second monosaccharide the OH bonded to C-1 must have been up and the OH bonded to C-4 of the third monosacchardide is also up. All three structures are digestible by humans.

71. Anthracite has the greatest percentage of carbon and the smallest percentage of oxygen in its composition. This means that anthracite has the greatest fuel value of the four types of coal.

72. *Advantages*

    1. Complete combustion produces only water.
    2. Hydrogen has a higher fuel value than any other fuel.

    Disadvantages

    1. Low density results in low yields of energy per milliliter of liquid. This low heat-to-volume ratio is one factor limiting the use of liquid hydrogen for internal combustion engines.
    2. It has a very low boiling point. This makes it difficult and expensive both to liquefy and to store as a liquid.
    3. It is highly flammable.

73. The greater the carbon content and the lower the percentage of oxygen, the greater the fuel value. The order of fuel values is anthracite > bituminous > subbituminous > lignite.

74. Coal is not a compound. It is a mixture of solid carbon and impurities. Since coal is a mixture, it does not have a definite composition, so an empirical formula calculation from one sample of coal would be different from another sample. Determining an empirical formula is not possible because coal is not a pure substance.

75. Using Appendix 4 at the end of the text, we can calculate ΔH for each of these processes.

    Calculating the fuel value for reaction 1

$$H_2(g) + CO(g) + O_2(g) \rightarrow H_2O(g) + CO_2(g)$$

$$\Delta H^{\circ}_{rxn} = \left[\Delta H^{\circ}_{f,H_2O} + \Delta H^{\circ}_{f,CO_2}\right] - \left[\Delta H^{\circ}_{f,H_2} + \Delta H^{\circ}_{f,CO} + \Delta H^{\circ}_{f,O_2}\right]$$

From Appendix 4, the heats of formation (ΔH°_f) for reactants and products are as follows:

$H_2O(g) = -241.8$ kJ/mol
$CO(g) = -110.5$ kJ/mol
$CO_2(g) = -393.5$ kJ/mol
$H_2(g) = 0$ kJ/mol
$O_2(g) = 0$ kJ/mol

$$\begin{aligned}\Delta H^{\circ}_{rxn} &= [(-241.8\text{ kJ}) + (-393.5\text{ kJ})] - [(-110.5\text{ kJ}) \\ &\quad + (0.0\text{ kJ}) + (0.0\text{ kJ})] \\ &= [(-635.3\text{ kJ})] - [(-110.5\text{ kJ})] \\ &= -635.3\text{ kJ} + 110.\ 5\text{ kJ} \\ &= -524.8\text{ kJ}\end{aligned}$$

In this reaction, 1 mole of H_2 gas and 1 mole of CO gas react with excess O_2 to produce −524.8 kJ of heat energy.

We next determine how much heat energy is produced per gram of the combined reactants H_2 and CO.

1 mol $H_2 = 2$ g
1 mol CO = 28 g

There are a total of 30 g of reactants in the H_2/CO mixture. The mixture produces 524.8 kJ of heat energy.

To obtain a fuel value, divide the heat produced (524.8 kJ) by the 30 g of the mixture to be combusted.

$$\text{fuel value} = \frac{524.8\text{ kJ}}{30\text{ g}} = 17.5\text{ kJ/g}$$

Calculating the fuel value for reactions 2a and 2b:

$$2\,H_2(g) + CO(g) \rightarrow CH_3OH(l) \qquad (2a)$$

$$CH_3O(l) + \tfrac{3}{2}\,O_2(g) \rightarrow 2\,H_2O(g) + CO_2(g) \qquad (2b)$$

First, we can make use of Hess's law to combine reactions 2a and 2b to get a single reaction. Remember, both reactions and the heat associated with reactions are additive.

$$\begin{array}{c} 2\,H_2(g) + CO(g) \rightarrow CH_3OH(l) \\ \underline{CH_3O(l) + \tfrac{3}{2}\,O_2(g) \rightarrow 2\,H_2O(g) + CO_2(g)} \\ 2\,H_2(g) + CO(g) + CH_3OH(l) + \tfrac{3}{2}\,O_2(g) \\ \rightarrow CH_3OH(l) + 2\,H_2O(g) + CO_2(g)\end{array}$$

Since $CH_3OH(l)$ is present on both sides of the equation, it can be cancelled out.

The final equation is

$$2\,H_2(g) + CO(g) + \tfrac{3}{2}\,O_2(g) \rightarrow 2\,H_2O(g) + CO_2(g)$$

Now we can calculate ΔH for this reaction using data from Appendix 4. The ΔH°_f for reactants and products are as follows:

$H_2O(g) = -241.8$ kJ/mol
$CO(g) = -110.5$ kJ/mol
$CO_2(g) = -393.5$ kJ/mol
$H_2(g) = 0$ kJ/mol
$O_2(g) = 0$ kJ/mol

$$\Delta H^{\circ}_{rxn} = [2(-241.8\text{ kJ}) + (-393.5\text{ kJ})] - [(-110.5\text{ kJ}) + 2(0.0\text{ kJ}) + \tfrac{3}{2}(0.0\text{ kJ})]$$
$$= [(-483.6\text{ kJ} + (-393.5\text{ kJ})] - [(-110.5)]$$
$$= -877.1\text{ kJ} + 110.5\text{ kJ}$$
$$= -766.6\text{ kJ}$$

So 766.6 kJ of heat energy is released when 2 moles of H_2 reacts with 1 mole of CO and excess O_2.

1 mol H_2 = 2 g, so 2 mol H_2 = 4 g

1 mol CO = 28 g

1 mol of CO and 2 mol H_2 correspond to a total mass of 32 g of reactants (28 g CO and 4 g H_2) being combusted.

The fuel value for the reactants in this process is

$$\text{fuel value} = \frac{766.6\text{ kJ}}{32\text{ g}} = 24.0\text{ kJ/g}$$

This two-step process produces more heat energy per gram (24.0 kJ/g) than does the one-step process (17.5 kJ/g).

77. Carboxylic acids are more soluble in water than are aldehydes with the same number of carbon atoms because carboxylic acids have two possible sites (both oxygens of the carboxyl) for hydrogen bonds to form, while aldehydes have only one site (the carbonyl oxygen).

78. Methylamine (CH_3NH_2) contains a single carbon atom, whereas butylamine ($CH_3CH_2CH_2CH_2NH_2$) contains a 4-carbon hydrocarbon chain.

    Because its nonpolar portion is larger, the butylamine is much less polar than the methylamine molecule and we expect it to be less soluble in water.

    The more polar methylamine (CH_3NH_2) should be more soluble in water than the less polar butylamine ($CH_3CH_2CH_2CH_2NH_2$).

79. Ethane, CH_3CH_3, should have a higher fuel value than ethanol, CH_3CH_2OH, because the oxygen present in ethanol adds to the molar mass with no appreciable increase to the fuel value.

80. The molar mass of acetic acid (CH_3COOH) is 60 g/mol (32 g are due to the two oxygen atoms).

    The molar mass of ethanal (CH_3CHO) is 44 g/mol (16 g is due to the oxygen atom).

    The carbon and hydrogen content of both compounds is 28 g/mol. Since oxygen adds no significant fuel value to a substance, the extra oxygen content of acetic acid decreases its fuel value and ethanal has a higher fuel value than acetic acid.

81. Propanoic acid is a 3-carbon carboxylic acid. The propanoate ion is formed by removing the acidic hydrogen (the H bonded to the OH of the carboxyl) from propanoic acid.

    There are two resonance forms of both propanoic acid and the propanoate ion. The Lewis structures are as follows:

```
    H   H  :O:                  H   H  :Ö:⁻
    |   |   ||                  |   |   |
H — C — C — C — Ö — H   ↔   H — C — C — C = Ö — H
    |   |   |                   |   |   |      +
    H   H   H                   H   H   H
```

Propanoic acid

```
    H   H  :O:                  H   H  :Ö:⁻
    |   |   ||                  |   |   |
H — C — C — C — Ö:⁻     ↔   H — C — C — C = Ö
    |   |   |                   |   |   |
    H   H   H                   H   H   H
```

Propanoate ion

83. An amine is a nitrogen atom bonded to one, two, or three hydrocarbon groups, or an ammonia molecule with one, two, or three hydrogen atoms replaced with hydrocarbon chains.

    To name an amine:

    1. Identify the longest chain directly bonded to the nitrogen atom.
    2. Name this chain by changing the *ane* suffix to *yl* and adding the word amine (all one word). This is the base amine name.
    3. If any of the hydrocarbon chains are the same, use the prefixes *di* to indicate that two identical hydrocarbon chains are bonded to the nitrogen atom and *tri* to indicate three identical hydrocarbon chains bonded to the nitrogen. For example, two CH_3 groups would be dimethyl and three CH_3CH_2 groups would be triethyl.
    4. The complete name will include the name of each hydrocarbon chain bonded to the nitrogen atom with a prefix in front of it (if there are two or three identical chains). The last chain named should be the longest chain. Lastly there are no spaces between the names of the different chains.
    5. A more correct naming system (not described in the text) is to name each groups using the letter N (to indicate the hydrocarbon chain is bonded to the nitrogen atom) in front of each chain. If there are two identical groups use N,N. An N is not used for the longest chain.

a.

```
                    CH3
                     |
     N    or  CH3CH2 — N — CHCH3
                ↗      |      ↖
        ethyl group   CH3   isopropyl group
                       ↑
                  methyl group
```

The longest chain is the isopropyl group.

The name of this compound is ethylmethylisopropylamine or N-ethyl-N-methylisopropylamine.

b. or

$$\begin{array}{c} CH_3 \\ | \\ CH_3-N-CHCH_3 \\ | \\ CH_3 \end{array}$$

isopropyl group

The longest chain is an isopropyl group. In addition, there are two methyl groups bonded to the nitrogen atom.

The name of this compound is dimethylisopropylamine, or N,N-dimethylisopropylamine.

c. or

$$\begin{array}{c} CH_3 \quad H \\ | \quad\;\; | \\ CH_3CH_2CH-N-CH_3 \end{array}$$

sec-butyl group

The longest chain is a *sec*-butyl group. There is also a methyl group bonded to the nitrogen atom.

The name of this compound is methyl-*sec*-butylamine, or N-methyl-*sec*-butylamine.

d. or

$$\begin{array}{c} CH_3 \qquad H \\ | \qquad\quad | \\ CH_3CHCH_2-N-CH_3 \end{array}$$

← methyl group

This is an isobutyl group

The longest chain is an isobutyl group. In addition, there is a methyl group bonded to the nitrogen atom.

The name of this compound is methylisobutylamine, or N-methylisobutylamine.

85. Use the heat of formation data in Appendix 4 and the formula

$$\Delta H^o_{rxn} = \left[\sum n\Delta H^o_{f,products}\right] - \left[\sum m\Delta H^o_{f,reactants}\right]$$

The reaction is

$$4\ CH_3NH_2(g) + 2\ H_2O(l) \rightarrow 3\ CH_4(g) + CO_2(g) + 4\ NH_3(g)$$

From Appendix 5, the standard heats of formation (ΔH^o_f) are as follows:

$CH_3NH_2(g) = -23.0$ kJ/mol
$H_2O(l) = -285.8$ kJ/mol
$CH_4(g) = -74.8$ kJ/mol
$CO_2(g) = -393.5$ kJ/mol
$NH_3(g) = -46.1$ kJ/mol

$$\begin{aligned} \Delta H^o_{rxn} &= (3\Delta H^o_{f,CH_4} + \Delta H^o_{f,CO_2} + \Delta H^o_{f,NH_3}) \\ &\quad - (4\Delta H^o_{f,CH_3NH_2} + 2\Delta H^o_{f,H_2O}) \\ &= [3\text{ mol}(-74.8\text{ kJ/mol}) + 1\text{ mol}(-393.5\text{ kJ/mol}) \\ &\quad + 4\text{ mol}(-46.1\text{ kJ/mol})] - [4\text{ mol}(-23.0\text{ kJ/mol}) \\ &\quad + 2\text{ mol}(-285.8\text{ kJ/mol})] \\ &= [(-224.4\text{ kJ}) + (-393.5\text{ kJ}) + (-184.4\text{ kJ})] \\ &\quad -[(-92.0\text{ kJ}) + (-571.6\text{ kJ})] \\ &= [-802.3\text{ kJ}] - [-663.6\text{ kJ}] \\ &= -138.7\text{ kJ} \end{aligned}$$

87. Combustion analysis is based on the assumption that the combustion is complete. If an excess of oxygen is not used, the combustion could be incomplete and the results would not be reproducible.

88. First of all, the starting sample may not contain any oxygen. Secondly, if the sample does contain oxygen, there may not be enough oxygen present in the sample to convert all of the carbon into carbon dioxide and the hydrogen into water vapor (or liquid water). Lastly, lack of sufficient oxygen leads to incomplete combustion. This results in the formation of carbon monoxide instead of carbon dioxide.

89. A molecular formula tells us the actual number of atoms of each element in a substance. An empirical formula is the smallest whole number ratio of the atoms of each element in a substance. A molecular formula must be a whole number multiple of the empirical formula.

90. The two pieces of information needed to determine a molecular formula from the results of an elemental analysis of an organic compound are:

    1. The empirical formula, which can be calculated from the analysis
    2. The molar mass, which must be determined from a different experiment

91. This sample contains only the elements carbon and hydrogen. We need to determine how many moles of carbon and hydrogen are present in the sample. We can calculate how many moles are in the sample if we know the mass of CO_2 and H_2O produced when the sample is combusted.

    To convert mass of CO_2 in milligrams to moles of C:

    1. Convert milligrams of CO_2 to moles of CO_2 (divide by the molar mass, 44.0 g/mol). Note that we have to convert milligrams to grams first (1 g = 1000 mg).
    2. Convert moles of CO_2 to moles of C (from the formula we see that 1 mole CO_2 = 1 mole C).

Performing the calculation:

$$\text{moles of C} = 440\text{mg } CO_2 \times \frac{1\text{ g } CO_2}{1000\text{ mg } CO_2}$$

$$\times \frac{1\text{ mol } CO_2}{44.0\text{ g } CO_2} \times \frac{1\text{ mol C}}{1\text{ mol } CO_2} = 0.0100\text{ mol C}$$

To convert milligrams of H_2O to moles of H:

1. Convert milligrams of H_2O to moles of H_2O (divide by the molar mass, 18.0 g/mol). Note that we have to convert milligrams to grams first (1 g = 1000 mg).
2. Convert moles of H_2O to moles of H (from the formula we see that 1 mole H_2O = 2 moles H).

Performing the calculation:

$$\text{moles of H} = 135\text{ mg H}_2\text{O} \times \frac{1\text{ g H}_2\text{O}}{1000\text{ mg H}_2\text{O}}$$

$$\times \frac{1\text{ mol H}_2\text{O}}{18.0\text{ g H}_2\text{O}} \times \frac{2\text{ mol H}}{1\text{ mol H}_2\text{O}} = 0.0150\text{ mol H}$$

To determine the empirical formula, first write a formula using the number of moles calculated: $C_{0.0100}H_{0.0150}$.

Next divide each subscript by the smallest number (0.0100):

$$C_{\frac{0.0100}{0.0100}}H_{\frac{0.0150}{0.0100}} = C_{1.00}H_{1.50}$$

We need to multiply each subscript by 2 to get whole numbers.

$$C_{1.00\times2}H_{1.50\times2} = C_2H_3$$

The empirical formula of this substance is C_2H_3.

To find the molecular formula from the empirical formula (C_2H_3) and the molar mass, use the following the procedure:

1. Determine the molar mass of the empirical formula. In this case

$$2\text{ C} \times 12.0 = 24.0\text{ g}$$

$$3\text{ H} \times 1.0 = \underline{3.0\text{ g}}$$

$$27.0\text{ g/mol}$$

2. Divide the molar mass of the compound by the molar mass of the empirical formula.

$$\frac{\text{molar mass of compound}}{\text{molar mass of empirical formula}}$$

The problem gives the molar mass of the compound as 270 g/mol. We calculated the molar mass of the empirical formula to be 27.0 g/mol.

$$\frac{270\text{ g/mol}}{27.0\text{ g/mol}} = 10$$

Note: The molar mass of the compound must be a whole number multiple of the molar mass of the empirical formula. In this case it is 10.

3. Multiply each subscript in the empirical formula by the whole number calculated in step 2 (10).

$$C_{2\times10}H_{3\times10} = C_{20}H_{30}$$

The molecular formula of this substance is $C_{20}H_{30}$.

93. Information given:

The sample combusted contains the elements C, H, and O.
The mass of the sample is 192 mg (0.192 g).
The mass of CO_2 produced is 528 mg (0.528 g).
The mass of H_2O produced is 216 mg H_2O (0.216 g).

Convert 528 mg of CO_2 to moles of C:

$$\text{moles of C} = 0.528\text{ g CO}_2 \times \frac{1\text{ mol CO}_2}{44.0\text{ g CO}_2}$$

$$\times \frac{1\text{ mol C}}{1\text{ mol CO}_2} = 0.0120\text{ mol C}$$

Convert 216 mg of H_2O to moles of H:

$$\text{moles of H} = 0.216\text{ g H}_2\text{O} \times \frac{1\text{ mol H}_2\text{O}}{18.0\text{ g H}_2\text{O}}$$

$$\times \frac{2\text{ mol H}}{1\text{ mol H}_2\text{O}} = 0.0240\text{ mol H}$$

Next we need to determine the mass of oxygen in the sample. To do this convert moles of C and H to the mass of each element.

$$\text{mass of C} = 0.0120\text{ mol C} \times \frac{12.0\text{ g C}}{1\text{ mol C}} = 0.144\text{ g C}$$

$$\text{mass of hydrogen} = 0.0240\text{ mol H} \times \frac{1.008\text{ g H}}{1\text{ mol H}}$$

$$= 0.0242\text{ g H}$$

Now we can determine the mass of oxygen in the sample.

mass of sample = 0.192 g
mass of C = 0.144 g
mass of H = 0.0242 g

$$\text{mass of O} = \text{mass of sample} - \text{mass of C} - \text{mass of H}$$

$$= 0.192\text{ g} - 0.144\text{ g} - 0.0242\text{ g}$$

$$= 0.0238\text{ g}$$

The sample contains 0.0238 g of oxygen.

Next determine the number of moles of oxygen in the sample.

$$0.0238\text{ g O} \times \frac{1\text{ mol g O}}{16.0\text{ g O}} = 0.00149\text{ mol O}$$

Now we are able to determine the empirical formula of methylheptenone, using the moles calculated for each elements:

C = 0.0120 mol
H = 0.0240 mol
O = 0.00149 mol

$$C_{0.0120}H_{0.0240}O_{0.00149}$$

Next divide each subscript by the smallest number (0.00149).

$$C_{\frac{0.0120}{0.00149}}H_{\frac{0.0240}{0.00149}}O_{\frac{0.00149}{0.00149}} = C_{8.05}H_{16.1}O_1, \text{ or } C_8H_{16}O_1$$

Note that the carbon and hydrogen subscripts can be rounded to whole numbers.

The empirical formula for methylheptenone is $C_8H_{16}O$.

95. Structural isomers are substances that have the same molecular formula, but a different arrangement of atoms within the molecule.

Geometric isomers have an identical structure in terms of linkage (i.e., the order in which atoms are bonded to each other), but the relative orientation of atoms differs.

Geometric isomers can arise because of the lack of free rotation associated with a carbon–carbon double bond (C=C). When different groups (or atoms) are bonded to the carbon atoms of double bonds, two possible arrangements exist. These two structures are geometric isomers.

For example, let's look at 2-butene. The condensed structural formula is

$$CH_3CH = CHCH_3$$
$$\uparrow$$
double bond

Let's draw a more detailed structure, showing the approximate bond angles. Remember these structures are planar with respect to the carbon-carbon double bond, so this part of the molecule is flat.

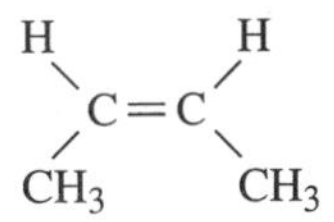

```
H         H          H         CH3
 \       /            \       /
  C = C                C = C
 /       \            /       \
CH3       CH3       CH3        H
```

Both H atoms are on one side of the double bond (*cis* isomer).

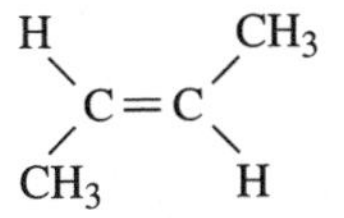

The H atoms are on opposite sides of the double bond (*trans* isomer).

96. All of the carbon-carbon bonds are single bonds so they are free to rotate relative to each other. Geometric isomers are not possible because the atoms do not have fixed orientations.

97. No. The general formula for both an alkene and a cycloalkane (a cyclic alkane) is C_nH_{2n}. Since they both have the same molecular formula, combustion analysis will yield the same results for both compounds. The percentage composition and the empirical formulas derived from the combustion analysis will be identical for both compounds.

98. No. The general formula for both an alkyne and a cycloalkene (cyclic alkene) is C_nH_{2n-2}. Since they both have the same general formula, the mass ratios of the elements are the same and the combustion analysis for both is identical.

99. Both groups bonded to C-1 are hydrogen atoms. A requirement for *cis-trans* isomerism is that each carbon that forms the double bond must be bonded to two different groups. For example, let's look at 1-butene. The condensed structural formula is $CH_2 = CHCH_2CH_3$.

A more detailed structure is

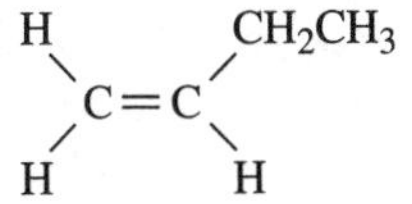

```
H         CH2CH3
 \       /
  C = C
 /       \
H         H
```

Both groups bonded to C-1 are hydrogen atoms. The atoms bonded to C-2 are a hydrogen atom and an ethyl (CH_3CH_2) group. 1-Butene cannot have geometric isomers because both carbon atoms of the double bond must have different groups to have *cis-trans* isomers.

100. Alkynes contain a carbon-carbon triple bond. This allows each carbon atom of the triple bond to form only one more bond (carbon always forms four bonds in stable substances).

In an alkyne the groups bonded to the carbons in the triple bond have a bond angle of 180 degrees (linear), so there is only one possible arrangement of atoms and no possibility for *cis-trans* isomerism.

An example is 2-butyne

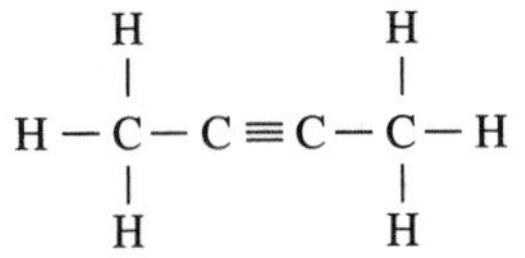

```
    H           H
    |           |
H - C - C ≡ C - C - H
    |           |
    H           H
```

101. a. Propene has a 3-carbon chain with a double bond.
Structural formula: $CH_2{=}CHCH_3$
The Lewis structure is

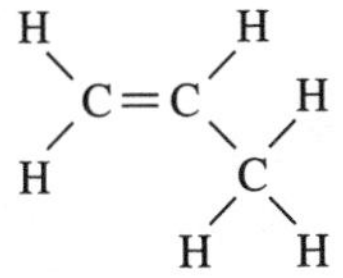

```
H       H
 \     /
  C = C    H
 /     \  /
H       C
       / \
      H   H
```

b. Propyne has a 3-carbon chain with a triple bond.
Structural formula: $CH_3 - C \equiv CH$
The Lewis structure is

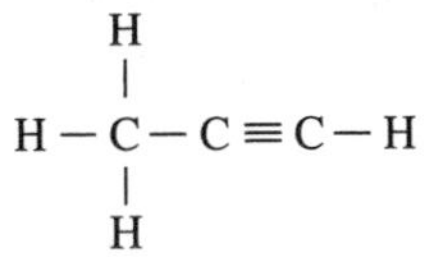

```
    H
    |
H - C - C ≡ C - H
    |
    H
```

103. a. *cis*-3-Heptene has a 7-carbon chain with a double bond between C-3 and C-4. The prefix *cis* means

that the H atoms bonded to C-3 and C-4 are on the same side of the double bond.

The condensed structural formula is

$$\begin{array}{ccccc} & H & & H & \\ & & C=C & & \\ & CH_2 & & CH_2CH_2CH_3 & \\ CH_3 & & & & \end{array}$$

b. *cis*-1,2-Dichloro-1-propene has a 3-carbon chain. The carbon–carbon double bond is between C-1 and C-2. C-1 and C-2 also have a single chlorine atom bonded to each of them. The prefix *cis* tells us that both chlorine atoms are on the same side of the double bond.

The condensed structural formula is

$$\begin{array}{ccc} Cl & & Cl \\ & C=C & \\ H & & CH_3 \end{array}$$

c. *trans*-4-Methyl-2-hexene has a 6-carbon chain. There is a double bond between C-2 and C-3 and a methyl group bonded to C-4. In the *trans* isomer the H atoms bonded to C-2 and C-3 are on opposite sides of the double bond.

The condensed structural formula is

$$\begin{array}{cccc} & & CH_3 & \\ & & | & \\ H & & CH & \\ & C=C & & CH_2CH_3 \\ CH_3 & & H & \end{array}$$

105. a. or $CH_3CH_2CH_2CH{=}CHCH_3$

This is a 6-carbon chain with a carbon–carbon double bond between C-2 and C-3. The geometry is *trans*; that is, the carbons forming the hexene chain lie on opposite sides of the double bond.

The name of this substance is *trans* 2-hexene.

b. or

$$\begin{array}{l} CH_3 \\ \quad CH-C\equiv C-CH_3 \\ CH_2CH_3 \end{array}$$

The longest chain contains six carbon atoms. There is a carbon–carbon triple bond between C-2 and C-3, and a methyl group bonded to C-4.

The name of this substance is 4-methyl-2-hexyne.

CHAPTER 13

Entropy and Free Energy and Fueling the Human Engine

1. The heat of solution for calcium chloride dissolved in water is quite exothermic. This heat makes it more effective than NaCl in melting ice on a sidewalk. When sodium chloride dissolves in water, its heat of solution is close to zero (that is, it is not noticeably exothermic or endothermic).

   Calcium chloride produces three ions per formula unit compared to only two ions per formula unit for sodium chloride. This extra ion makes the total particle molality (the number of ions produced per formula unit × solute molality) greater for calcium chloride. However, the nearly twice the molar mass of $CaCl_2$ (111 g/mol) when compared to NaCl (58.5 g/mol) results in NaCl being more effective the lower the freezing point on a per gram basis. (1 gram of NaCl lowers the freezing point of ice more than 1 g of $CaCl_2$.)

2. The heat of solution depends on:
   (1) The amount of energy required to overcome the attractive forces between solute only particles and solvent only particles. This process is endothermic.
   (2) The amount of energy released when solute and solvent particle interact. This process is exothermic.

3. The endothermic terms in Equation 13.1 are:

   $\Delta H_{\text{H bond}}$. This corresponds to the energy required to overcome the hydrogen bonds between water molecules. Since it requires energy, this process is endothermic and ΔH is positive.

   ΔH_{ionic}. This term corresponds to the energy required to break apart the ions in an ionic compound. This process requires energy so it is endothermic and ΔH_{ionic} is also positive.

4. The more positive the heat of solution, the lower the solubility of an ionic compound in water. In general, a process with a positive (endothermic) heat term is unlikely to be spontaneous.

5. If more energy is released when ions and water dipoles interact than the sum of the energies required to separate the ions in an ionic compound and to overcome the hydrogen bonding between water molecules, then it increases the probability that the heat of solution will be exothermic.

6. $$\Delta H_{\text{hydration}} = \Delta H_{\text{H bonds}} + \Delta H_{\text{ion-dipole}}$$

   The more positive the charge on the ion, the greater the amount of energy released by the ion–dipole attraction between the ion and the water molecules. We would expect the ion–dipole interaction to be more exothermic for Al^{3+} than for Mg^{2+}, so the heat of hydration for Al^{3+} ions will be more exothermic (negative) than the heat of hydration for Mg^{2+} ions.

7. The chloride ions (Cl^-) are significantly smaller than nitrate ions (NO_3^-). The smaller the ion size, the closer the ions can get to the water molecules and the larger the ion–dipole interaction. This results in the chloride ions having a more exothermic heat of hydration. The heat of hydration of NO_3^- ions should be less exothermic than the heat of hydration of Cl^- ions.

8. The larger the ion charge, the greater the attraction between the ion and the water dipole and the more exothermic the process. The smaller the ion size, the closer the ions can get to the water molecules. This again increases the attraction between the ion and the water dipole and results in a more exothermic process. Both higher absolute ion charge and smaller ion size lead to a more exothermic heat of hydration.

9. For NH_4NO_3, $\Delta H_{soln} = +26$ kJ/mol

   We can determine how much heat is absorbed or released when each sample is dissolved in water by multiplying the heat of solution times the number of moles of the substance dissolved.

$$\text{heat} = 0.30\,\text{mole } NH_4NO_3 \times \frac{+26\,\text{kJ}}{\text{mol}} = +7.8\,\text{kJ}$$

When 0.30 mol of NH_4NO_3 dissolves, 7.8 kJ of heat are absorbed from the solution, making the solution colder.

For $MgSO_4$, $\Delta H_{soln} = -91$ kJ/mol

$$\text{heat} = 0.10\,\text{mol } M_gSO_4 \times \frac{-91\,\text{kJ}}{\text{mol}} = -9.1\,\text{kJ}$$

When 0.10 mol of $MgSO_4$ dissolves, 9.1 kJ of heat are released to the solution, making the solution hotter.

The total heat change when the two processes are combined is

$$+7.8\,\text{kJ} + (-9.1\,\text{kJ}) = -1.3\,\text{kJ}$$

The temperature increases because the overall process releases 1.3 kJ of heat (that is, it is very slightly exothermic).

11. The temperature change of any solution can be calculated using the following formula:

$$\text{heat} = \text{mole} \times \text{molar heat capacity} \times \text{change in temperature}\ (q = nC\Delta T)$$

In this problem, we are given the mass of solution and we are told to assume that the molar heat capacity of the solution is the same as that of water.

Mass of solution = 1 kg of water + 16 g of $MgSO_4$ = 1016 g
$\Delta H_{solution}$ for $MgSO_4 = -91$ kJ/mol
Molar heat capacity of water = 75.3 J/mol · °C

To determine the heat released (the process is exothermic since the heat of solution is negative), convert the mass of $MgSO_4$ to moles and then multiply the number of moles of $MgSO_4$ by the heat of solution.

$$\text{moles of } MgSO_4 = 16\,\text{g } MgSO_4 \times \frac{1\,\text{mol } MgSO_4}{120.4\,\text{g } MgSO_4} = 0.133\,\text{mol } MgSO_4$$

$$\text{heat released} = 0.133\,\text{mol } MgSO_4 \times \frac{-91\,\text{kJ}}{\text{mol}} = -12.1\,\text{kJ}$$

Dissolving 16 g of $MgSO_4$ will release 12.1 kJ of heat, which can be used to heat the solution.

We know the amount of heat (q) generated in this process. All we need to do is to solve for ΔT. Since the molar heat capacity has units of J/mol · °C while the calculated heat is in kJ, we need to convert −12.1 kJ into −12,100 J (1 kJ = 1000 J). Since the dissolving process is exothermic, the solution absorbs heat energy and the temperature of the solution must increase. Since we are looking at the temperature change of the surroundings, the sign of the energy change must be positive. We need to change the sign of q (from negative to positive) when we do the following calculations:

$$q = nC\Delta T \qquad \text{so} \qquad \Delta T = \frac{q}{nC}$$

However, before we can do this, we need to determine the number of moles of water in the solution.

$$\text{moles of } H_2O = 1000\,\text{g } H_2O \times \frac{1\,\text{mol } H_2O}{18.0\,\text{g } H_2O} = 55.6\,\text{mol } H_2O$$

Since the solution also contains some dissolved $MgSO_4$, we need to add the number of moles $MgSO_4$ to the number of moles of water to get the total number of moles in the solution.

$$\text{Total moles} = 55.6 + 0.133 = 55.733$$

Substituting in the second equation, we get

$$\Delta T = \frac{q}{nC} = \frac{12{,}100\,\text{J}}{(55.733\,\text{mol})(75.3\,\text{J/mol}\cdot{}^\circ\text{C})} = 2.88^\circ\text{C}$$

The temperature of the solution increases by 2.88°C, so the temperature increases from 20.0°C to 22.88°C.

To the correct number of significant figures, the final temperature should be 22.9°C.

13. From the information given, we can calculate the amount of heat generated when 10.00 g of sodium acetate dissolves in 1.000 L of water (1000 mL = 1000 g). Information given:

   The molar heat capacity of the solution is 75.3 J/mol · °C.

   The initial temperature of the water is 20.00°C.

   The final temperature of the solution is 19.6°C.

   10.00 grams of sodium acetate (CH_3CO_2Na) are dissolved in the water.

The amount of heat associated with this process can be calculated using:

$$q = nC\Delta T$$

n = moles of water + moles of sodium acetate
C = 75.3 J/mol · °C
$\Delta T = 19.60^\circ\text{C} - 20.00^\circ\text{C} = -0.40^\circ\text{C}$

Since the temperature of the solution decreases, the heat associated with the process must be positive (endothermic).

We need to change the sign on q in order to have the correct sign for the heat associated with the dissolving process.

If we know the total moles in solution, we can solve for q.

Next, let's calculate the number of moles of both the water and the sodium acetate present in the solution.

$$\text{moles of } CH_3CO_2Na = 10.00\,g\ CH_3CO_2Na \times \frac{1 mol\ CH_3CO_2Na}{82.0\,g\ CH_3CO_2Na} = 0.122\,mol\ CH_3CO_2Na$$

$$\text{moles of } H_2O = 1000\,g\ H_2O \times \frac{1 mol\ H_2O}{18.0\,g\ H_2O} = 55.6\,mol\ H_2O$$

Total number of moles in the solution $= 55.6 + 0.122 = 55.722$

Now we are ready to calculate the heat absorbed by this process. $q = nC\Delta T = (55.722\,mol)(75.3\,J/mol \cdot °C)(0.40°C) = 1678\,J$

The system absorbs 1700 J (the answer should be reported to 2 significant figures) of heat energy from the water (thereby decreasing the temperature of the water).

The problem asks us to find the molar heat of solution of sodium acetate ($NaCH_3CO_2$). We can determine this quantity by dividing the heat generated by this process (1700 J) by the moles of sodium acetate dissolved—we have already calculated the number of moles of sodium acetate dissolved. Performing this calculation yields:

$$\text{Heat of solution of sodium acetate} = \frac{1700\,J}{0.122\,mol} = 13{,}934\,J/mol \text{ or } 14\,kJ/mol \text{ (answer rounded to 2 significant figures)}$$

15. Only the relationship described in part a is true.

We can use the equation $\Delta H_{solution} = \Delta H_{H\ Bond} + \Delta H_{ionic} + \Delta H_{ion\text{-}dipole}$ to determine the numerical value (or in this question the sign) of $\Delta H_{solution}$.

$\Delta H_{H\text{-}Bond}$ is the energy required to overcome the hydrogen bonds between water molecules. It must be a positive quantity (endothermic).

ΔH_{ionic} is related to the lattice energy. The lattice energy is the amount of energy released when positive and negative ions combine to form an ionic solid. This quantity is negative (exothermic). ΔH_{ionic} is the energy required to separate an ionic compound into its individual ions. Chemically, it is the reverse reaction when compared to the reaction for the lattice energy. It corresponds to $\Delta H_{lattice} = -U$ (lattice energy). It also must be a positive quantity (endothermic).

$\Delta H_{ion\text{-}dipole}$ is the energy released when an ion interacts with the dipole (of water molecules in this case). It is an exothermic process (negative quantity).

a. $\Delta H_{H\ Bond} + \Delta H_{ionic} > \Delta H_{ion\text{-}dipole}$

We can see that if $\Delta H_{H\ Bond} + \Delta H_{ionic} > \Delta H_{ion\text{-}dipole}$, then the sum of the three terms must be greater than zero, resulting in a positive value for $\Delta H_{solution}$.

b. $\Delta H_{H\ Bond} + \Delta H_{ionic} < \Delta H_{ion\text{-}dipole}$

In this example the sum of the two positive terms ($\Delta H_{H\ Bond}$ and ΔH_{ionic}) is less than the value of the negative term. $\Delta H_{solution}$ must be a negative quantity.

c. $\Delta H_{H\ Bond} + \Delta H_{ionic} = \Delta H_{ion\text{-}dipole}$

In this example, the sum of the three quantities must equal zero.

17. The lattice energy can be calculated using the equation

$$\Delta H_{solution} = -U + \Delta H_{hydration}$$

Solving for the lattice energy, we get

$$U = \Delta H_{hydration} - \Delta H_{solution}$$

For LiCl

$\Delta H_{solution} = +17\,kJ/mol \quad \Delta H_{hydration} = -823\,kJ/mol$

$$U = -823\,kJ/mol - 17\,kJ/mol$$

$$U = -840\,kJ/mol$$

For NaCl

$\Delta H_{solution} = +47\,kJ/mol \quad \Delta H_{hydration} = -723\,kJ/mol$

$$U = -723\,kJ/mol - 47\,kJ/mol$$

$$U = -770\,kJ/mol$$

For KCl

$\Delta H_{solution} = +52\,kJ/mol \quad \Delta H_{hydration} = -649\,kJ/mol$

$$U = -649\,kJ/mol - 52\,kJ/mol$$

$$U = -701\,kJ/mol$$

19. The enthalpy of solution ($\Delta H_{solution}$) can be calculated using the equation

$$\Delta H_{solution} = \Delta H_{hydration} - U$$

We need to determine the heat of solution, $\Delta H_{solution}$.

For NaF

$\Delta H_{hydration} = -841\,kJ/mol \qquad U = -914\,kJ/mol$

$$\Delta H_{solution} = -841\,kJ/mol - (-914\,kJ/mol)$$

$$\Delta H_{solution} = +73\,kJ/mol$$

For NaCl

$\Delta H_{hydration} = -723\,kJ/mole \qquad U = -770\,kJ/mol$

$$\Delta H_{solution} = -723\,kJ/mol - (-770\,kJ/mol)$$

$$\Delta H_{solution} = +47\,kJ/mol$$

For NaBr

$$\Delta H_{\text{hydration}} = -694\,\text{kJ/mol} \qquad U = -728\,\text{kJ/mol}$$

$$\Delta H_{\text{solution}} = -694\,\text{kJ/mol} - (-728\,\text{kJ/mol})$$

$$\Delta H_{\text{solution}} = +34\,\text{kJ/mol}$$

Note: The values for $\Delta H_{\text{solution}}$ calculated in this last part of question 19 do not agree with the value given in problem 18.

21. The entropy of a group of smaller molecules is greater than the entropy of the larger molecule formed when the smaller ones combine. Therefore, the entropy change in this process is negative; that is, the entropy decreases.

22. A gas has significantly higher absolute molar entropy when compared to the liquid or solid phases of the same substance. This is due to the lack of order associated with the gaseous state.

23. Dissolving a gas in a liquid results in a decrease in entropy.
    A substance in the gaseous state is very disordered (high entropy). When it dissolves in a liquid it becomes much more ordered. This is due to the decrease in volume to move in and so less freedom to move in the dissolved state than in the pure gaseous state.

24. If we consider the system to be the Drano and the surroundings to be the water, then:

    ΔS_{system} is positive. (Dissolving a solid in a liquid results in the solid having higher entropy).
    $\Delta S_{\text{surroundings}}$ is positive. (The process is exothermic, so heat is transferred to the surroundings, thus increasing the entropy).

25. If we consider the system to be the ice cubes and the surroundings to be the lemonade, then:

    ΔS_{system} is positive. The ice cubes must absorb heat energy to melt. Alternatively, the change from solid to liquid results in the liquid water having higher entropy than the ice cubes.
    $\Delta S_{\text{surroundings}}$ is negative. The surroundings supply heat energy to the ice, thus decreasing the temperature of the surroundings and lowering the entropy.

26. Fullerenes are expected to have a higher absolute molar entropy than diamond.
    Fullerenes are discrete molecules that contain upwards of 60 carbon atoms arranged in the shape of a geodesic dome. This structure results in many more molecular motions than in diamond. The bonding in diamond is classified as network covalent. That is, there is an extended structure containing each carbon atom being bonded to four other carbon atoms. This structure contains many more than 60 carbon atoms. However, the atoms in the diamond structure are held rigidly in place resulting in diamond having a very small absolute molar entropy.

27. The superfluid state is more ordered than the ordinary liquid state. The entropy decreases in changing from ordinary ^{3}He to the superfluid form of ^{3}He.

    ^{3}He (ordinary liquid) → ^{3}He (superfluid liquid)
    less ordered (higher entropy) → more ordered (lower entropy)

28. Reversing a process changes the sign of the entropy change (ΔS).

29. The sulfur atom being slightly larger than an oxygen atom results in $H_2S(g)$ having a slightly larger absolute molar entropy than $H_2O(g)$. The greater standard molar entropy for H_2S is due to the greater number of fundamental particle (protons, neutrons, and electrons) in the sulfur atom.
    Dipole moments in the gas phase should have no effect on the absolute molar entropy because the average kinetic energies of the molecules are much greater than the forces of attractions between molecules. That is, gas molecules do not tend to stick together.

30. S_8 contains eight atoms per molecule. This allows more internal motions leading to a greater absolute molar entropy as compared to the S_2 molecule.

31. Choices *a* and *c* are possible, choice *b* is not possible.
    If a process is possible, that means that the signs for ΔS_{system} and $\Delta S_{\text{surroundings}}$ could result in the given sign for $\Delta S_{\text{universe}}$.

    $$\Delta S_{\text{universe}} = \Delta S_{\text{system}} + \Delta S_{\text{surroundings}}$$

    Keep in mind the above equation when determining if the combinations listed in the problem are possible.
    a. If $\Delta S_{\text{surroundings}}$ is more positive than ΔS_{system} is negative, then $\Delta S_{\text{universe}}$ has to be a positive quantity. This process would be spontaneous.
    b. If ΔS_{system} and $\Delta S_{\text{surroundings}}$ are both negative, then $\Delta S_{\text{universe}}$ has to be a negative quantity. This combination cannot give a positive value for $\Delta S_{\text{universe}}$.
    c. If ΔS_{system} is more negative than $\Delta S_{\text{surroundings}}$ is positive, then $\Delta S_{\text{universe}}$ must be a negative quantity.

33. The entropy of many substances increases with increasing complexity of the substances. We can use the molar masses of similar substances to determine complexity. The greater the molar mass, the higher the absolute molar entropy. Another method to determine complexity is to determine the number of atoms in one molecule of the substance. The more atoms in a molecule, the greater the complexity and the higher the absolute molar

entropy. Lastly, we need to remember that solids tend to have lower entropy than liquids, and liquids have considerably lower entropy than gases.

a. $CH_4(g) < CF_4(g) < CCl_4(g)$

The molar masses are

$CH_4 = 16$ g/mol
$CF_4 = 88$ g/mol
$CCl_4 = 154$ g/mol

Each substance contains five atoms per molecule.

b. $CH_3OH(l) < C_2H_5OH(l) < C_3H_7OH(l)$

The molar masses are

$CH_3OH = 32$ g/mol (6 atoms)
$C_2H_5OH = 46$ g/mol (9 atoms)
$C_3H_7OH = 60$ g/mol (12 atoms)

c. $HF(g) < H_2O(g) < NH_3(g)$

The molar masses are

$HF = 20$ g/mol (2 atoms)
$H_2O = 18$ g/mol (3 atoms)
$NH_3 = 17$ g/mol (4 atoms)

In this case, each substance has the about the same molar mass. The number of atoms per molecule will be more important in determining the complexity of each substance.

The actual values are 193.8 J/°C for HF, 188.7 J/°C for H_2O, and 192.3 J/°C for NH_3.

35. a. Entropy decreases; ΔS is negative
A brick wall is more ordered than a pile of bricks. In this case the system goes from a less ordered (higher entropy state, loose bricks) to a more ordered (lower entropy state, the brick wall). The entropy change ΔS ($S_{final} - S_{initial}$) must be negative because the initial state (random pile of bricks) has a higher entropy than the final state (the brick wall).

b. Entropy decreases; ΔS is negative.
The single pile is more ordered (the final state, lower entropy) than a yard full of leaves (the initial state, higher entropy). The entropy change is negative because the initial state has a higher entropy than the final state.

c. Entropy decreases; ΔS is negative.
The entropy decreases (ΔS is negative) because $AgCl(s)$ is more ordered (the final state, lower entropy) than the $Ag^+(aq)$ and $Cl^-(aq)$ ions dissolved in water (the initial state, higher entropy).

d. Entropy increases; ΔS is positive.

$$Zn(s) + 2\,HCl(aq) \rightarrow H_2(g) + ZnCl_2(aq)$$

The products (final state) have higher entropy because 1 mole of gas (H_2) and 1 mole of an aqueous species ($ZnCl_2$) have higher entropy than 1 mole of solid (Zn) and 2 moles of an aqueous substance (HCl).

37. If we know the absolute molar entropy of substances, then ΔS can be calculated for any process using the equation

$$\Delta S_{reaction} = \sum n S^o_{products} - \sum m S^o_{reactants}$$

a. $N_2(g) + O_2(g) \rightarrow 2\,NO(g)$

From the Appendix, the absolute molar entropies (S°) are

$N_2(g) = 191.5$ J/mol · K
$O_2(g) = 205.0$ J/mol · K
$NO(g) = 210.7$ J/mol · K

$$\begin{aligned}\Delta S &= [(2\,\text{mol}) \times (S^o_{NO})] - [(1\,\text{mol}) \\ &\quad \times (S^o_{N_2}) + (1\,\text{mol}) \times (S^o_{O_2})] \\ &= [(2\,\text{mol}) \times (210.7\,\text{J/mol}\cdot\text{K})] - [(1\,\text{mol}) \\ &\quad \times (191.5\,\text{J/mol}\cdot\text{K}) + (1\,\text{mol}) \\ &\quad \times (205.0\,\text{J/mol}\cdot\text{K})] \\ &= (421.4\,\text{J/K}) - (191.5\,\text{J/K} + 205.0\,\text{J/K}) \\ &= 421.4\,\text{J/K} - 396.5\,\text{J/K} \\ &= +24.9\,\text{J/K}, \quad \text{or} + 24.9\,\text{J/°C}\end{aligned}$$

b. $2\,NO(g) + O_2(g) \rightarrow 2\,NO_2(g)$

From the Appendix, the absolute molar entropies are

$NO(g) = 210.7$ J/mol · K
$O_2(g) = 205.0$ J/mol · K
$NO_2(g) = 240.0$ J/mol · K

$$\begin{aligned}\Delta S &= [(2\,\text{mol}) \times (S^o_{NO_2})] - [(2\,\text{mol}) \\ &\quad \times (S^o_{NO}) + (1\,\text{mol}) \times (S^o_{O_2})] \\ &= [(2\,\text{mol}) \times (240.0\,\text{J/mol}\cdot\text{K})] - [(2\,\text{mol}) \\ &\quad \times (210.6\,\text{J/mol}\cdot\text{K} + 1\,\text{mol}) \\ &= (480.0\,\text{J/K}) - (421.4\,\text{J/K} + 205.0\,\text{J/K}) \\ &= \times(205.0\,\text{J/mol}\cdot\text{K})]480.0\,\text{J/K} - 626.4\,\text{J/K} \\ &= -146.4\,\text{J/K}, \quad \text{or} - 146.4\,\text{J/°C}\end{aligned}$$

c. $NO(g) + \frac{1}{2}O_2(g) \rightarrow NO_2(g)$

From the Appendix, the absolute molar entropies are

$NO(g) = 210.7$ J/mol · K
$O_2(g) = 205.0$ J/mol · K
$NO_2(g) = 240.0$ J/mol · K

$$\Delta S = [(1\text{ mol}) \times (S^{\circ}_{NO_2})] - [(1\text{ mol}) \times (S^{\circ}_{NO}) + (\tfrac{1}{2}\text{ mol}) \times (S^{\circ}_{O_2})]$$

$$= [(1\text{ mol}) \times (240.0\text{ J/mol}\cdot\text{K})] - [(1\text{ mol}) \times (210.6\text{ J/mol}\cdot\text{K}) + (\tfrac{1}{2}\text{ mol}) \times (205.0\text{ J/mol}\cdot\text{K})]$$

$$= (240.0\text{ J/K}) - (210.7\text{ J/K} + 102.5\text{ J/K})$$

$$= 240.0\text{ J/K} - 313.2\text{ J/K}$$

$$= -73.2\text{ J/K}, \quad \text{or } -73.2\text{ J/°C}$$

d.

$$2\,NO_2(g) \rightarrow N_2O_4(g)$$

From the Appendix, the absolute molar entropies are
$NO_2(g) = 240.0$ J/mol · K
$N_2O_4(g) = 304.2$ J/mol · K

$$\Delta S = (1\text{ mol}) \times (S^{\circ}_{N_2O_4}) - (2\text{ mol}) \times (S^{\circ}_{NO_2})$$

$$= (1\text{ mol}) \times (304.2\text{ J/mol}\cdot\text{K}) - (2\text{ mol}) \times (240.0\text{ J/mol}\cdot\text{K})$$

$$= (304.2\text{ J/K}) - (480.0\text{ J/K})$$

$$= -175.8\text{ J/K}, \quad \text{or } -175.8\text{ J/°C}$$

39.

$$Cl(g) + O_3(g) \rightarrow ClO(g) + O_2(g)$$

$$\Delta S = +19.9\text{ J/K}$$

$$\Delta S = [(1\text{ mol}) \times (S^{\circ}_{O_2}) + (1\text{ mol}) \times (S^{\circ}_{ClO})] - [(1\text{ mol}) \times (S^{\circ}_{Cl}) + (1\text{ mole}) \times (S^{\circ}_{O_3})]$$

From the Appendix, the absolute molar entropies are
$Cl(g) = 165.2$ J/mol · K
$O_2(g) = 205.0$ J/mol · K
$O_3(g) = 238.8$ J/mol · K

We know that $\Delta S_{reaction} = +19.9$ J/K.
Substituting this value gives

$$19.9\text{ J/K} = [(1\text{ mol}) \times (S^{\circ}_{O_2}) + (1\text{ mol}) \times (S^{\circ}_{ClO})] - [(1\text{ mol}) \times (S^{\circ}_{Cl}) + (1\text{ mol}) \times (S^{\circ}_{O_3})]$$

Next, substitute the remaining S° values and solve for S° of $ClO(g)$.

$$19.9\text{ J/K} = [(1\text{ mol}) \times (205.0\text{ J/mol}\cdot\text{K}) + (1\text{ mol}) \times (S^{\circ}_{ClO})] - (1\text{ mol}) \times (165.2\text{ J/mol}\cdot\text{K}) + (1\text{ mol}) \times (238.8\text{ J/mol}\cdot\text{K})$$

$$19.9\text{ J/K} = (205.0\text{ J/K} + 1\text{ mol}) \times (S^{\circ}_{ClO}) - (165.2\text{ J/K} + 238.8\text{ J/K})$$

$$19.9\text{ J/K} = 205.0\text{ J/K} + 1\text{ mol} \times (S^{\circ}_{ClO}) - 404.0\text{ J/K}$$

Collecting the numerical terms:

$$19.9\text{ J/K} - 205.0\text{ J/K} + 404.0\text{ J/K} = (1\text{ mol}) \times (S^{\circ}_{ClO})$$

$$(1\text{ mol}) \times (S^{\circ}_{ClO}) = 218.9\text{ J/K}$$

Dividing both sides by 1 mole yields

$$S^{\circ}_{ClO} = 218.8\text{ J/mol}\cdot\text{K}, \quad \text{or } 218.9\text{ J/mole}\cdot\text{°C}$$

41. No, all exothermic reactions are not spontaneous. The thermodynamic quantity that determines spontaneity is called the free energy (ΔG°), and it depends on both the enthalpy change (ΔH°) and the entropy change (ΔS°).

$$\Delta G^{\circ} = \Delta H^{\circ} - T\Delta S^{\circ}$$

For a reaction to be spontaneous ΔG° must be negative.
A negative ΔH° or a positive ΔS° will favor a reaction being spontaneous.
Many exothermic processes are spontaneous but not all of them.

42.

$$\Delta G^{\circ} = \Delta H^{\circ} - T\Delta S^{\circ}$$

For a reaction to be spontaneous, ΔG° must be negative.
If ΔH° is positive and ΔS° is positive the reaction will not be spontaneous until the $T\Delta S$ term is greater than the ΔH term. Since ΔH and ΔS are not very temperature dependent, these quantities can assumed to be constant during the given temperature change. As the temperature increases the $T\Delta S$ term gets significantly larger, but the ΔH term remains nearly the same. When the temperature becomes high enough, the $T\Delta S$ term becomes greater than the ΔH term. When this happens ΔG becomes negative.

43. For process that is always spontaneous, no matter what the temperature, ΔG° must always be negative. This will occur if ΔH° is negative and ΔS° is positive. Any process that meets this requirement will be spontaneous at all temperatures. The combustion of hydrocarbons is both exothermic (negative ΔH°) and results in an increase in entropy (ΔS° is positive), so the combustion of hydrocarbons is spontaneous at all temperatures. (However, this does not mean that the reaction occurs very quickly at all temperatures. In fact, at lower temperatures this process occurs very, very slowly.)
Another process that would be spontaneous at all temperatures is rolling a ball down a hill.

44. A process that is never spontaneous must have a positive ΔH° value and a negative ΔS° value, so that ΔG° is always a positive quantity.
Going up to the top of a mountain is a process that is never spontaneous. It requires some form of energy to be used for this to occur.

Most combustion processes are always spontaneous. If one reverses a combustion process that is spontaneous under all conditions, then it should never be spontaneous. We can consider the following reaction.

$$6\,CO_2(g) + 6\,H_2O(g) \rightarrow C_6\,H_{12}O_6(S) + 6\,O_2(g)$$

We know that this reaction must be endothermic (because the reverse reaction is a combustion reaction, which are always exothermic) and there is also a decrease in entropy (12 moles of gaseous reactants against only 6 moles of gaseous products).

A positive ΔH^o value and a negative ΔS^o value have to result in a positive ΔG^o value.

45. $$CO_2(s) \rightarrow CO_2(g)$$

In the sublimation process, ΔS^o must be positive because the substance changes from the solid state to the gaseous state.

ΔH^o is positive, since we know it requires heat energy to sublime any substance. (Sublimation can be considered to be a combination of melting and boiling.)

ΔG^o must be negative because we know that dry ice spontaneously sublimes at room temperature.

47. Spontaneous processes occur on their own without outside intervention. There are two thermodynamic quantities that favor a process being spontaneous: an increase in entropy (ΔS is positive) and a decrease in enthalpy (ΔH is negative).
    a. Spontaneous. This process occurs on its own. As gas particles spread out through a room, they become less ordered (or disorder increases). This means that the entropy increases. An increase in entropy favors a process being spontaneous
    b. Nonspontaneous. Clocks do not fix themselves. To fix a clock one has to input energy and work.
    c. Spontaneous. We know that metallic iron reacts with oxygen and water vapor in the air to form rust.
    d. Spontaneous (above 0°C). Ice melts in tap water because the water temperature is above the freezing point.

49. For NaBr

$$NaBr \rightarrow Na^+(aq) + Br^-(aq)$$

We are given the following information:

$\Delta H^o = -1$ kJ/mol
$\Delta S^o = 57$/J/mol · K

We can determine ΔG from the equation

$$\Delta G^o = \Delta H^o - T\Delta S^o$$

Before substituting these values into the equation, we need to make sure that ΔH^o and ΔS^o have the same energy units. ΔH^o is reported in units of kJ, while ΔS^o is reported in units of J.

We need to convert one of them. Let's convert ΔS^o to kJ/mol · K.

$$57\,\text{J/mol} \cdot \text{K} = 0.057\,\text{kJ/mol} \cdot \text{K}\ (1\,\text{kJ} = 1000\,\text{J})$$

Additionally the temperature must be in kelvins (K = C° + 273).

$$25°\text{C} = 298\,\text{K}$$

We can now calculate ΔG^o:

$$\begin{aligned}\Delta G^o &= \Delta H^o - T\Delta S^o\\ &= (-1\,\text{kJ/mol}) - (298\,\text{K})(0.057\,\text{kJ/mol} \cdot \text{K})\\ &= (-1\,\text{kJ/mol}) - (17.0\,\text{kJ/mol})\\ &= -18.0\,\text{kJ/mol}\end{aligned}$$

ΔG^o is negative as we expect. (Since NaBr is soluble in water, this is a spontaneous process.)

For NaI

$$NaI \rightarrow Na^+(aq) + I^-(aq)$$

We are given the following information:

$\Delta H^o = -7$ kJ/mol
$\Delta S^o = 74$/J/mol · K

We can determine ΔG from the equation

$$\Delta G^o = \Delta H^o - T\Delta S^o$$

The temperature must be in kelvins, and we need to match units for ΔH^o and ΔS^o. Let's convert ΔS^o to kJ/mol · K.

$$\Delta S^o = 74\,\text{J/mol} \cdot \text{K} = 0.074\,\text{kJ/mol} \cdot \text{K}$$

We can now calculate ΔG^o.

$$\begin{aligned}\Delta G^o &= \Delta H^o - T\Delta S^o\\ &= (-7\,\text{kJ/mol}) - (298\,\text{K})(0.074\,\text{kJ/mol} \cdot \text{K})\\ &= (-7\,\text{kJ/mol}) - (22.1.0\,\text{kJ/mol})\\ &= -29.1\,\text{kJ/mol}\end{aligned}$$

When 1 mole of NaI dissolves, the free energy is lowered by 29.1 kJ/mol. ΔG^o is negative as we expect. (Since NaI is soluble in water, this a spontaneous process.)

51. a. We are given the balanced equation

$$H_2O(g) + C(s) \rightarrow H_2(g) + CO(g)$$

and we are asked to find the value of ΔG^o for this process.

We can calculate this quantity by looking up the ΔG_f^o for each substance involved in this reaction in

Appendix 4. Then we can calculate ΔG° using the equation

$$\Delta G^o_{reaction} = \sum n \Delta G^o_{f,products} - \sum m \Delta G^o_{f,reactants}$$

From Appendix 4, we obtain the following values for ΔG^o_f

$H_2O(g) = -228.6$ kJ/mol
$C(s) = 0.0$ kJ/mol
$H_2(g) = 0.0$ kJ/mol
$CO(g) = -137.2$ kJ/mol

Substituting in the equation we get

$$\begin{aligned}\Delta G^o_{reaction} &= (1 \text{ mol} \times \Delta G^o_{f,H_2} + 1 \text{ mol} \times \Delta G^o_{f,CO})\\ &\quad - (1 \text{ mol} \times \Delta G^o_{f,H_2O} + 1 \text{ mol} \times \Delta G^o_{f,C})\\ &= (1 \text{ mol}) \times (0.0 \text{ kJ/mol} + 1 \text{ mol})\\ &\quad \times (-137.2 \text{ kJ/mol})\\ &\quad - (1 \text{ mol}) \times (-228.6 \text{ kJ/mol} + 1 \text{ mol})\\ &\quad \times (0.0 \text{ kJ/mol})\\ &= (-137.2 \text{ kJ}) - (-228.6 \text{ kJ}) = +91.4 \text{ kJ}\end{aligned}$$

This reaction is not spontaneous at room temperature.

b. To determine over what temperature range this reaction is spontaneous we can use the formula

$$\Delta G^o = \Delta H^o - T\Delta S^o$$

If we can find the temperature when $\Delta G^o = 0$, then we know the temperature at which the reaction changes from spontaneous to nonspontaneous. Depending on the signs of ΔH^o and ΔS^o, we can then determine whether this reaction will be spontaneous at a temperature higher or lower than this calculated temperature.

We need to determine both ΔH^o and ΔS^o for this reaction. We can do this using the following formulas:

$$\Delta H^o_{reaction} = \sum n \Delta H^o_{f,products} - \sum m \Delta H^o_{f,reactants}$$

$$\Delta S^o_{reaction} = \sum n S^o_{products} - \sum m S^o_{reactants}$$

Again, we need to look up values of ΔH^o_f and S^o in Appendix 4.

Note: for free energy and enthalpy we can only measure the change in these quantities (ΔG^o and ΔH^o). For the entropy, we can measure absolute values (S^o).

From Appendix 4, we obtain the following values for ΔH^o_f:

$H_2O(g) = -241.8$ kJ/mol
$C(s) = 0.0$ kJ/mol
$H_2(g) = 0.0$ kJ/mol
$CO(g) = -110.5$ kJ/mol

and for S^o:

$H_2O(g) = 188.7$ J/mol
$C(s) = 5.7$ J/mol
$H_2(g) = 130.6$ J/mol
$CO(g) = -197.7$ J/mol

Substituting into the equation for ΔH^o we get

$$\begin{aligned}\Delta H^o_{reaction} &= (1 \text{ mol} \times \Delta H^o_{f,H_2} + 1 \text{ mol} \times \Delta H^o_{f,CO})\\ &\quad - (1 \text{ mol} \times \Delta H^o_{f,H_2O} + 1 \text{ mol} \times \Delta H^o_{f,C})\\ &= [(1 \text{ mol}) \times (0.0 \text{ kJ/mol}) + (1 \text{ mol})\\ &\quad \times (-110.5 \text{ kJ/mol})]\\ &\quad - [(1 \text{ mol}) \times (-241.8 \text{ kJ/mol})\\ &\quad + (1 \text{ mol}) \times (0.0 \text{ kJ/mol})]\\ &= (-110.5 \text{ kJ}) - (-241.8 \text{ kJ})\\ &= +131.3 \text{ kJ}\end{aligned}$$

Substituting into the equation for ΔS^o we get

$$\begin{aligned}\Delta S^o_{reaction} &= (1 \text{ mol} \times S^o_{H_2} + 1 \text{ mol} \times S^o_{CO})\\ &\quad - (1 \text{ mol} \times S^o_{H_2O} + 1 \text{ mol} \times S^o_{C})\\ &= [(1 \text{ mol}) \times (130.6 \text{ J/mol}) + (1 \text{ mol})\\ &\quad \times (197.7 \text{ J/mol})] - [(1 \text{ mol})\\ &\quad \times (188.7 \text{ J/mol}) + (1 \text{ mol}) \times (5.7 \text{ J/mol})]\\ &= (130.6 \text{ J/K} + 197.7 \text{ J/K})\\ &\quad - (188.7 \text{ J/K} + 5.7 \text{ J/K})\\ &= 328.3 \text{ J/K} - 194.4 \text{ J/K}\\ &= +133.9 \text{ J/K}\end{aligned}$$

Let's determine the temperature at which this reaction has a $\Delta G^o = 0$.

$$\begin{aligned}\Delta G^o &= \Delta H^o - T\Delta S^o\\ 0 &= \Delta H^o - T\Delta S^o\\ \Delta H^o &= T\Delta S^o\\ T &= \Delta H^o/\Delta S^o\end{aligned}$$

$\Delta H^o = 131.3$ kJ $\quad \Delta S^o = 133.9$ J/K $= 0.1339$ kJ/K

$$\begin{aligned}T &= \frac{\Delta H^o}{\Delta S^o}\\ &= \frac{131.3 \text{ kJ}}{0.1339 \text{ kJ/K}} = 981 \text{ K}\end{aligned}$$

When ΔH^o and ΔS^o are both positive, we should expect the reaction to be spontaneous at higher temperatures (see Table 13.2).

This reaction is spontaneous at temperatures above 981 K and it is not spontaneous at temperatures below 981 K.

53. In this problem, we need to calculate ΔH^o and ΔS^o for the reaction

$$SO_2(g) + 2\ H_2S(g) \rightarrow 3\ S(s) + 2\ H_2O(g)$$

To do this we need to look up the pertinent data in Appendix 4.
Values for S^o:
$SO_2(g) = 248.2$ J/mol · K
$S(s) = 32.1$ J/mol · K
$H_2S(g) = 205.6$ J/mol · K
$H_2O(g) = 188.7$ J/mol · K
Values for ΔH_f^o:
$SO_2(g) = -296.8$ J/mol
$S(s) = 0.0$ J/mol
$H_2S(g) = -20.2$ J/mol
$H_2O(g) = -241.8$ J/mol
Substituting these values we obtain:

$$\begin{aligned}\Delta H^o &= [(2 \text{ mol } H_2O) \times (-241.8 \text{ kJ/mol}) + (3 \text{ mol S}) \\ &\quad \times (0.0 \text{ kJ/mol})] - (1 \text{ mol } SO_2) \\ &\quad \times [(-296.8 \text{ kJ/mol} + (2 \text{ mol } H_2S) \\ &\quad \times (-20.2 \text{ kJ/mol})] \\ &= (-483.6 \text{ kJ} + 0.0 \text{ kJ}) \\ &\quad -[(-296.8 \text{ kJ}) + (-40.4 \text{ kJ})] \\ &= (-483.6 \text{ kJ}) - (-337.2 \text{ kJ}) \\ &= -146.4 \text{ kJ}\end{aligned}$$

$$\begin{aligned}\Delta S^o &= [(2 \text{ mol } H_2O) \times (188.7 \text{ J/mol} \cdot \text{K}) + (3 \text{ mol S}) \\ &\quad \times (32.1 \text{ J/mol} \cdot \text{K})] - [(1 \text{ mol } SO_2) \\ &\quad \times [(248.2 \text{ J/mol} \cdot \text{K}) + (2 \text{ mol } H_2S) \\ &\quad \times (205.6 \text{ J/mol} \cdot \text{K})] \\ &= (377.4 \text{ J/K} + 96.3 \text{ J/K}) \\ &\quad - (248.2 \text{ J/K} + 411.2 \text{ J/K}) \\ &= (473.7 \text{ J/K}) - (659.4 \text{ J/K}) \\ &= -185.7 \text{ J/K}\end{aligned}$$

ΔH^o is negative and ΔS^o is negative.

This reaction should be spontaneous at lower temperatures. We can determine the highest temperature at which this reaction is spontaneous by setting $\Delta G^o = 0$ and solving for the temperature in the formula

$$\Delta G^o = \Delta H^o - T\Delta S^o$$

Setting $\Delta G^o = 0$ gives us

$$0 = \Delta H^o - T\Delta S^o \qquad \text{or} \qquad \Delta H^o = T\Delta S^o$$

Solving for T, we get

$$T = \frac{\Delta H^o}{\Delta S^o}$$

The temperature here corresponds to the highest temperature at which this reaction is spontaneous.

Note: Make sure the energy units of ΔH^o and ΔS^o are the same.

$$\Delta S^o = -185.7 \text{ J/K} = -0.1857 \text{ kJ/K}$$

$$T = \frac{\Delta H^o}{\Delta S^o} = \frac{-146.4 \text{ kJ}}{-0.1857 \text{ kJ/K}} = 788.4 \text{ K}$$

This reaction is spontaneous at all temperatures below 788.4 K.

55. In each part we need to look up the values for ΔG_f^o in Appendix 4 and then calculate ΔG_{rxn}^o from the equation

$$\Delta G_{reaction}^o = \sum n\Delta G_{f,products}^o - \sum m\Delta G_{f,reactants}^o$$

a. $$N_2(g) + O_2(g) \rightarrow 2\ NO(g)$$

Values for ΔG_f^o:
$NO = 86.6$ kJ/mol
$N_2 = 0.0$ kJ/mol
$O_2 = 0.0$ kJ/mol

$$\begin{aligned}\Delta G_{rxn}^o &= (2 \text{ mol} \times \Delta G_{f,NO}^o) \\ &\quad - (1 \text{ mol} \times \Delta G_{f,N_2}^o + 1 \text{ mol} \times \Delta G_{f,O_2}^o) \\ &= (2 \text{ mol}) \times (86.6 \text{ kJ/mol}) \\ &\quad - [(1 \text{ mol}) \times (0.0 \text{ kJ/mol}) \\ &\quad + (1 \text{ mol}) \times (0.0 \text{ kJ/mol})] \\ &= 173.2 \text{ kJ} - 0.0 \text{ kJ} \\ &= 173.2 \text{ kJ}\end{aligned}$$

b. $$2\ NO(g) + O_2(g) \rightarrow 2\ NO_2(g)$$

Values for ΔG_f^o:
$NO = 86.6$ kJ/mol
$O_2 = 0.0$ kJ/mol
$NO_2 = 51.3$ kJ/mol

$$\begin{aligned}\Delta G_{rxn}^o &= (2 \text{ mol} \times \Delta G_{f,NO_2}^o) \\ &\quad - (1 \text{ mol} \times \Delta G_{f,O_2}^o + 2 \text{ mol} \times \Delta G_{f,NO}^o) \\ &= (2 \text{ mol}) \times (51.3 \text{ kJ/mol}) \\ &\quad - [(1 \text{ mol}) \times (0.0 \text{ kJ/mol}) \\ &\quad + (2 \text{ mol}) \times (86.6 \text{ kJ/mol})] \\ &= 102.6 \text{ kJ} - 173.2 \text{ kJ} \\ &= -70.6 \text{ kJ}\end{aligned}$$

c. $$NO(g) + \frac{1}{2}O_2(g) \rightarrow NO_2(g)$$

ΔG^o_f NO = 86.6 kJ/mol
ΔG^o_f O_2 = 0.0 kJ/mol
ΔG^o_f NO_2 = 51.3 kJ/mol

$$\begin{aligned}\Delta G^o_{rxn} &= (1 \text{ mol} \times \Delta G^o_{f,NO_2}) \\ &\quad - (\tfrac{1}{2} \text{ mol} \times \Delta G^o_{f,O_2} + 1 \text{ mol} \times \Delta G^o_{f,NO}) \\ &= (1 \text{ mol}) \times (51.3 \text{ kJ/mol}) \\ &\quad - [(\tfrac{1}{2} \text{ mol}) \times (0.0 \text{ kJ/mol}) \\ &\quad + (1 \text{ mol}) \times (86.6 \text{ kJ/mol})] \\ &= 51.3 \text{ kJ} - 86.6 \text{ kJ} \\ &= -35.3 \text{ kJ}\end{aligned}$$

Note: All of the quantities in this reaction are one half those found in part *b*. We should expect the ΔG^o_{rxn} value to be half of that calculated in part *b*. This is exactly what we find (−35.3 kJ is half of −70.6 kJ).

d. $$2\ NO_2(g) \rightarrow N_2O_4(g)$$

Values for ΔG^o_f:
NO_2 = 51.3 kJ/mol
N_2O_4 = 97.8 kJ/mol

$$\begin{aligned}\Delta G^o_{rxn} &= (1 \text{ mol} \times \Delta G^o_{f,N_2O_4}) - (2 \text{ mol} \times \Delta G^o_{f,NO_2}) \\ &= (1 \text{ mol}) \times (97.8 \text{ kJ/mol}) \\ &\quad - (2 \text{ mol}) \times (51.3 \text{ kJ/mol}) \\ &= 97.8 \text{ kJ} - 102.6 \text{ kJ} \\ &= -4.8 \text{ kJ}\end{aligned}$$

57. To answer this question, we need to know the signs of ΔH^o_{rxn} and ΔS^o_{rxn} for each process. The best way to determine this is to use Appendix 4 to calulate ΔH^o_{rxn} and ΔS^o_{rxn} for each reaction.

For reaction 55a

$$N_2(g) + O_2(g) \rightarrow 2\ NO(g)$$

$NO(g)$ ΔH^o_f = 90.3 kJ/mol S^o = 210.7 J/mol · K
$O_2(g)$ ΔH^o_f = 0.0 kJ/mol S^o = 205.0 J/mol · K
$N_2(g)$ ΔH^o_f = 0.0 kJ/mol S^o = 191.5 J/mol · K

Note: The units of moles will cancel out in the calculation below and have thus been omitted.

$$\begin{aligned}\Delta H^o_{rxn} &= [(2)(90.3 \text{ kJ}] - [(1)(0.0 \text{ kJ}) + (1)(0.0 \text{ kJ})] \\ &= +180.6 \text{ kJ}\end{aligned}$$

$$\begin{aligned}\Delta S^o_{rxn} &= [(2)(210.7 \text{ J/K})] - [(1)(191.5 \text{ J/K}) \\ &\quad + (1)(205.0 \text{ J/K})] \\ &= 421.4 \text{ J/K} - (191.5 \text{ J/K} + 205.0 \text{ J/K}) \\ &= +24.9 \text{ J/K}\end{aligned}$$

ΔH^o_{rxn} and ΔS^o_{rxn} are both positive, so this reaction is spontaneous only at higher temperatures.

For reaction 55b

$$2\ NO(g) + O_2(g) \rightarrow 2\ NO_2(g)$$

$NO_2(g)$ ΔH^o_f = 33.2 kJ/mol S^o = 240.0 J/mol · K
$O_2(g)$ ΔH^o_f = 0.0 kJ/mol S^o = 205.0 J/mol · K
$NO(g)$ ΔH^o_f = 90.3 kJ/mol S^o = 210.7 J/mol · K

$$\begin{aligned}\Delta H^o_{rxn} &= (2)(33.2 \text{ kJ}) - [(2)(90.3 \text{ kJ}) + (1)(0.0 \text{ kJ})] \\ &= 66.4 \text{ kJ} - 180.6 \text{ kJ} = -114.2 \text{ kJ}\end{aligned}$$

$$\begin{aligned}\Delta S^o_{rxn} &= (2)(240.0 \text{ J/K}) - [(2)(210.7 \text{ J/K}) \\ &\quad + (1)(205.0 \text{ J/K})] \\ &= 480.0 \text{ J/K} - (421.4 \text{ J/K} + 205.0 \text{ J/K}) \\ &= 480.0 \text{ J/K} - 626.4 \text{ J/K} = -146.4 \text{ J/K}\end{aligned}$$

Both ΔH^o_{rxn} and ΔS^o_{rxn} are negative, so this reaction will be spontaneous at lower temperatures only.

For reaction 55c

$$NO(g) + \frac{1}{2}\ O_2(g) \rightarrow NO_2(g)$$

This is the same reaction as in part *a* of this question. The signs of ΔH^o_{rxn} and ΔS^o_{rxn} will be the same is in part *a*. Since only half the quantities are reacted in part *c*, the values calculated for ΔH^o_{rxn} and ΔS^o_{rxn} should also be halved.

$$\Delta H^o_{rxn} = \frac{1}{2}(180.6 \text{ kJ}) = 90.3 \text{ kJ}$$

$$\Delta S^o_{rxn} = \frac{1}{2}(24.8 \text{ J/K}) = 12.4 \text{ J/K}$$

ΔH^o_{rxn} and ΔS^o_{rxn} are both positive, so this reaction is spontaneous only at higher temperatures.

For reaction 55d

$$2\ NO_2(g) \rightarrow N_2O_4(g)$$

$NO_2(g)$ ΔH^o_f = 33.2 kJ/mol S^o = 240.0 J/mol · K
$N_2O_4(g)$ ΔH^o_f = 9.2 kJ/mol S^o = 304.2 J/mol · K

$$\begin{aligned}\Delta H^o_{rxn} &= (1)(9.2 \text{ kJ}) - (2)(33.2 \text{ kJ}) \\ &= 9.2 \text{ kJ} - 66.4 \text{ kJ} = -57.2 \text{ kJ}\end{aligned}$$

$$\begin{aligned}\Delta S^o_{rxn} &= (1)(304.2 \text{ J/K}) - (2)(240.0 \text{ J/K}) \\ &= 304.2 \text{ J/K} - 480.0 \text{ J/K} = -175.8 \text{ J/K}\end{aligned}$$

Both ΔH^o_{rxn} and ΔS^o_{rxn} are negative, so this reaction will be spontaneous at lower temperatures only.

a. Reactions 55a and 55c will be spontaneous at high temperatures.
b. Reactions 55b and 55d will be spontaneous at low temperatures.
c. None of the reactions will be spontaneous at all temperatures.

59. Enantiomers are molecules whose structures are mirror images of one another and are not identical. For example, your left and right hand are enantiomers; they are mirror images of one another and they are not identical as is evident when you try to put your left glove on your right hand.

60. The term chiral refers to the potential for forming enantiomers. An atom that has four different groups bonded tetrahedrally can be chiral. For example, the following molecule contains a chiral carbon atom.

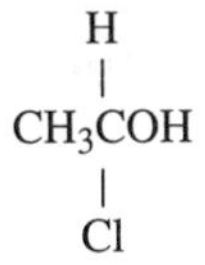

```
   H
   |
CH3COH
   |
   Cl
```

The central carbon is bonded to four different groups: CH_3, OH, Cl, and H.

Mirror images of substances that are chiral are enantiomers (isomers) of each other. These types of isomers are called stereoisomers, and are distinguished by designating them as either left-handed or right-handed structures.

61. A racemic mixture is one that contains a 50:50 mixture of enantiomers.

62. An unsaturated fatty acid has one or more double bonds in its hydrocarbon structure, a saturated fatty acid does not.

63. (*S*)-amino acids are structures in which the substituent groups bonded to the chiral atom trace a counterclockwise path going from the highest priority group to the lowest.

Note: The easiest way to determine the correct *R* and *S* orientations is by using an organic model kit.

For example, let's look at the following structure of 2-butanol. Carbon-2 of this structure is a chiral center. We view the structure such that the group with lowest priority (in this case hydrogen) is directed away from the viewer. In terms of a two-dimensional structure (that you see on paper), the hydrogen should be on a horizontal bond (directed back). To determine the *R* and *S* configuration in the structures below, the viewer looks at the molecule from an angle that lines up the H atom directly behind the C atom.

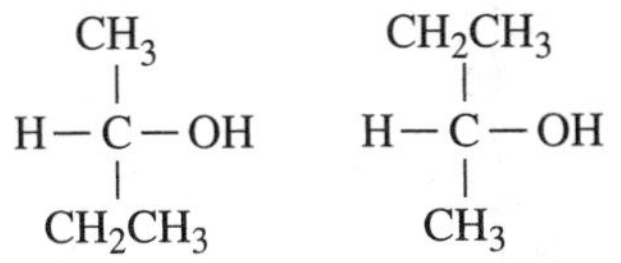

```
   CH3         CH2CH3
    |            |
H—C—OH       H—C—OH
    |            |
  CH2CH3        CH3
```

In order of priority, we designate the groups

a—OH
b—CH_3CH_2
c—CH_3

In the structure on the left, the arrangement of OH, CH_3CH_2, and CH_3 groups is clockwise. This is the *R* isomer. It is named (*R*)-2-butanol.

In the structure on the right, the arrangement of OH, CH_3CH_2, and CH_3 groups is counterclockwise. This is the *S* isomer. It is named (*S*)-2-butanol. *S*-amino acids are the forms that make up the proteins in our bodies.

64. The only amino acid that does not contain a chiral center (i.e., has no enantiomers) is glycine.

65. In an α-amino acid the amino group must be bonded to the same carbon atom as the carboxyl group (COOH), as in the following structure:

```
R  R
|  |
N—C—COOH
|  |
R  R
```

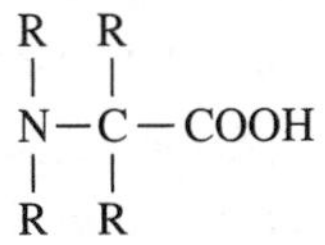

where R can be a hydrogen atom or any other group.

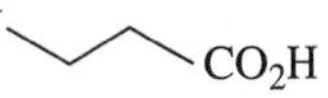

a. H_2N (skeletal) CO_2H or $H_2N—CH_2CH_2CO_2H$

Structure *a* is not an α-amino acid because there are two carbon atoms between the amino group and the carboxyl group.

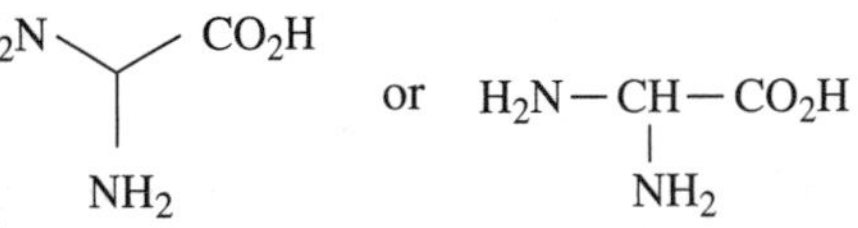

b. H_2N (skeletal) CO_2H, NH_2 or

```
H2N—CH—CO2H
     |
    NH2
```

Structure *b* is an α-amino acid because the amino group and the carboxyl group are bonded to the same carbon atom.

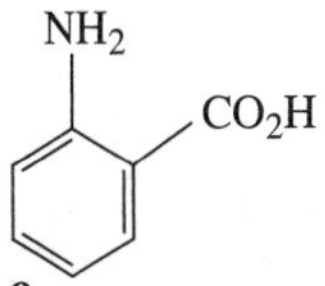

c. (benzene ring with NH_2 and CO_2H)

Structure *c* is not an α-amino acid because there are two carbon atoms between the amino group and the carboxyl group.

Structures *a* and *c* are not α-amino acids.

66. **a.** (skeletal: O, NH_2) or

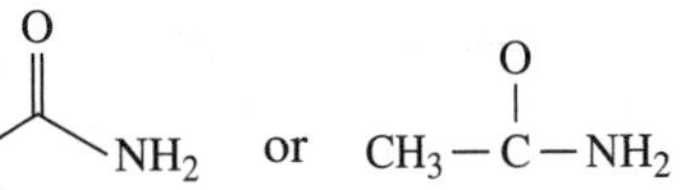

```
      O
      ||
CH3—C—NH2
```

Structure *a* is not an amino acid, because it does not contain a carboxyl group.

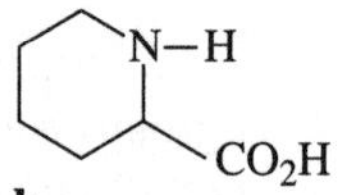

```
   N–H
      CO2H
b.
```

Structure *b* is an α-amino acid because the amino group (the nitrogen in the ring structure) and the carboxyl group are bonded to the same carbon atom.

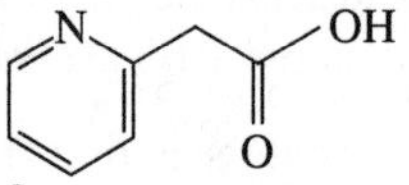

```
 N        OH
       O
c.
```

Structure *c* is not an α-amino acid because there are two carbon atoms between the amino group (the nitrogen in the ring structure) and the carboxyl group.

67. In each reaction the OH of the carboxyl group and one of the hydrogen atoms from the amine group combine to form water. The remainder of the molecules slide together to form a peptide link.

a. alanine and serine

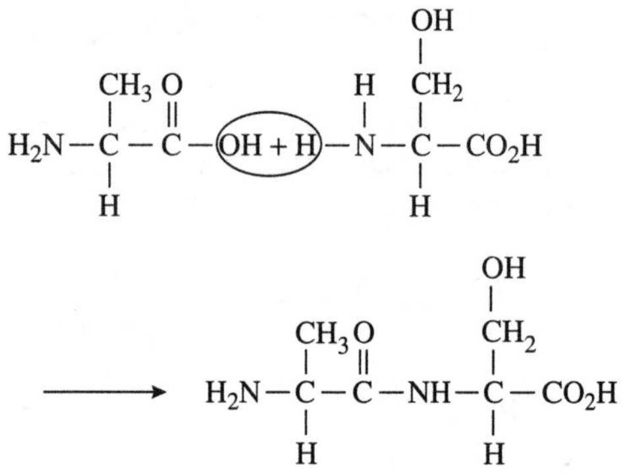

```
                               OH
                               |
       CH3 O              H   CH2
       |   ||             |   |
H2N – C – C –(OH + H)– N – C – CO2H
       |                      |
       H                      H

                              OH
                              |
            CH3 O             CH2
            |   ||            |
—→   H2N – C – C – NH – C – CO2H
            |                 |
            H                 H
```

b. alanine and phenylalanine

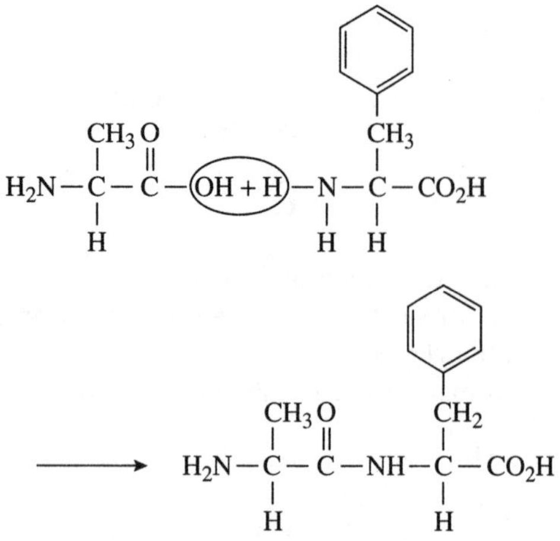

```
                            (C6H5)
                              |
       CH3 O                 CH3
       |   ||                 |
H2N – C – C –(OH + H)– N – C – CO2H
       |                 |    |
       H                 H    H

                            (C6H5)
                              |
            CH3 O            CH2
            |   ||            |
—→   H2N – C – C – NH – C – CO2H
            |                 |
            H                 H
```

c. alanine and valine

```
                            CH3  CH3
                               \ /
       CH3 O                   CH
       |   ||                  |
H2N – C – C –(OH + H)– N – C – CO2H
       |                 |    |
       H                 H    H

                            CH3  CH3
                               \ /
            CH3 O              CH
            |   ||             |
—→   H2N – C – C – NH – C – CO2H
            |                  |
            H                  H
```

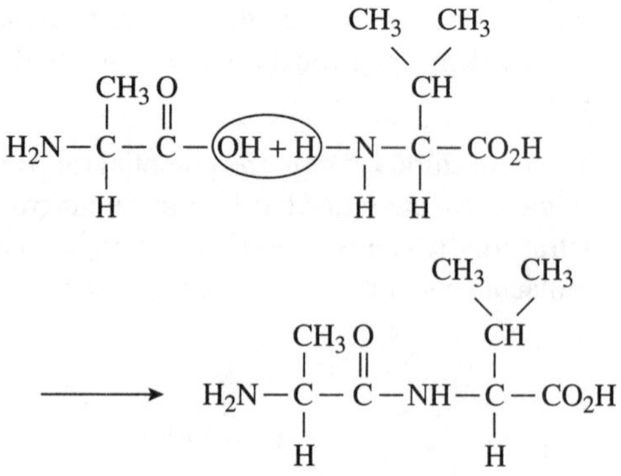

d. methionine + alanine + glycine

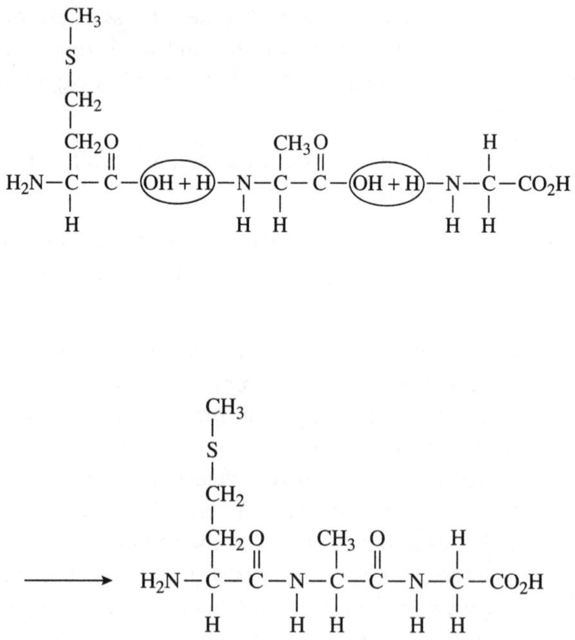

```
       CH3
       |
       S
       |
       CH2
       |
       CH2 O                 CH3 O                        H
       |   ||                |   ||                       |
H2N – C – C –(OH + H)– N – C – C –(OH + H)– N – C – CO2H
       |                 |   |                 |   |
       H                 H   H                 H   H

            CH3
            |
            S
            |
            CH2
            |
            CH2 O     CH3 O      H
            |   ||    |   ||     |
—→   H2N – C – C – N – C – C – N – C – CO2H
            |        |   |       |   |
            H        H   H       H   H
```

e. methionine + valine + alanine

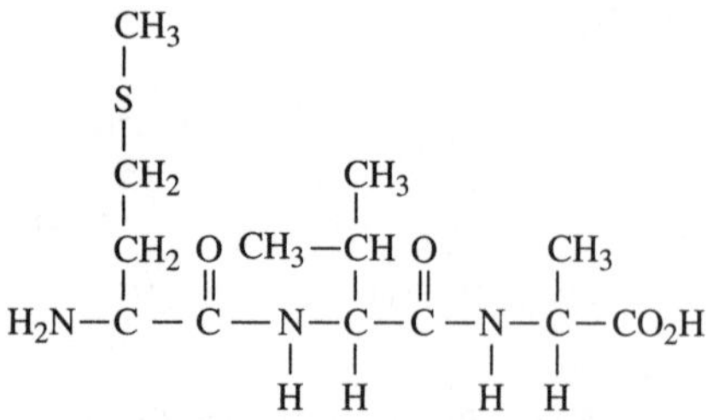

```
       CH3
       |
       S
       |
       CH2        CH3
       |          |
       CH2 O CH3–CH O        CH3
       |   ||     |  ||       |
H2N – C – C – N – C – C – N – C – CO2H
              |   |       |   |
              H   H       H   H
```

f. serine + glycine + tyrosine

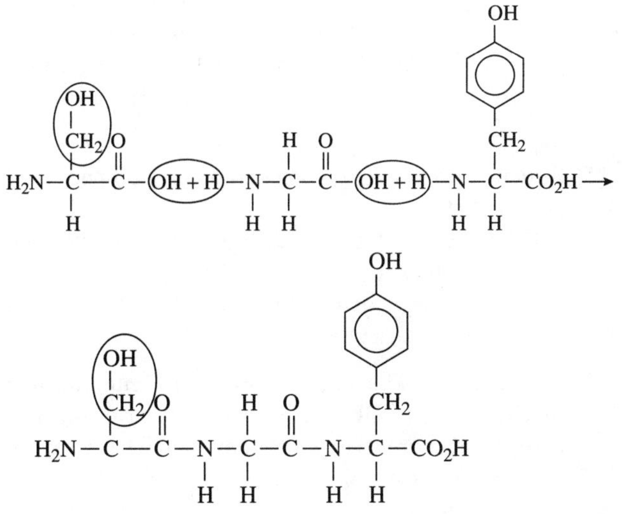

69. The five amino acids are tyrosine, glycine, glycine, valine, and methionine.

To determine the identity of the amino acids we need to split the peptide between the C and N of each amide linkage, $\overset{O}{\overset{\|}{C}}-N$.
The line structure of this compound is

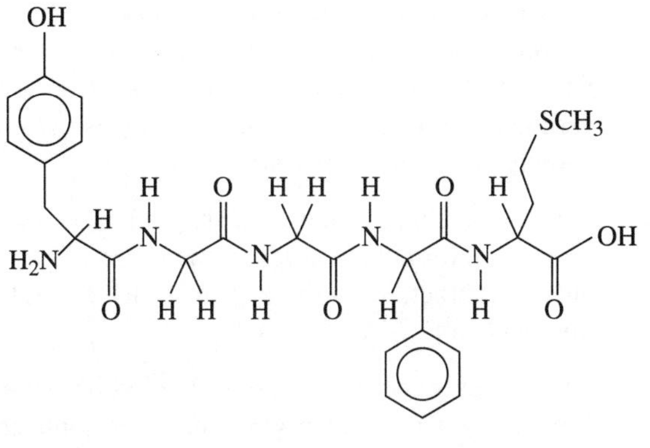

Drawing in all the atoms gives us:

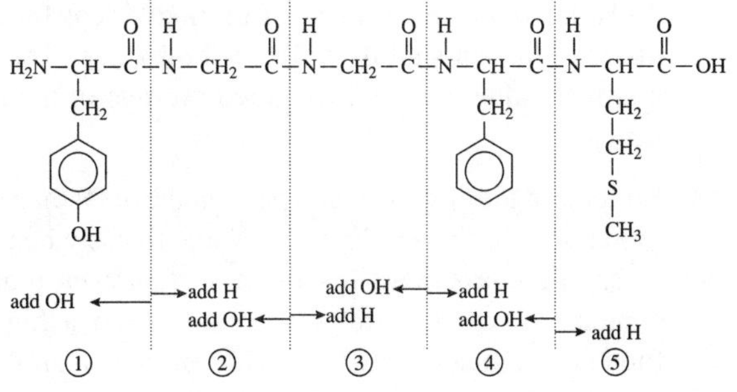

We can divide this compound into amino acids by splitting the amide linkages (carbonyls bonded to a nitrogen), as shown by the vertical lines. Add an OH group to the carbon of the carbonyl and a hydrogen atom to the nitrogen.

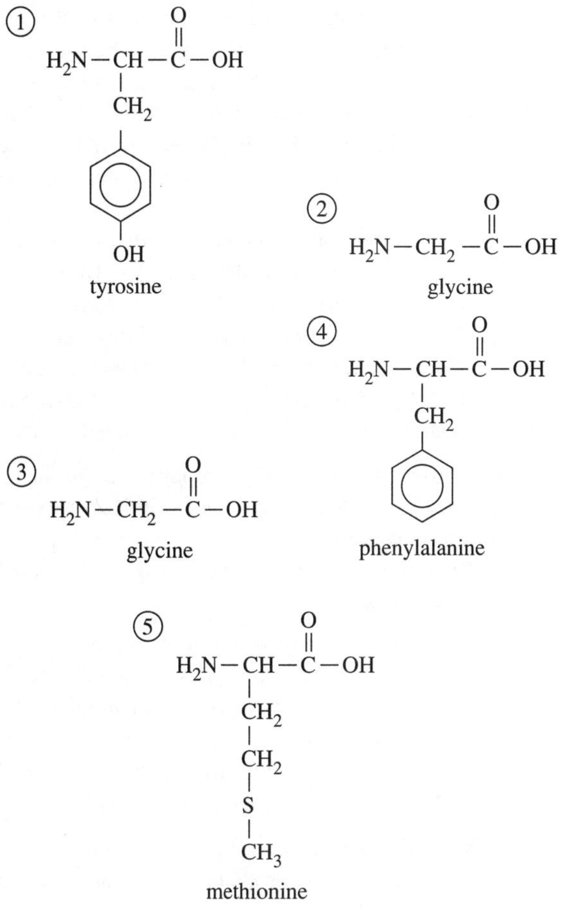

71. Objects *a*, *b*, and *d* are chiral.
73. They are all (*S*) sterioisomers.

Rules for the *R-S* System

1. Assign a priority to each of the groups bonded to the carbon atom. Label these groups as *a*, *b*, *c*, and *d*, so that *a* has the highest priority and *d* the lowest priority.

   Use the following rules to determine priority.

   i. Priority is determined by the atomic number of the atom that is directly bonded to the carbon atom.
   ii. The atom that has the highest atomic number is given the highest priority (letter *a*). The atom that has the lowest atomic number is given the lowest priority (letter *d*).
   iii. If two identical atoms are bonded to the carbon atom, then use the next set of atoms that are directly bonded to the atom. Continue this process until an *a*, *b*, *c*, *d* order has been established.

Using this rule, if both an ethyl and a methyl group are bonded to a carbon atom, the ethyl would be given the higher priority because it contains a carbon–carbon bond (both of these atoms have an atomic number of 6). The methyl group contains only hydrogen atoms bonded to the carbon atom (atomic numbers 6 and 1). The atomic number combination 6 and 6 of the ethyl group is higher than the 6 and 1 combination of the methyl.

2. Next rotate the molecule such that the group with the lowest priority is directed away from your eye, that is, directly behind the central atom.
3. Trace a path from *a* to *b* to *c*. If this tracing is in a clockwise direction, the stereoisomer is designated as (*R*). If this tracing is in a counterclockwise direction the stereoisomer is designated as (*S*).

The easiest way to see the orientation of all the groups is to use a model kit. It is very difficult to see these orientations in your head or on a piece of paper.

a. The order of priority for the groups in this substance is as follows:

a. NH_2	nitrogen (atomic number 7)
b. COOH	carbon (atomic number 6), then oxygen (atomic number 8)
c. $CH_2C_6H_5$	carbon (atomic number 6), then another carbon (atomic number 6)
d. H	hydrogen (atomic number 1)

Note: COOH has a higher priority than $CH_2C_6H_5$. Each group has a carbon bonded to the central atom. In COOH the carbon is bonded to an oxygen and in $CH_2C_6H_5$ the carbon is bonded to another carbon. Oxygen has a higher atomic number, so it has the higher priority.

This is an (*S*) stereoisomer.

b. The order of priority for the groups in this substance is as follows:

a. NH_2	nitrogen (atomic number 7)
b. COOH	carbon (atomic number 6), then oxygen (atomic number 8)
c. $CH_2CH_2SCH_3$	carbon (atomic number 6), then another carbon (atomic number 6)
d. H	hydrogen (atomic number 1)

See the explanation in part *a* on how to differentiate between *b* and *c*.

This is an (*S*) stereoisomer.

c. The order of priority for the groups in this substance is as follows:

a. NH	nitrogen (atomic number 7)
b. COOH	carbon (atomic number 6), then oxygen (atomic number 8)
c. C-C	carbon (atomic number 6), then another carbon (atomic number 6)
d. H	hydrogen (atomic number 1)

See the explanation in part *a* on how to differentiate between *b* and *c*.

This is an (*S*) stereoisomer.

75. Structure *a* has only carbon-carbon single bonds, so it is saturated.

    Structure *b* contains one carbon-carbon double bond, so it is unsaturated.

    Structure *c* contains four carbon-carbon double bonds, so it is unsaturated.

77. Both leucine and isoleucine have a molecular formula of $C_5H_{10}NH_2COOH$. Since they both have the same carbon and hydrogen content, they should have approximately the same fuel value.

78. Fats contain a greater percentage of carbon and hydrogen and less oxygen as compared with carbohydrates and proteins. This greater carbon and hydrogen content increases the fuel value of fats.

    The typical formula of a fat is $C_{57}H_{110}O_6$ (about 1 oxygen atom for every 9.5 carbon atoms).

    The typical formula of a carbohydrate is $C_6H_{12}O_6$ or $C_{12}H_{22}O_{11}$ (about 1 oxygen atom for each carbon atom).

    Proteins are combinations of amino acids. A typical formula of an amino acid is $C_2H_4NO_2$ (glycine) or $C_{11}H_{11}O_2O_2$ (tryptophan). In proteins there is from 1 carbon atom for each oxygen to maybe as many as 5.5 carbons for each oxygen atom.

79. Glucose, mannose, and galactose all have the same molecular formula. However, there are slight differences in structure so these isomers have similar, but not identical, fuel values.

80. Glucose exists in three forms: the cyclic hemiacetal forms, α-glucose and β-glucose, and the linear form (aldohexose form). These three forms exist as an equilibrium mixture in which there is very little ($< 1\%$) in the aldohexose form. Each of the individual forms is likely to have a slightly different fuel value. The normal fuel value will be a weighted average of the three forms.

81. Butter is made up of a greater percentage of fat than most foods, so it has a very high fuel value to use when necessary. Fats have more than twice the fuel value (caloric content) of carbohydrates or proteins, so a high-fat food is an efficient way to provide energy on an Arctic expedition.

82. Starch has a slightly higher fuel value than glucose because it has a greater carbon and hydrogen content (or

a smaller oxygen content) on a percentage basis. When sugar units combine to form complex monosaccharides such as starch, water molecules are removed. This removes oxygen atoms from the new substance but does not remove carbon atoms. This results in an increased carbon content.

83. Running at 5 mph corresponds to running a mile every 12 minutes and requires 480–600 Calories/hour.
    The fuel value of fat is 38 kJ/g.
    The number of calories corresponding to a pound of fat is

$$1 \text{ pound} \times \frac{454 \text{ g}}{1 \text{ pound}} \times \frac{38 \text{ kJ}}{\text{g}} = 17{,}252 \text{ kJ}$$

Converting from kJ to Cal (remember 1 Cal = 1000 cal = 1 Kcal)

$$17{,}252 \text{ kJ} \times \frac{1 \text{ Cal}}{4.184 \text{ kJ}} = 4123 \text{ Cal}$$

One pound of fat contains about 4123 Cal.
If in 1 hour you burn 480 Calories per hour, the time required to burn 4123 Cal is

$$4123 \text{ Cal} \times \frac{1 \text{ hr}}{480 \text{ Cal}} = 8.6 \text{ hr}$$

Next, determine how many miles you would run in 8.6 hours (running at 5 mph).

$$8.6 \text{ hr} \times 5 \text{ miles/hr} = 43 \text{ miles}$$

If you burn 600 Calories per hour, the time required to burn a pound of fat is

$$4123 \text{ Cal} \times \frac{1 \text{ hr}}{600 \text{ Cal}} = 6.9 \text{ hr}$$

Next, determine how many miles you would run in 6.9 hours (running at 5 mph).

$$6.9 \text{ hr} \times 5 \text{ miles/hr} = 34.5 \text{ miles}$$

An average person needs to run between 34.5 miles and 43 miles to use all the energy provided by a pound of fat.

85. Fat has a fuel value of 38 kJ/g; that is, 1 g of fat will produce 38 kJ of energy. If all of this energy is used to heat water, we can then use the equation

$$q = (\text{moles of } H_2O)(\text{molar heat capacity of water})$$
$$(\Delta T) = nc_p\Delta T$$

$q = 38$ kJ
$\Delta T = 33°C$

molar heat capacity of water is 75.3 J/mol · °C
We can solve the above equation for moles of H_2O:

$$\text{moles of } H_2O = \frac{q}{(\text{molar heat capacity})(\Delta T)}$$

Before substituting, we see that the units of q are in kJ, while the molar heat capacity has an energy unit of J. We need to convert one of these energy units so they are the same.

$$38 \text{ kJ} = 38{,}000 \text{ J}$$

Now we can substitute the data and calculate how many moles of water can be heated by 1 g of water.

$$\text{moles of } H_2O = \frac{q}{(\text{molar heat capacity})(\Delta T)}$$
$$= \frac{38{,}000 \text{ J}}{(75.3 \text{ J/mole} \cdot °C)(33°C)}$$
$$= 15.3 \text{ mol } H_2O$$

Lastly, convert moles of water to mass in grams by multiplying by the molar mass.

$$\text{mass of } H_2O = 15.3 \text{ mol } H_2O \times \frac{18.0 \text{ g } H_2O}{\text{mole } H_2O}$$
$$= 275 \text{ g } H_2O$$

87. Step 1 in glycolysis is the conversion of glucose into glucose 6-phosphate. This reaction requires energy. The needed energy is provided by the breakdown of an ATP molecule. The overall energy change for this process results in a $\Delta G°$ of −17.7 KJ (see text, page 651).
    Step 2 converts glucose 6-phosphate into fructose 6-phosphate. This is a molecular rearrangement. One should expect that a molecular rearrangement should occur with a small energy change.

88. Since it requires energy to convert ADP into ATP, this process is only possible if other steps provide this energy. The exergonic steps allow for the net formation of 2 ATP molecules in the overall glycolysis process. The net result is there is a cache of stored energy which can be used at some later time.

89. The reactions along with their corresponding $\Delta G°$ values are additive. That is, we can use Hess's law to calculate the free energy change of a multi-step process.

90. For steps 3 and 7 (Figure 13.18), there is an increase in entropy.

91. The missing product is NH_3.
    In this reaction, an amide is converted into a carboxylic acid and ammonia by the addition of water across the amide linkage. An OH group from water bonds to the carbonyl carbon and a H from water bonds to the nitrogen.

The reaction with complete structures is

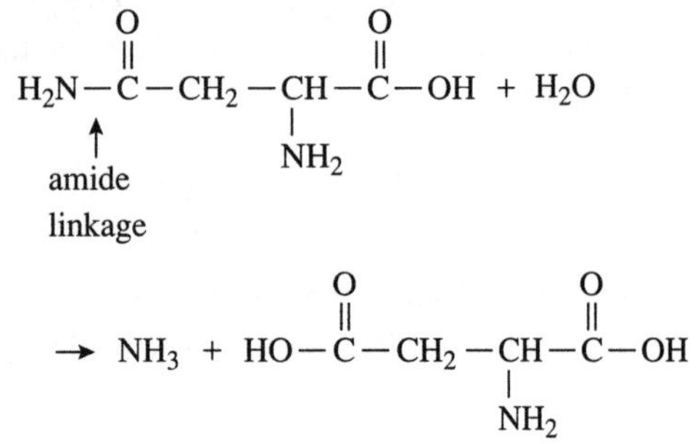

$$H_2N-\overset{\overset{\displaystyle O}{\|}}{C}-CH_2-\underset{\underset{\displaystyle NH_2}{|}}{CH}-\overset{\overset{\displaystyle O}{\|}}{C}-OH + H_2O$$

↑ amide linkage

$$\rightarrow NH_3 + HO-\overset{\overset{\displaystyle O}{\|}}{C}-CH_2-\underset{\underset{\displaystyle NH_2}{|}}{CH}-\overset{\overset{\displaystyle O}{\|}}{C}-OH$$

The molecule will split at the amide linkage marked by the vertical arrow.

93. The reaction is

$$\text{maltose} + H_2O \rightarrow 2\ \text{glucose}$$

For this reaction

$$\Delta G^o_{rxn} = (2\ \text{mol}\ \times \Delta G^o_{f,\text{glucose}}) - (1\ \text{mol}\ \times \Delta G^o_{f,\text{maltose}} + 1\ \text{mol} \times \Delta G^o_{f,H_2O})$$

The problem listed the free energies of formation (ΔG_f) as:

glucose $= -1274.5$ kJ/mol
maltose $= -2246.6$ kJ/mol
$H_2O = -285.8$ kJ/mol

Substituting these values, we get

$$\Delta G^o_{rxn} = (2\ \text{mol}) \times (-1274.5\text{kJ/mol}) - [(1\ \text{mol}) \times (-2246.6\ \text{kJ/mol}) + (1\ \text{mol}) \times (-285.8\ \text{kJ/mol})]$$
$$= -16.6\ \text{kJ}$$

The free energy change is -16.6 kJ.

95. The three kinds of subunits used to form DNA are

1. A phosphate group
2. A five-carbon cyclic sugar called deoxyribose
3. A nitrogen-containing base

The two components that form the "backbone" are the phosphate group and the sugar.

96. A codon is a sequence of three nucleotides that codes for a specific amino acid. There are 20 amino acids. If the codon were made up of only two nucleotides, it would not be possible to have a unique codon for each amino acid since there are only 16 possible combinations made up of two nucleotides.

97. The strands of the double helix of DNA are held together by hydrogen bonds.

98. Base pairing refers to the hydrogen bonding that occurs between two specific pairs of bases that hold the two strands of a DNA molecule together.
   In DNA the complementary base pairs are

adenine and thymine
guanine and cytosine

99. Adenosine is a combination of adenine (a base) and deoxyribose. Adding a phosphate unit from (H_3PO_4) to C-5 of the sugar results in the formation of adenosine 5′-monophosphate.
   The resulting structure is

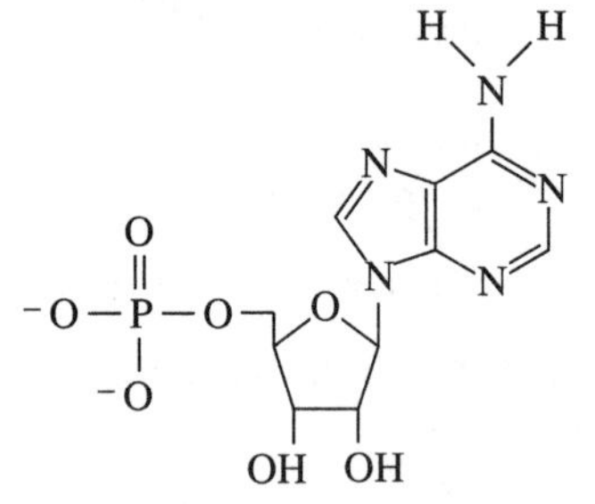

H H N N N O ‖ $^-$O—P—O O N N | $^-$O OH OH

adenosine 5′-monophosphate

The number of H atoms on the phosphate group depends on the pH.

101. The complementary base pairs in DNA are

Thymine (T) and adenine (A)
Cytosine (C) and guanine (G)

The original sequence is T-C-G-G-T-A.
   Using the above base pairings we get the complementary sequence A-G-C-C-A-T.

CHAPTER 14 Chemical Kinetics and Air Pollution

1. The average rate is the rate of a reaction averaged over a specific time interval. The instantaneous rate is the rate at any particular moment during a chemical reaction.

2. Yes. Since the average rate is determined using a specific time interval, by decreasing the time to smaller intervals, one can, in the limit, find a time interval that equals an instantaneous rate. The average rate is also equal to the instantaneous rate if we choose a time interval where the rate is constant.

3. The average rate of a chemical reaction usually changes over time because the concentrations of the reactants and products change as a reaction takes place.

4. Yes. The instantaneous rate of a chemical reaction changes over time because the concentrations of the reactants and products change as a reaction takes place.

5. a. $-\dfrac{\Delta[NH_3]}{\Delta t} = \dfrac{\Delta[H^+]}{\Delta t} = \dfrac{\Delta[NO_2^-]}{\Delta t}$

   b. $\dfrac{\Delta[NO_2^-]}{\Delta t} = -\dfrac{2}{3}\dfrac{\Delta[O_2]}{\Delta t}$

   c. $-\dfrac{\Delta[NH_3]}{\Delta t} = -\dfrac{2}{3}\dfrac{\Delta[O_2]}{\Delta t}$

7. a. $-2\dfrac{\Delta[H_2O_2]}{\Delta t} = \dfrac{\Delta[OH]}{\Delta t}$

   b. $-\dfrac{\Delta[ClO]}{\Delta t} = -\dfrac{\Delta[O_2]}{\Delta t} = \dfrac{\Delta[ClO_3]}{\Delta t}$

   c. $-2\dfrac{\Delta[N_2O_5]}{\Delta t} = -2\dfrac{\Delta[H_2O]}{\Delta t} = \dfrac{\Delta[HONO_2]}{\Delta t}$

9. In the graph N_2O must be represented by the line that indicates decreasing concentration over time since it is the only reactant. Of the two lines that indicate increasing concentration over time, the uppermost line must represent N_2 since it is formed twice as fast as O_2. The remaining line represents O_2.

11. $SO_2 + 3\,CO \rightarrow 2\,CO_2 + COS$

   a. $\dfrac{\Delta[CO_2]}{\Delta t} = -\dfrac{2}{3}\dfrac{\Delta[CO]}{\Delta t}$

   b. $\dfrac{\Delta[COS]}{\Delta t} = -\dfrac{\Delta[SO_2]}{\Delta t}$

   c. $-\dfrac{\Delta[CO]}{\Delta t} = -3\dfrac{\Delta[SO_2]}{\Delta t}$

13. a. $2\,ClO \rightarrow Cl_2 + O_2$

$$\frac{\Delta[Cl_2]}{\Delta t} = -\frac{2.95 \times 10^6\ M\ ClO}{s} \times \frac{1\text{ mole } Cl_2}{-2\text{ mole ClO}}$$
$$= 1.48 \times 10^6\ M\ Cl_2/s$$
$$\frac{\Delta[O_2]}{\Delta t} = -\frac{2.95 \times 10^6\ M\ ClO}{s} \times \frac{1\text{ mole } O_2}{-2\text{ mole ClO}}$$
$$= 1.48 \times 10^6\ M\ O_2/s$$

   b. $ClO + O_3 \rightarrow O_2 + ClO_2$

$$\frac{\Delta[O_2]}{\Delta t} = -\frac{9.03 \times 10^3\ M\ ClO}{s} \times \frac{\text{mole } O_2}{-\text{mole ClO}}$$
$$= 9.03 \times 10^3\ M\ O_2/s$$
$$\frac{\Delta[ClO_2]}{\Delta t} = -\frac{9.03 \times 10^3\ M\ ClO}{s} \times \frac{\text{mole } ClO_2}{-\text{mole ClO}}$$
$$= 9.03 \times 10^3\ M\ ClO_2/s$$

15. $NO_2^- + O_3 \rightarrow NO_3^- + O_2$

   Average rate$_{0-100\,\mu s}$

$$\frac{\Delta[O_3]}{\Delta t} = \frac{[O_3]_{100} - [O_3]_0}{100\,\mu s - 0} = -1.37 \times 10^{-5}\ M\ O_3/\mu s$$

Average rate$_{200-300\,\mu s}$

$$\frac{\Delta[O_3]}{\Delta t} = \frac{[O_3]_{300} - [O_3]_{200}}{300\,\mu s - 200\,\mu s}$$

$$= -5.50 \times 10^{-6}\,M\,O_3/\mu s$$

17. $2\,ClO \rightarrow Cl_2O_2$

Using the balanced equation and the data given, the concentration of Cl_2O_2 was calculated at each time given. The balanced equation tells us that for every two moles of ClO that react, one mole of Cl_2O_2 forms. At time = 1s, the concentration of ClO has dropped from 2.60×10^{11} to 1.08×10^{11}.

$$2.60 \times 10^{11} - 1.08 \times 10^{11} = 1.52 \times 10^{11}$$

$$\frac{(1.52 \times 10^{11}\text{ molecules ClO})(1\text{ molecule }Cl_2O_2)}{cm^3(2\text{ molecules ClO})}$$

$$= 7.60 \times 10^{10}\text{ molecules }Cl_2O_2/cm^3$$

The concentrations of Cl_2O_2 at the other times were obtained in a similar manner.

Time(s)	[ClO]	[Cl_2O_2]
0	2.60×10^{11}	0
1	1.08×10^{11}	7.60×10^{10}
2	6.83×10^{10}	9.59×10^{10}
3	4.99×10^{10}	1.05×10^{11}
4	3.93×10^{10}	1.10×10^{11}
5	3.24×10^{10}	1.14×10^{11}
6	2.76×10^{10}	1.16×10^{11}

Following the method used in Sample Exercise 14.3 of the textbook, the instantaneous rates of ClO and Cl_2O_2 at 1 second can be obtained. As in the sample exercise, the tangent line was drawn, the convenient lines on the tangent line were picked, and the slope was calculated in the standard mathematical way. The calculations below used estimated concentration and time data. Your estimates may be different.

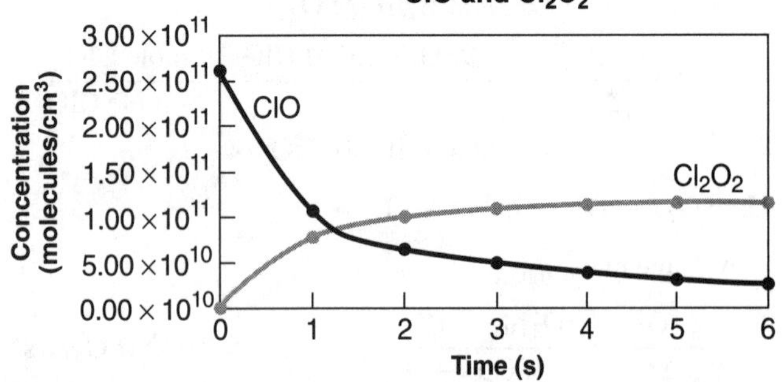

For ClO:

$$\text{Instantaneous rate at 1 second} = \frac{\Delta y}{\Delta x}$$

$$= \frac{[ClO]_2 - [ClO]_0}{2\,s - 0\,s}$$

$$= \frac{1.0 \times 10^{10} - 2.0 \times 10^{11}}{2 - 0}$$

$$= -9.5 \times 10^{10}\text{ molecules/cm}^3 \cdot s$$

For Cl_2O_2:

$$\text{Instantaneous rate at 1 second} = \frac{\Delta y}{\Delta x}$$

$$= \frac{[Cl_2O_2]_2 - [ClO]_0}{2\,s - 0\,s}$$

$$= \frac{1.25 \times 10^{11} - 3.0 \times 10^{10}}{2 - 0}$$

$$= 4.75 \times 10^{10}\text{ molecules/cm}^3 \cdot s$$

19. Two different reactions could have the same reactants, but it is unlikely that two reactions would ever have the same order of reactants and the same value of the rate constant.

20. Since the units of rate a rate law expression are conventionally M/s, reactions of various orders must have rate constants with different units in order to ensure that the units on both sides of the expression agree. Let's look at the units that k should have to give units of M/s for the rate.

 If the reaction is zero order

$$\text{rate} = k$$

In order for the rate units to be M/s the units of k for a zero-order reaction must be M/s.

 For a first-order reaction

$$\text{rate} = k[X]$$

The units for the equation are as follows:

$$M/s = kM$$

Solving for k

$$k = 1/s$$

The units of k for a first-order reaction must be s^{-1}.

21. Yes. The half-life, $t_{1/2}$, is a measure of time and hence a half-life should always be measured in units of time.

22. No. The half-life of a first-order reaction is independent of the concentration of the reactants and inversely proportional to the rate constant for that particular reaction.

$$t_{1/2} = \frac{\ln 2}{k} = \frac{0.693}{k}$$

23. Doubling the initial concentration of a second-order reactant will decrease the half-life by one half. This is easy to see from the equation for second-order half-lives.

$$t_{1/2} = \frac{1}{k[A]_0}$$

24. Yes. Since these reactions have the same value of k and they are both first order, they must have equal half-lives.

25. a. Rate $= k[A][B]$ This reaction is first-order with respect to A, first-order with respect to B, and second-order overall.
    b. Rate $= k[A]^2[B]$ This reaction is second-order with respect to A, first-order with respect to B, and third-order overall.
    c. Rate $= k[A][B]^3$ This reaction is first-order with respect to A, third-order with respect to B, and fourth-order overall.

27. a. Rate $= k[O][NO_2]$ The units of the rate constant will be $M^{-1}\,s^{-1}$.
    b. Rate $= k[NO]^2[Cl_2]$ The units of the rate constant will be $M^{-2}\,s^{-1}$.
    c. Rate $= k[CHCl_3][Cl_2]^{1/2}$ The units of the rate constant will be $M^{-1/2}\,s^{-1}$.

29. a. Rate $= k[BrO]$
    b. Rate $= k[BrO]^2$
    c. Rate $= k[BrO]$
    d. Rate $= k[BrO]^0 = k$

31. From the data given we know that the reaction is second-order overall. This leaves many possibilities for the orders of the individual reactants. In order to determine if the reaction is first order in each reactant we need to know order of one of the individual reactants. This can be obtained by measuring the reaction rate in two kinetics experiments, in which the concentration of one of the reactants is varied and the concentration of the other reactant is held constant.

33. a. Rate $= k[NO_2][O_3]$
    b. Rate $= k[NO_2][O_3]$
    Rate $= (1.93 \times 10^4\ M^{-1}\,s^{-1})(1.8 \times 10^{-8} M)(1.4 \times 10^{-7} M)$
    Rate $= 4.9 \times 10^5\ M/s$
    c. The rate of appearance of NO_2 will be $4.9 \times 10^5\ M/s$.
    d. Since the reaction is first-order with respect to O_3, if the concentration of O_3 is doubled, the rate of the reaction will be doubled if all are conditions are kept constant.

35. Since all of the reactions involve two reactants, the concentrations of all reactants are the same, and each given k has the same units, the reaction with the largest value of k will yield the highest rate. Reaction C is the fastest since it has the largest value of k of these four second-order reactions.

37. Assuming that the changes made were not accompanied by any other changes, the rate law for this reaction would be

$$\text{rate} = k[NO]^1[NO_2]^1[H_2O]^0$$

This reaction would be considered pseudo-second-order since the concentration of H_2O is essentially constant.

39. $2\,ClO_2(g) + 2\,OH^-(aq) \rightarrow ClO_3^-(aq) + ClO_2^-(aq) + H_2O(l)$

By inspection of Experiments 1 and 2, we see that if the $[OH^-]$ is held constant and the $[ClO_2]$ is tripled, the rate of the reaction triples. This indicates that the reaction is first order in ClO_2.

By inspection of Experiments 2 and 3, we see that if the $[ClO_2]$ is held constant and the $[OH^-]$ is tripled, the rate of the reaction triples. This indicates that the reaction is first order in OH^-.

The rate law for this reaction is

$$\text{rate} = k[ClO_2][OH^-]$$

The rate constant can be obtained by substituting the data from Experiment 1 in the preceeding equation

$$0.0248\ M/s = k(0.060\ M)(0.030\ M)$$

$$k = 1.4 \times 10^1\ M^{-1}\,s^{-1}$$

41. $H_2 + 2\,NO \rightarrow 2\,H_2O + N_2$

By inspection of Experiments 1 and 2, we see that if the $[H_2]$ is held constant and the $[NO]$ is doubled, the rate of the reaction quadruples. This indicates that the reaction is second order in NO.

By inspection of Experiments 3 and 4, we see that if the $[NO]$ is held constant and the $[H_2]$ is doubled, the rate of the reaction doubles. This indicates that the reaction is first order in H_2.

The rate law for this reaction is

$$\text{rate} = k[H_2][NO]^2$$

We can determine k by substituting the data from Experiment 1 in the preceeding equation

$$0.0248\ M/s = k(0.212\ M)(0.136\ M)^2$$

$$k = 6.32\ M^{-2}\,s^{-1}$$

43. $Cl_2O_2 \rightarrow 2\ ClO$

Since the reaction is first order, we know that a plot of ln [Cl_2O_2] versus time will yield a straight line and the slope of the line will equal $-k$.

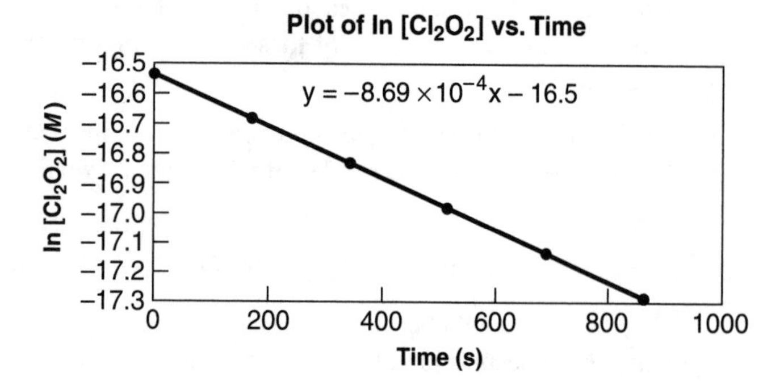

The rate constant has a value of $8.69 \times 10^{-4}\ \mu s^{-1}$.

$$\text{slope} = \frac{\Delta y}{\Delta x} = \frac{-17.283 - -16.534}{862 - 0}$$

$$= -8.69 \times 10^{-4}\ s^{-1}$$

$$k = 8.69 \times 10^{-4}\ s^{-1}$$

The slope can also be obtained by performing a linear regression of the data in the plot.

45. $2\ ClO(g) \rightarrow Cl_2O_2$

Using the data given the following plot was obtained.

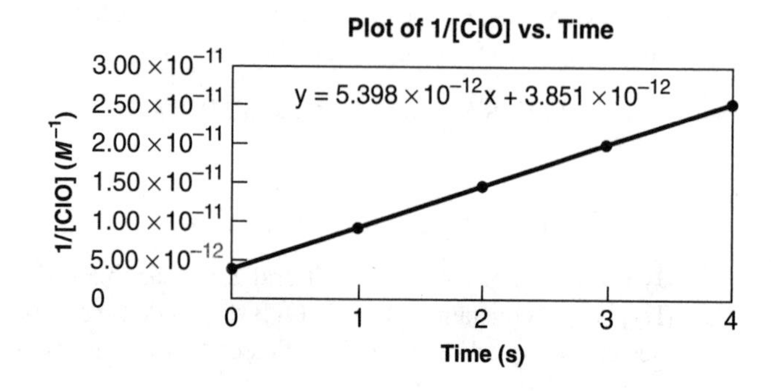

For a second-order reaction, a plot of inverse concentration versus time gives a straight line whose slope is k.

$$\text{slope} = \frac{\Delta y}{\Delta x} = \frac{2.54 \times 10^{-11} - 3.85 \times 10^{-12}}{4 - 0}$$

$$= 5.39 \times 10^{-12}\ M^{-1}\ s^{-1}$$

$$k = 5.39 \times 10^{-12}\ M^{-1}\ s^{-1}$$

The slope can also be obtained by performing a linear regression on the data in the plot. This method yields a slope equal to $5.40 \times 10^{-12}\ M^{-1}\ s^{-1}$ and hence a $k = 5.40 \times 10^{-12}\ M^{-1}\ s^{-1}$.

47. $2\ HNO_2(aq) \rightarrow NO(g) + NO_2(g) + H_2)(l)$

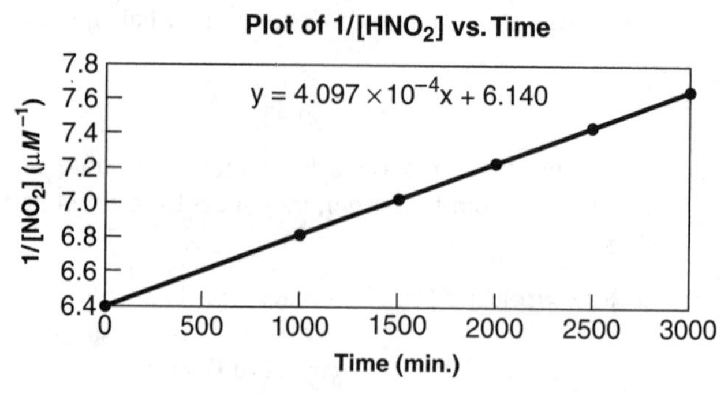

Since a plot of inverse concentration versus time is linear, the reaction is second order.

$$\text{rate} = k[HNO_2]^2$$

Slope is equal to k for an inverse concentration versus time plot of second-order data.

$$\text{slope} = \frac{\Delta y}{\Delta x} = \frac{7.639 - 6.410}{3000 - 0}$$

$$= 4.097 \times 10^{-4}\ \mu M^{-1}\ \text{min}^{-1}$$

$$k = 4.097 \times 10^{-4}\ \mu M^{-1}\ s^{-1}$$

A linear regression of the data in the plot yields the same value of k.

49. A plot of ln [NH_3] versus time is linear; therefore the reaction is first order in NH_3. Such a plot yields a slope equal to $-k$.

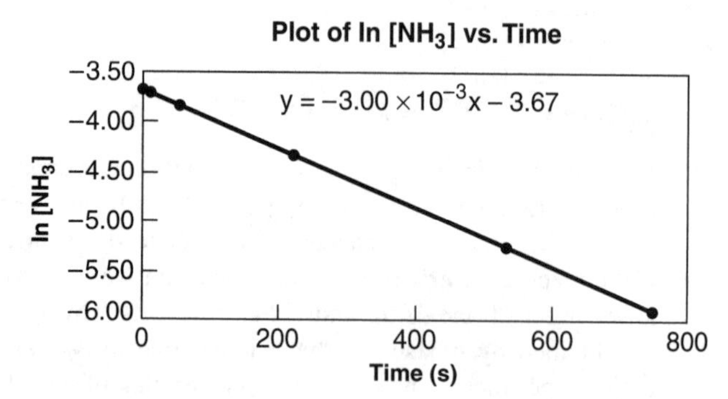

The rate law is

$$\text{rate} = k[NH_3]$$

$$\text{slope} = \frac{\Delta y}{\Delta x} = \frac{-5.90 - (-3.67)}{746 - 0} = -2.99 \times 10^{-3}\ s^{-1}$$

$$k = 2.99 \times 10^{-3}\ s^{-1}$$

A linear regression of the data in the plot yields a slope equal to $-3.00 \times 10^{-3}\ s^{-1}$ and a k equal to $3.00 \times 10^{-3}\ s^{-1}$.

51. a. Since the plot of ln $[N_2O]$ versus time is linear, the reaction is first order and the rate law is

$$\text{rate} = k[N_2O]$$

Our initial amount of N_2O is 100 (100%) and the final is 6.25 (6.25%).

$$\ln\frac{[N_2O]}{[N_2O]_0} = -kt$$

$$\ln[N_2O] - \ln[N_2O]_0 = -kt$$

b.

$$\begin{aligned}
\ln[N_2O] &= -kt + \ln[N_2O]_0 \\
\ln 6.25 &= -kt + \ln 100 \\
\ln 6.25 - \ln 100 &= -kt \\
\ln 100 - \ln 6.25 &= kt \\
\ln(100/6.25) &= kt \\
\ln 16 &= kt \\
\frac{\ln 16}{k} &= t
\end{aligned}$$

$$\text{number of half-lives} = \frac{t}{t_{1/2}}$$

$$\text{number of half-lives} = \frac{(\ln 16)/k}{(\ln 2)/k}$$

$$\text{number of half-lives} = 4$$

53. a. Since the plot of $1/[C_4H_6]$ versus time was linear, the reaction is second order and the rate law is

$$\text{rate} = k[C_4H_6]^2$$

b. The original amount would be considered 100%. The number of half-lives necessary to reach 3.1% of the original amount would be five half-lives.

$$\frac{100\%}{2^5} = 3.125\%$$

55. To calculate the half-life of a second-order reaction we use the formula

$$t_{1/2} = \frac{1}{k[A]_0}$$

In Problem 45 we obtained a value of 5.40×10^{-12} $M^{-1}\,s^{-1}$ for the rate constant and were given a value of 2.60×10^{11} M for the initial concentration.

$$\begin{aligned}
t_{1/2} &= \frac{1}{k[A]_0} \\
&= \frac{1}{(5.40 \times 10^{-12}\,M^{-1}\,s^{-1})(2.60 \times 10^{11}\,M)} \\
&= 7.12 \times 10^{-1}\,s
\end{aligned}$$

57. $C_{12}H_{22}O_{11}(aq) + H_2O(l) \rightarrow 2\,C_6H_{12}O_6(aq)$

The following plot of ln $[C_{12}H_{22}O_{11}]$ versus time is linear.

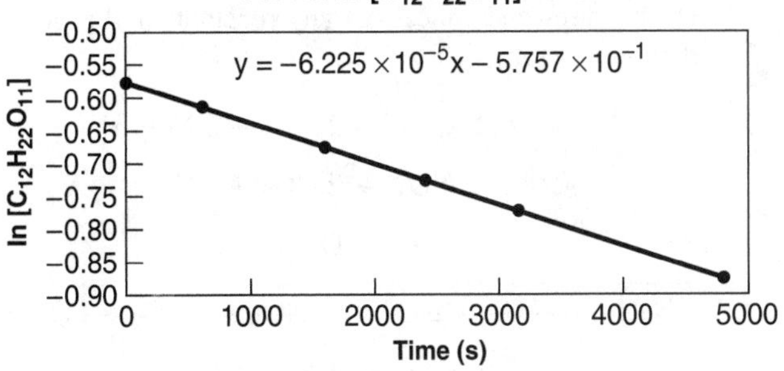

Since water is both a solvent and a reactant, we say the reaction is pseudo-first-order in sucrose. For the rate law we can write

$$\text{rate} = k[C_{12}H_{22}O_{11}][H_2O]$$

Since the concentration of water remains essentially constant (55.5 M) we can say

$$\text{rate} = k'[C_{12}H_{22}O_{11}]$$

where $k' = k[H_2O]$. The second-order rate constant is k and the pseudo-first-order rate constant is k'.

Slope equals $-k'$ in the plot we obtained.

$$\begin{aligned}
\text{slope} &= \frac{\Delta y}{\Delta x} = \frac{(-0.875 - -0.576)}{4800 - 0} \\
&= -6.23 \times 10^{-5}\,s^{-1} \\
k' &= 6.23 \times 10^{-5}\,s^{-1} \\
k' &= k[H_2O] \\
6.23 \times 10^{-5}\,s^{-1} &= k\,55.1\,M \\
k &= 1.12 \times 10^{-6}\,M^{-1}\,s^{-1}
\end{aligned}$$

59. These reactions cannot proceed by the same mechanism since they have very different rate laws. Based solely on the reactants, the reactions could have "similar" mechanisms. Only reactions with similar rate laws can have similar mechanisms. Since the mechanism and the rate law are intimately related, these two reactions cannot have similar mechanisms because their rate laws are so different. The common chemical between the reactions is first order for one reaction and second order for the other.

60. Yes, it is possible for these reactions to have similar mechanisms since their rate laws and reactants are similar.

61. a. Rate = $k[SO_2Cl_2]$, unimolecular
b. Rate = $k[NO_2][CO]$, bimolecular
c. Rate = $k[NO_2]^2$, bimolecular

63. In order to determine the overall reaction we simply sum up the steps and cancel out any reactants and products that are the same.

$$2\,N_2O_5 \rightarrow 2\,NO_3 + 2\,NO_2$$
$$2\,NO_3 \rightarrow 2\,NO_2 + 2\,O$$
$$2\,O \rightarrow O_2$$
$$2\,N_2O_5 + \cancel{2\,NO_3} + \cancel{2\,O} \rightarrow \cancel{2\,NO_3} + 4\,NO_2 + \cancel{2\,O} + O_2$$
$$2\,N_2O_5 \rightarrow 4\,NO_2 + O_2$$

65. If step 2 were the rate-determining step then the rate law would be

$$\text{rate} = k[O][N_2]$$

We are told that the step needed to break oxygen molecules into oxygen atoms is fast and reversible. We can use this information to derive an expression for [O] that we can substitute into the rate law

$$O_2 \rightleftharpoons 2\,O$$
$$k_f[O_2] = k_r[O]^2$$
$$\frac{k_f[O_2]}{k_r} = [O]^2$$
$$\frac{k_f^{1/2}[O_2]^{1/2}}{k_r^{1/2}} = [O]$$

Substituting back into the rate law we obtain

$$\text{rate} = k[O][N_2] = \frac{k\,k_f^{1/2}[O_2]^{1/2}[N_2]}{k_r^{1/2}} = k_{\text{overall}}\,[O_2]^{1/2}[N_2]$$

This matches the experimentally determined rate law. Step 2 must be the rate-determining step.

67. If the first step was the r.d.s.:

$$\text{rate} = k[NO][Cl_2]$$

since the rate law for the overall reaction is equal to the rate law for the r.d.s.

If the second step was the r.d.s.:

$$\text{rate} = k_2[NOCl_2][NO]$$

We must substitute in the $NOCl_2$ since it is not a reactant.

$$k_{1f}[NO][Cl_2] = k_{1r}[NOCl_2]$$
$$\frac{k_{1f}}{k_{1r}}[NO][Cl_2] = [NOCl_2]$$

Substituting in for $[NOCl_2]$ we obtain

$$\text{rate} = \frac{k_2 k_{1f}}{k_{1r}}[NO]^2[Cl_2]$$
$$\text{rate} = k_{\text{overall}}[NO]^2[Cl_2]$$

If this was the correct mechanism, then the first step must be the r.d.s since it is the only r.d.s. that yields a rate law that agrees with the experimental data, first order in both NO and Cl_2.

69. a. The first step in the mechanism is the rate-determining step, so the rate law for the reaction is equal to the rate law for this step. For the proposed mechanism,

$$\text{rate} = k[NO_2]$$

This mechanism matches the experimental rate law for the photochemical decomposition of NO_2.

b. The second step of the mechanism is the rate-determining step. The rate law for the reaction is equal to the rate law for this step.

$$\text{rate} = k_2[N_2O_4]$$

Since N_2O_4 is not a reactant, we must substitute for it by assuming the previous step is fast and reversible. We then obtain

$$k_{1f}[NO_2]^2 = k_{1r}[N_2O_4]$$

solving for $[N_2O_4]$ we obtain

$$[N_2O_4] = \frac{k_{1f}}{k_{1r}}[NO_2]^2$$

Substituting into the rate law we obtain:

$$\text{rate} = k_2\frac{k_{1f}}{k_{1r}}[NO_2]^2$$
$$\text{rate} = k_{\text{overall}}[NO_2]^2$$

This mechanism matches the experimental rate law for the thermal decomposition of NO_2.

c. The first step in the mechanism is the rate-determing step, so the rate law for the reaction is equal to the rate low for this step. For the proposed mechanism, rate $= k[NO_2]^2$.

This mechanism matches the experimental rate law for the thermal decomposition of NO_2.

71. a. False. The sign of the enthalpy of a reaction is not necessarily a good predictor of rate. For example, an exothermic reaction with a relatively high E_a could be considered slow.

b. True. Reactions with positive values of ΔG are likely to be slow since they are nonspontaneous in the forward direction and spontaneous in the reverse direction. This means the rate of the reaction in the reverse direction is faster than the rate of the reaction in the forward direction.

c. False. The sign of the enthalpy change of a reaction is not necessarily a good predictor of rate. For example, an endothermic reaction could have a relatively small E_a and proceed more quickly than an exothermic reaction with a large E_a.

d. False. The sign of the entropy change of a reaction is not necessarily a good predictor or rate.

72. a. False. The spontaneity of a reaction does not tell us about the rate of the reaction.

b. False. Endothermic reactions with high activation energies can be slow.

c. False. The sign of the entropy change is not an absolute predictor of the reaction rate.

d. False. Any reaction at a relatively low temperature would be expected to be relatively slow since a smaller percentage of the reacting species possess the necessary energy to react.

73. The order of a reaction is determined by its mechanism, the route by which reactants change into products. Rate is temperature dependent since it depends on how often molecules collide with the correct orientation to yield an effective collision. As the temperature of a reaction is varied, movements of the molecules change. These changes affect their ability to react in a given time period. The order will not change since the mechanism will not change as the temperature is varied.

74. An exothermic reaction is a reaction where energy is released because the reactants have more energy than the products. For this type of reaction, the E_a is lower for than the E_a for the reverse reaction. The energy of the activated complex is the same for both forward and reverse reactions, but the E_a for the reverse reaction is larger in an exothermic reaction because the reverse reaction begins at a lower energy level. See Figure 14.17 of the text.

75. $k = Ae^{-E_a/RT}$

The Arrhenius equation shows that k is affected by changes in temperature. By examing the equation we can see that the reaction with $E_a = 150$ kJ/mol will have the larger percent increase in k, and hence the larger increase in rate as the temperature is raised.

76. Changing the temperature will yield the more accurate rate constant. The addition of a catalyst will not allow us to determine k for the uncatalyzed reaction. The rate and the ate constant for the catalyzed reaction will be larger than those of the uncatalyzed reaction at the same temperature. Using the data from several temperatures, one should be able to obtain an Arrhenius plot that will allow the calculation of k at the desired temperature. Such plots should not mix catalyzed and uncatalyzed data. The data for catalyzed and uncatalyzed reactions cannot be used in the same plot because they have different values of E_a. The Arrhenius equation is not capable of dealing with two activation energies.

77.

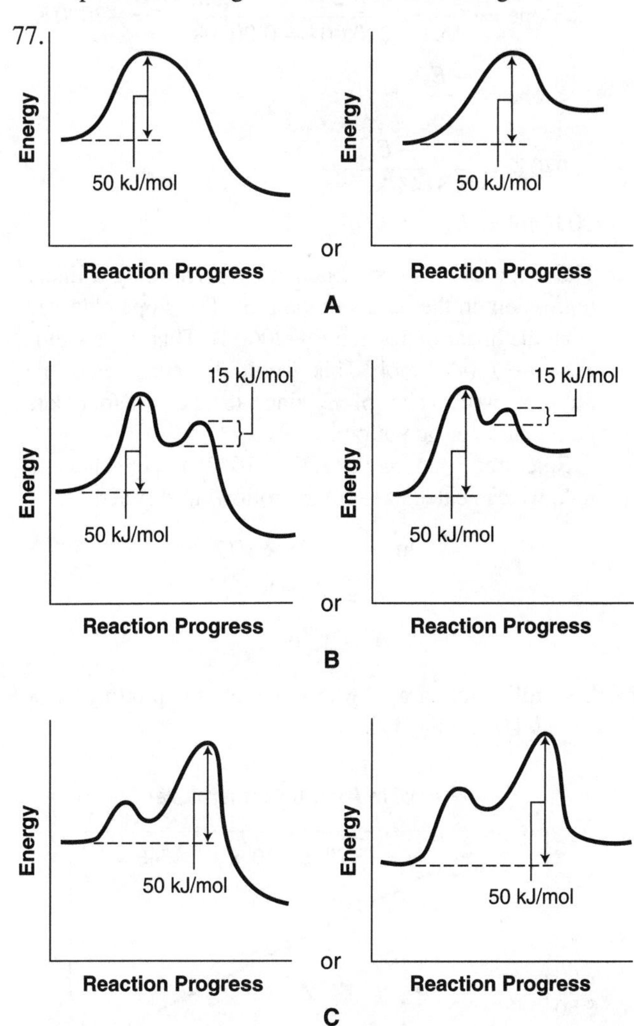

79. The reaction profile indicates that there are two activation energies. This eliminates b, the one-step mechanism. Since the first activation energy is lower than the second, we expect the first step to be relatively fast. The best fit for the reaction profile is c.

81. The following graph was obtained by plotting ln k versus $1/T$ for the data given:

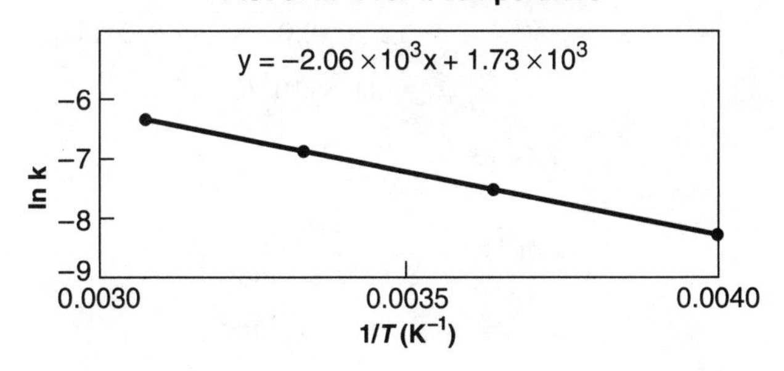

We can calculate E_a since the slope equals E_a/R for a plot of ln k versus $1/T$. The slope can be obtained by using two points in the plot.

$$\text{Slope} = \frac{\Delta y}{\Delta x} = \frac{(-8.2396 - -6.3368)}{(0.004 - 0.00308)} = -2070\,\text{K}$$

$$\text{Slope} = \frac{-E_a}{R}$$

$$-2070\,\text{K} = \frac{-E_a}{8.314\,\text{J/K}\cdot\text{mol}}$$

$$17200\,\text{J/mol} = E_a$$

The slope can also be obtained by performing a linear regression on the points in the plot. The slope obtained from the linear regression is −2060 K. This slope yields $E_a = -17100$ J/mol. This would be considered the more accurate value of E_a since it takes all four data points into consideration.

Since the y-intercept (1.73×10^{-3}) is equivalent to lnA, we can calculate A in the following manner:

$$\ln \text{A} = 1.73 \times 10^{-3}$$

$$\text{A} = e^{1.73\times10^{-3}}$$

$$\text{A} = 1.00$$

83. The following graph was obtained by plotting ln k versus $1/T$ for the data given:

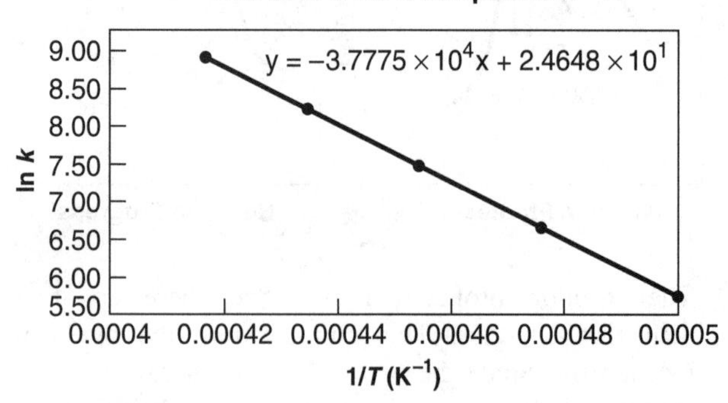

a. We can calculate E_a since the slope equals $-E_a/R$ for a plot of ln k versus $1/T$. The slope can be obtained by using two points in the plot.

$$\text{slope} = \frac{\Delta y}{\Delta x} = \frac{(5.76 - 8.91)}{(0.0005 - 0.000417)}$$

$$= -3.80 \times 10^4\,\text{K}$$

$$\text{slope} = \frac{-E}{R}$$

$$-3.80 \times 10^4\,\text{K} = \frac{-E_a}{8.314\,\text{J/K}\cdot\text{mol}}$$

$$E_a = 3.16 \times 10^5\,\text{J/mol}$$

The slope can also be obtained by performing a linear regression on the points in the plot. The slope from the linear regression is −37800 K. This slope yields $E_a = 3.14 \times 10^5$ J/mol. This would be considered the more accurate value of E_a since it takes all five data points into consideration.

b. Since the y-intercept (24.648) is equivalent to ln A, we can calculate A in the following manner:

$$e^{24.648} = \text{A} = 5.06 \times 10^{10}$$

c. By plugging into the natural logarithm version of the Arrhenius equation we can solve for k at 300 K.

$$\ln k = -\left(\frac{E_a}{R}\right)\left(\frac{1}{T}\right) + \ln A$$

$$\ln k = -\left(\frac{314,000\,\text{J/mol}}{8.314\,\text{J/K}\cdot\text{mol}}\right)\left(\frac{1}{300\,\text{K}}\right) + 24.648$$

$$\ln k = -101$$

$$k = 1.05 \times 10^{-44} M^{-1/2}\text{s}^{-1}$$

85. The following graph was obtained by plotting ln k versus $1/T$ for the data given:

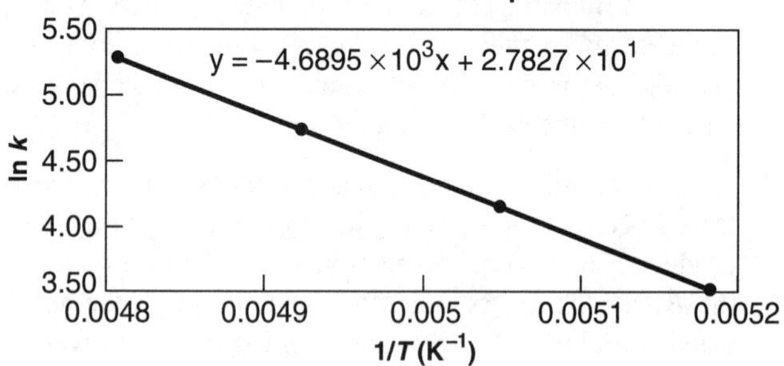

We can calculate E_a since the slope equals $-E_a/R$ for a plot of ln k versus $1/T$. The slope can be calculated by using two points in the plot.

$$\text{slope} = \frac{\Delta y}{\Delta x} = \frac{(3.53 - 5.28)}{(0.005181 - 0.004808)} = -5170\,\text{K}$$

$$\text{slope} = \frac{-E_a}{R}$$

$$-5170\,\text{K} = \frac{-E_a}{8.314\,\text{J/K}\cdot\text{mol}}$$

$$E_a = 4.30 \times 10^4\,\text{J/mol}$$

The slope can also be obtained by performing a linear regression on the points in the plot. The slope from the linear regression is −4690 K. This slope yields $E_a = 3.9 \times 10^4$ J/mol. This would be considered the more accurate value of E_a since it takes all four data points into consideration.

The linear regression also yields a y-intercept which is equivalent to ln A for a plot of ln k versus $1/T$.

$$\ln A = 27.827$$
$$A = e^{27.827}$$
$$A = 1.22 \times 10^{12}$$

87. Yes. Since a catalyst lowers the E_a of a reaction, this yields a larger value of k in the Arrhenius equation $k = Ae^{-E_a/RT}$. A larger value of k results in a faster rate.

88. If the addition of a catalyst changes the mechanism, the reaction will have a different rate-determining step, and hence a different rate law. Catalysts work by lowering the activation energy of the overall reaction. This is accomplished by providing a different pathway (i.e., mechanism) for the reaction. For example, the uncatalyzed reaction to convert *cis*-2-butene to *trans*-2-butene has a rate law of rate $= k[cis\text{-2-butene}]$ while the I_2 catalyzed reaction has a rate law of rate $= k[cis\text{-2-butene}][I_2]^{1/2}$.

89. Yes. A catalyst lowers the E_a for a reaction. If it is lowered for the forward reaction it must be lowered for the reverse. This point will be important when we discuss Le Châtelier's principle in the next chapter.

90. No. Since CO is consumed during the course of the reaction and not produced, it cannot be a catalyst.

91. A bimolecular reaction will obey pseudo-first-order kinetics when the concentration of one reactant is sufficiently high that it remains essentially constant during the course of the reaction.

92. The concentration of a homogeneous catalyst is not found in the rate law for the reaction it catalyzes since it is not found in the overall balanced equation for the reaction.

93. NO is a catalyst for the reaction since it is both consumed and produced during the course of the reaction. N_2O could not be considered a catalyst in this reaction since it is consumed and not regenerated during the course of the reaction.

95. $k = Ae^{-E_a/RT}$

By substituting the given values into the Arrhenius equation we obtain the following values of k:

$O_3 + O \rightarrow O_2 + O_2$

$$k = (8.0 \times 10^{-12}\ \text{cm}^3/\text{molecule} \cdot \text{s})e^{-\left(\frac{1.71\times10^7\ \text{J/mol}}{(8.314\ \text{J/K}\cdot\text{mol})(298\text{K})}\right)}$$

$$k = 8.0 \times 10^{-15}\ \text{cm}^3\text{molecules}^{-1}\ \text{s}^{-1}$$

$O_3 + Cl \rightarrow ClO + O_2$

$$k = (2.9 \times 10^{-11}\ \text{cm}^3/\text{molecule} \cdot \text{s})e^{-\left(\frac{2.16\times10^3\ \text{J/mol}}{(8.314\ \text{J/K}\cdot\text{mol})(298\text{K})}\right)}$$

$$k = 1.2 \times 10^{-11}\ \text{cm}^3\text{molecules}^{-1}\ \text{s}^{-1}$$

Since the reaction with chlorine radicals has a larger value of k, this reaction will proceed faster than the reaction with oxygen radicals.

CHAPTER 15 Chemical Equilibrium and Why Smog Persists

1. At equilibrium the rate of the forward reaction is equal to the rate of the reverse reaction.

2. There are many possible examples of dynamic equilibrium. For example, an unopened bottle of soda likely contains many reactions at equilibrium. There is an equilibrium between the solid and gaseous phases of water.

$$H_2O(l) \rightleftharpoons H_2O(g)$$

If you used a bottle of vinegar today, there was an equilibrium between the acetic acid and its conjugate base.

$$HC_2H_3O_2(aq) + H_2O(l) \rightleftharpoons C_2H_3O_2^-(aq) + H_3O^+(aq)$$

3. No. The concentrations of the reactants and products are still changing at 20 μs.

4. No. Equilibrium is established when the components of the mass action expression equal the equilibrium constant. The mass action expression for the reaction $aA + bB \rightleftharpoons cC + dD$ is

$$K = \frac{[C]^c[D]^d}{[A]^a[B]^b}$$

Equilibrium is established when the mass action expression for a particular reaction equals the equilibrium constant for that particular reaction.

5.

$^{14}N_2$:	28 g/mol	Original reactant.
$^{14}N^{15}N$:	29 g/mol	Possibly formed by: $^{15}N_2O + {}^{14}N^{15}NO \rightarrow {}^{15}N_2 + {}^{14}N^{15}N + O_2$ or $2^{14}N^{15}NO \rightarrow 2^{14}N^{15}N + O_2$
$^{15}N_2$:	30 g/mol	Possibly formed by: $2^{15}N_2O \rightarrow 2^{15}N_2 + O_2$
$^{16}O_2$:	32 g/mol	Original reactant.
$^{14}N_2O$:	44 g/mol	Possibly formed by: $2^{14}N^{15}N + O_2 \rightarrow 2^{14}N^{15}NO$ or $^{15}N_2 + {}^{14}N^{15}N + O_2 \rightarrow {}^{15}N_2O + {}^{14}N^{15}NO$
$^{15}N_2O$:	46 g/mol	Original reactant.

7. Since the reaction is at equilibrium, then

$$k_f[A] = k_r[B]$$

Rearranging we obtain

$$\frac{k_f}{k_r} = \frac{[B]}{[A]}$$

Since $k_f > k_r$ we know that the equilibrium expression above will have an equilibrium constant greater than 1.

8. One form of the Arrhenius expression is $k = e^{-Ea/(RT)}$. From this we see that a reaction with a relatively small E_a will have a relatively large k. This means that for this reaction k_f will be greater than k_r and hence k_f/k_r, which is equal to K, will have a value greater than 1.

9. In conditions where $Q < K$, reactions will proceed toward products. Increasing the amount of products and decreasing the amount of reactants will increase the value of Q so that it approaches K.

10. K is the ratio of rate constants for the forward and reverse reactions, k_f and k_r.

$$K = \frac{k_f}{k_r}$$

In order to have a large value of K with a small value of k_f, the value of k_r must be relatively small compared to k_f.

11. Since the reaction is at equilibrium, then

$$k_f[A] = k_r[B]$$

Rearranging we obtain

$$\frac{k_f}{k_r} = \frac{[B]}{[A]} = K$$

$$\frac{1.50 \times 10^{-2}\,s^{-1}}{4.50 \times 10^{-2}\,s^{-1}} = \frac{[B]}{[A]} = K$$

$$K = 3.33 \times 10^{-1}$$

13. a. $K = \frac{[N_2O_4]}{[N_2][O_2]^2}$

b. $K = \frac{[NO_2][N_2O]}{[NO]^3}$

c. $K = \frac{[N_2]^2[O_2]}{[N_2O]^2}$

15. Based on the graph, it appears that equilibrium has been established after five seconds. Estimating the concentrations at equilibrium and plugging them into the mass action expression we obtain

$$K_c = \frac{[N_2][O_2]^{1/2}}{[N_2O]} = \frac{[0.00032][0.00016]^{1/2}}{[7.10 \times 10^{-6}]} = 0.57$$

Estimated from the graph, $K_c = 0.55$. Note that your estimated concentrations and value of K_c may be slightly different.

17. $$K_c = \frac{[NO]^2}{[N_2][O_2]} = \frac{(3.1 \times 10^{-3})^2}{(3.3 \times 10^{-3})(5.8 \times 10^{-3})}$$

$$K_c = 5.0 \times 10^{-1}$$

19. Let's assume we have a 1-L reaction vessel, so the number of moles will be numerically equivalent to molarity.

	$H_2O(g)$ +	$CO(g)$	⇌	$H_2(g)$ +	$CO_2(g)$
I	0.015	0.015		0	0
C	$-x$	$-x$		$+x$	$+x$
E	$0.015 - x$	$0.015 - x$		x	x

We know that at equilibrium $[CO_2] = 0.0083\ M$; therefore $x = 0.0083$. Solving for the remaining equilibrium concentrations we obtain

$$K_c = \frac{[H_2][CO_2]}{[H_2O][CO]} = \frac{(8.3 \times 10^{-3})(8.3 \times 10^{-3})}{(6.7 \times 10^{-3})(6.7 \times 10^{-3})}$$

$$K_c = 1.53$$

21. $$K_c = 22 = \frac{[B]}{[A]}$$

$$Q = \frac{[B]}{[A]} = \frac{2.0}{0.1} = 20$$

$Q < K$, so the reaction will proceed toward products in order to achieve equilibrium.

23. $$K_{c(forward)} = \frac{[NO_2]^2}{[NO][NO_3]} \qquad K_{c(reverse)} = \frac{[NO][NO_3]}{[NO_2]^2}$$

$$K_{c(forward)} = \frac{1}{K_{c(reverse)}}$$

25. For the three reactions given:

$$2D \rightleftharpoons A + 2B \quad A + 2B \rightleftharpoons C \quad C \rightleftharpoons 2D$$

$$K_1 = \frac{[A][B]^2}{[D]^2} \quad K_2 = \frac{[C]}{[A][B]^2} = 3.3$$

$$K_3 = \frac{[D]^2}{[C]} = 0.041$$

Taking the reciprocal of K_2 and K_3 and multiplying these reciprocals will yield K_1.

$$K_1 = \frac{[A][B]^2}{[D]^2} = (K_2)^{-1}(K_3)^{-1} = \left(\frac{[A][B]^2}{[C]}\right)\left(\frac{[C]}{[D]^2}\right)$$

$$= \frac{[A][B]^2}{[D]^2}$$

$$K_1 = (K_2)^{-1}(K_3)^{-1} = (3.3)^{-1}(0.041)^{-1}$$

$$K_1 = 7.4$$

27. $$K = \frac{[SO_3]}{[SO_2][O_2]^{1/2}} \qquad K' = \frac{[SO_3]^2}{[SO_2]^2[O_2]}$$

$$K^2 = K' \quad \text{or} \quad K = (K')^{1/2}$$

29. $2\,SO_2(g) + O_2(g) \rightleftharpoons 2\,SO_3(g)$ $k_{Original} = \frac{[SO_3]^2}{[SO_2]^2[O_2]}$

a. $SO_2(g) + \frac{1}{2}O_2(g) \rightleftharpoons SO_3(g)$ $K_a = \frac{[SO_3]}{[SO_2][O_2]^{1/2}}$

The coefficients in this equation are halved compared to the original equation. Examining their mass action expressions we can see that K for this reaction, K_a, is equal to $(K_{Original})^{1/2}$ of the original reaction.

$$K_a = (K_{Original})^{1/2} = (2.4 \times 10^{-3})^{1/2} = 4.9 \times 10^{-2}$$

b. $2\,SO_3(g) \rightleftharpoons 2\,SO_2(g) + O_2(g)$ $K_b = \frac{[SO_2]^2[O_2]}{[SO_3]^2}$

This reaction is reversed compared to the original equation. Examining their mass action expressions we can see that K for this reaction, K_b, is equal to $(K_{Original})^{-1}$ of the original reaction.

$$K_b = (K_{Original})^{-1} = (2.4 \times 10^{-3})^{-1} = 4.2 \times 10^2$$

c. $SO_3(g) \rightleftharpoons SO_2(g) + \frac{1}{2}O_2(g)$ $K_c = \frac{[SO_2][O_2]^{1/2}}{[SO_3]}$

This reaction is reversed compared to the original equation. The coefficients in this equation are also half of those in the original equation. Examining their mass action expressions we can see that K for this reaction, K_c, is equal to $(K_{\text{Original}})^{-1/2}$ of the original reaction.

$$K_c = (K_{\text{Original}})^{-1/2} = (2.4 \times 10^{-3})^{-1/2} = 2.0 \times 10^1$$

31. According to the relationship, $K_p = K_c(RT)^{\Delta n}$, K_p will equal K_c when $\Delta n = 0$.

32. Yes. Since $K_p = K_c(RT)^{\Delta n}$
substituting 298 K for T and a positive value for Δn will always yield a K_p that is greater than the corresponding K_c.

33. a. Since $\Delta n = -1$, $K_c \neq K_p$.
    b. Since $\Delta n = 0$, $K_c = K_p$. (Note that Fe and FeO are solids!)
    c. Since $\Delta n = 0$, $K_c = K_p$.

35. $K_p = K_c(RT)^{\Delta n} = (4.95)[(0.08206\ \text{L} \cdot \text{atm/K} \cdot \text{mol})(598\ \text{K})]^{-1}$

$K_p = 1.01 \times 10^{-1}$

37. a.

$$K_p = \frac{P_C}{P_A P_B} = 1$$

$$Q = \frac{(1.0)}{(1.0)(1.0)} = 1$$

Since $Q = K$, the system is at equilibrium.

b. The temperature is 300 K.

$$K_p = K_c(RT)^{\Delta n}$$

$$1 = K_c[(0.08206\ \text{L} \cdot \text{atm/K} \cdot \text{mol})(300\ \text{K})]^{-1}$$

$$K_c = 25$$

$$K_c = \frac{[\text{C}]}{[\text{A}][\text{B}]}$$

$$Q = \frac{(1.0)}{(1.0)(1.0)} = 1$$

Since $Q < K_c$, the system is not at equilibrium. The reaction will proceed toward products in order to achieve equilibrium.

39. $K_p = K_c(RT)^{\Delta n} = (5 \times 10^5)[(0.08206\ \text{L} \cdot \text{atm/K} \cdot \text{mol})(298\ \text{K})]^{-1}$

$K_p = 2 \times 10^4$

41. $K_p = K_c(\text{RT})^{\Delta n}$

$$= (1.5 \times 10^{-3})[(0.08206\ \text{L} \cdot \text{atm/K} \cdot \text{mol})(298\ \text{K})]^0$$

$$K_p = 1.5 \times 10^{-3}\ (K_p = K_c \text{ since } \Delta n = 0)$$

$$K_p = \frac{(P_{NO})^2}{P_{N_2} P_{O_2}}$$

$$Q_p = \frac{(1.00 \times 10^{-3})^2}{(1{,}00 \times 10^{-3})(1.00 \times 10^{-3})} = 1$$

Since $Q_p > K_p$ the system is not at equilibrium. The reaction will proceed toward reactants in order to achieve equilibrium.

43. Yes. We can see this by looking at the formula $\Delta G^\circ = -RT \ln K$. This can be rearranged to obtain $K = e^{-\Delta G^\circ/(RT)}$.

In both forms we see that when K < 1, then $\Delta G^\circ > 0$.

44.

$$\Delta G^\circ = RT \ln K$$

If a reaction is in the gas phase, K is equal to K_p. K_p or K_c is used in the equation depending upon the phase of the reactants and products. K_p is only used when all species in a chemical reaction are gases, and K_c is used in all other cases. The K in the equation above is a thermodynamic equilibrium constant. A thermodynamic equilibrium constant has all gaseous species entered as the partial pressure of that particular gas and all species dissolved in liquids entered as molarities. That means that thermodynamic equilibrium constants can have both pressures and concentrations.

$$MnO_2(s) + 4\,H^+(aq) + 2\,Cl^-(aq) \rightleftharpoons Mn^{2+}(aq) + Cl_2(g) + 2\,H_2O(l)$$

For example, the thermodynamic equilibrium constant for the reaction above would be

$$K = \frac{[Mn^{2+}]P_{O_2}}{[H^+]^4[Cl^-]^2}$$

45.

$$\Delta G^\circ = -RT \ln K$$

$$\Delta G^\circ_{f,ClF} = -57.7\ \text{kJ/mol} = -RT \ln K$$

$$-5.77 \times 10^4\ \text{J/mol} = -(8.314\ \text{J/K} \cdot \text{mol}) \times (298\ \text{K}) \ln K$$

$$\ln K = 23.29$$

$$K = 1.30 \times 10^{10}$$

In the same way, we calculate K for the other two interhalogen compounds:

$$\Delta G^\circ_{f,ClBr} = -1.0\ \text{kJ/mol}; \qquad K = 1.49$$

$$\Delta G^\circ_{f,ICl} = -13.95\ \text{kJ/mol}; \qquad K = 279$$

The formation of $ClF(g)$ has the largest equilibrium constant of the three reactions. The more negative value of ΔG° results in the larger value for K.

47. $\Delta G^\circ = -RT \ln K$

$$\Delta G^\circ = 13.8\ \text{kJ/mol} = -RT \ln K$$

$$1.38 \times 10^4\ \text{J/mol} = -(8.314\ \text{J/K} \cdot \text{mol})(298\ \text{K}) \ln K$$

$$\ln K = -5.57$$

$$K = 3.81 \times 10^{-3}$$

49. a. To calculate ΔG° we need to know K for the reaction at 298 K. Equation 15.18 allows us to calculate this from the data at 973 K (700°C).

$$\ln\left(\frac{K_2}{K_1}\right) = \frac{-\Delta H^\circ}{R}\left(\frac{1}{T_2} - \frac{1}{T_1}\right) \qquad (15.18)$$

First, we need to calculate ΔH° using the values in Appendix 4.

$$\Delta H^\circ = \Sigma n \Delta H^\circ_{\text{f, prod}} - \Sigma m \Delta H^\circ_{\text{f, react}}$$
$$= (-241.8 \text{ kJ} + -110.5 \text{ kJ}) - (0 \text{ kJ} + -393.5 \text{ kJ})$$
$$= +41.2 \text{ kJ}$$

$$\ln\left(\frac{0.534}{K_1}\right) = \frac{-(4.12 \times 10^4 \text{ J/mol})}{8.314 \text{ J K}^{-1}\text{mol}^{-1}} \times \left(\frac{1}{973 \text{ K}} - \frac{1}{298 \text{ K}}\right)$$

Solving for K_1, we obtain $K_1 = 5.2 \times 10^{-6}$.

$$\Delta G^\circ = -RT \ln K$$
$$= -(8.314 \text{ J/K} \cdot \text{mol})(298 \text{ K}) \ln(5.2 \times 10^{-6})$$
$$= 3.01 \times 10^4 \text{ J/mol}$$

b.
$$\Delta G^\circ = \Sigma n \Delta G^\circ_{\text{f,prod}} - \Sigma m \Delta G^\circ_{\text{f, react}}$$
$$= (-228.6 \text{ kJ} + -137.2 \text{ kJ}) - (0 \text{ kJ} + -394.4 \text{ kJ})$$
$$= 28.6 \text{ kJ/mol} = 2.86 \times 10^4 \text{ J}$$

51. Adding the two equations in a Hess's law fashion we obtain ΔG° for the reaction we wish to calculate K_p.

$$N_2(g) + O_2(g) \rightleftharpoons \cancel{2NO(g)} \quad \Delta G^\circ = 173.2 \text{ kJ}$$
$$\cancel{2NO(g)} + O_2(g) \rightleftharpoons 2NO_2 \quad \Delta G^\circ = -69.7 \text{ kJ}$$
$$N_2(g) + 2O_2(g) \rightleftharpoons 2NO_2(g) \quad \Delta G^\circ = (173.2 \text{ kJ} + -69.7 \text{ kJ}) = 103.2 \text{ kJ}$$

$$\Delta G^\circ = RT \ln K$$

Recall that the value of K will be equal to K_p because the reaction involves only gas phase species.

$$\Delta G^\circ = RT \ln K_p$$
$$1.032 \times 10^5 \text{ J} = (8.314 \text{ J/K} \cdot \text{mol})(298 \text{ K}) \ln K_p$$
$$41.8 = \ln K_p$$
$$e^{41.8} = K_p$$
$$1.389 \times 10^{18} = K_p$$

53. No. Adding reactants could only change the value of Q, the reaction quotient. Note the value of Q may not change if the added reactants (e.g., solids) are not found in the expression for Q.

54.
$$2A \rightleftharpoons B$$

Consider the simple reaction above which $\text{Rate}_{\text{forward}} = K_f[A]^2$ and $\text{Rate}_{\text{reverse}} = k_r[B]$. If we increase [A], $\text{Rate}_{\text{forward}}$ will increase until equilibrium is reached. The rate of the reverse reaction also has to change as the [B] increases. Eventually, $\text{Rate}_{\text{forward}} = \text{Rate}_{\text{reverse}}$ and $K = k_f[B]/k_r[A]^2$. Does the rate of the reverse reaction change immediately and as much as the rate of the forward reaction? Clearly before equilibrium is reached, the rate of the forward reaction, must increase more than the rate of the reverse reaction; otherwise there wouldn't be any change in the equilibrium concentrations. So, both rates will change but the equilibrium constant will stay the same.

55.
$$Hb + 4CO(g) \rightleftharpoons Hb(CO)_4$$
$$Hb + 4O_2(g) \rightleftharpoons Hb(O_2)_4$$

A large increase in the partial pressure of O_2 drives the second reaction toward products. In terms of Le Châtelier's principle, having a very high partial pressure of O_2 shifts the second reaction toward products. As the second reaction shifts toward products, the amount of free Hb decreases. The decrease in [Hb] forces the first reaction to shift toward reactants, which decreases the $[Hb(CO)_4]$.

56. Yes. Equilibrium constants are temperature dependent, not pressure dependent.

57. Gas solubility is also an example of equilibrium. Increasing the partial pressure of a soluble gas drives the reaction toward products, which are dissolved gases in these cases. For example, we can think of the nitrogen in the atmosphere being involved in the following equilibrium in water samples:

$$N_2(g) \rightleftharpoons N_2(aq) \quad K = \frac{[N_2(g)]}{[N_2(aq)]}$$

As the partial pressure of N_2 is increased, this causes the reaction to shift toward products in order to reestablish equilibrium. The new state of equilibrium will have a higher concentration of dissolved N_2.

58. Adding an inert gas will only change the total pressure of the system—Dalton's law. The partial pressures of the gases found in the mass action expression do not change with the addition of an inert gas. Since the partial pressures of the gases in the mass action expression do not change, and since the position of equilibrium depends on the partial pressures of the gases in the mass action expression, the equilibrium is not upset.

59. When the reaction vessel is compressed to a smaller volume, pressure increases. The system will try to offset this pressure increase by shifting toward the side (reactants or products) containing fewer moles of gas.
   a. Toward reactants.
   b. Toward products.
   c. No effect (there are 2 moles of gas on each side).
   d. Toward products.

61. a. The addition of a species that appears in the mass action expression will cause the system to shift so

as to produce less of that species. O_3 is a reactant, so adding more of it will cause the reaction to shift toward products.

b. The addition of a species that appears in the mass action expression will cause the system to shift so as to reduce the amount of that species. O_2 is a product, so adding more of it will cause the reaction will shift toward reactants.

c. Compressing the mixture to one tenth of its initial volume will cause the reaction to shift toward the side having fewer moles of gas. The reaction will shift toward reactants since this side has two moles of gas and the products side has three moles of gas.

63. Decreasing the partial pressure of a species that appears in the mass action expression will cause the equilibrium to shift produce more of the removed species. O_2 is a reactant, so reducing its pressure will cause the reaction to shift toward reactants in order to offset the stress.

65. a. $PCl_5(g) \rightleftharpoons PCl_3(g) + Cl_2(g) \quad K_p = 23.6$ at 500 K

	PCl_5	PCl_3	Cl_2
I	0.560	0	0
C	$-x$	$+x$	$+x$
E	$0.560 - x$	x	x

$$K_p = \frac{(P_{PCl_3})(P_{Cl_2})}{(P_{PCl_5})}$$

$$\frac{(x)(x)}{(0.560 - x)} = 23.6$$

Solving the quadratic equation we obtain

$$x = 0.547 \quad \text{and} \quad x = -24.1$$

Since the negative value has no physical significance, we use $x = 0.547$ as our answer.

The equilibrium partial pressures are as follows:

$P_{Cl_2} = P_{PCl_3} = 0.547$ atm
$P_{PCl_5} = 0.013$ atm

b. Adding more Cl_2, a product, will shift the reaction toward reactants. The partial pressure of PCl_5 will increase and the partial pressure of PCl_3 will decrease.

67. $H_2O(g) + Cl_2O(g) \rightleftharpoons 2\,HOCl(g)$

$K_c = 0.0900$ at 298 K

	H_2O	Cl_2O	$HOCl$
I	0.00432	0.00432	0
C	$-x$	$-x$	$+2x$
E	$0.00432 - x$	$0.00432 - x$	$2x$

$$K_c = \frac{[HOCl]^2}{[H_2O][Cl_2O]}$$

$$\frac{(2x)^2}{(0.00432 - x)^2} = 0.0900$$

$$\frac{(2x)}{(0.00432 - x)} = 0.300$$

Simple algebra yields $x = 5.64 \times 10^{-4}$.

The equilibrium partial pressures are as follows:

$[H_2O] = 0.00376\,M$
$[Cl_2O] = 0.00376\,M$
$[HOCl] = 0.00113\,M$

69. $NO(g) + \frac{1}{2}O_2(g) \rightleftharpoons NO_2(g) \quad K_p = 2 \times 10^6$ at 298 K

$$K_p = \frac{(P_{NO_2})}{(P_{NO})(P_{O_2})^{1/2}}$$

$$\frac{(P_{NO_2})}{(P_{NO})(0.21)^{1/2}} = 2 \times 10^6$$

$$\frac{(P_{NO_2})}{(P_{NO})} = 9 \times 10^5$$

71. a. The flask initially contained pure $NO_2(g)$ at an unknown pressure. Let Y equal the initial partial pressure of $NO_2(g)$.

$2\,NO_2(g) \rightleftharpoons 2\,NO(g) + O_2(g)\ K_p = 158$ at 1000 K

	NO_2	NO	O_2
I	Y	0	0
C	$-2x$	$+2x$	$+x$
E	$Y-2x$	$2x$	x

We were told that the equilibrium partial pressure of $O_2(g)$ was 0.136 atm.

Since $P_{O_2} = 0.136$ atm, $x = 0.136$
$P_{NO} = 2x = 0.272$ atm

$$K_p = \frac{(P_{NO})^2(P_{O_2})}{(P_{NO_2})^2}$$

$$158 = \frac{(0.272)^2(0.136)}{(P_{NO_2})^2}$$

$$P_{NO_2} = 7.98 \times 10^{-3}$$

b. $P_{total} = P_{NO_2} + P_{NO} + P_{O_2} = 0.416$ atm

73. $N_2(g) + O_2(g) \rightleftharpoons 2\,NO(g)\ K_p = 0.050$ at 2473 K

	N_2	O_2	NO
I	0.79	0.21	0
C	$-x$	$-x$	$+2x$
E	$0.79-x$	$0.21-x$	$2x$

$$K_p = \frac{(P_{NO})^2}{(P_{N_2})(P_{O_2})}$$

$$\frac{(2x)^2}{(0.79 - x)(0.21 - x)} = 0.050$$

Solving for x using the quadratic equation or the method of successive approximations yields $x = 3.99 \times 10^{-2}$.

The partial pressures are as follows:

$P_{N_2} = 0.75$ atm
$P_{O_2} = 0.17$ atm
$P_{NO} = 0.08$ atm

75. $2\,H_2S(g) \rightleftharpoons 2\,H_2(g) + S_2(g)$ $K_c = 2.2 \times 10^{-4}$ at 1400 K

I	0.600	0	0
C	$-2x$	$+2x$	$+x$
E	0.600	$-2x$	x

$$K_c = \frac{[H_2]^2[S_2]}{[H_2S]^2}$$

$$\frac{(2x)^2(x)}{(0.600 - 2x)^2} = 2.2 \times 10^{-4}$$

Assume that $2x$ is small compared to 0.600, so that we can write

$$\frac{(2x)^2(x)}{0.600} = 2.2 \times 10^{-4}$$

Solving for x yields $x = 2.7 \times 10^{-2}$. Since this is less than 5% of the initial value (0.600 atm), our approximation is valid.

The concentrations at equilibrium are as follows:

$[H_2] = 5.2 \times 10^{-2}\,M$
$[S_2] = 2.7 \times 10^{-2}\,M$
$[H_2S] = 0.546\,M$

77. $CO(g) + Cl_2(g) \rightleftharpoons COCl_2(g)$ $K_c = 5.0$ at 600 K; $K_p = 0.10$ at 600 K

I	0.265	0.265	0
C	$-x$	$-x$	$+x$
E	$0.265-x$	$0.265-x$	x

$$K_p = K_c(RT)^{\Delta n}$$

$$K_p = 5.0[(0.0821\ \text{L}\cdot\text{atm/K}\cdot\text{mol})(600\ \text{K})]^{-1}$$

$$K_p = 0.10$$

$$K_p = \frac{P_{COCl_2}}{P_{CO}\,P_{Cl_2}}$$

$$\frac{(x)}{(0.265 - x)^2} = 0.10$$

Solving the quadratic equation yields $x = 0.00667$ and $x = 10.52$.

Since 10.52 atm is larger than the initial pressure of P_{CO} and P_{Cl_2}, $x = 0.00667$ is the usable solution.

The equilibrium partial pressures are as follows:

$P_{CO} = P_{Cl_2} = 0.258$ atm
$P_{COCl_2} = 6.67 \times 10^{-3}$ atm

79. $CO(g) + H_2O(g) \rightleftharpoons CO_2(g) + H_2(g)$ $K_c = 5.1$ at 700 K

I	0.050	0.050	0.050	0.050
C	$-x$	$-x$	$+x$	$+x$
E	$0.050 - x$	$0.050 - x$	$0.050 + x$	$0.050 + x$

$$K_c = \frac{[CO_2][H_2]}{[CO][H_2O]}$$

$$\frac{(0.050 + x)^2}{(0.050 - x)^2} = 5.1$$

$$\frac{(0.050 + x)}{(0.050 - x)} = 2.26$$

Solving for x yields $x = 1.9 \times 10^{-2}$.

The equilibrium concentrations are as follows:

$[CO] = [H_2O] = 0.031\,M$
$[CO_2] = [H_2] = 0.069\,M$

81. The reaction is exothermic. We can explain this in terms of Le Châtelier's principle. If the reaction was endothermic, increasing the temperature would shift the reaction toward products and the equilibrium constant would increase. The equilibrium constant for the reaction being discussed decreases with increasing temperature, so the reaction is exothermic.

82. Since an exothermic reaction can be considered as having heat for a product, an increase in temperature will shift the reaction toward reactants. The value of the equilibrium constant will decrease.

83. The reaction is exothermic. Only if heat is a product will decreasing the temperature yield an increase in the equilibrium constant.

84. Since the reaction is endothermic, an increase in temperature will shift the reaction toward products. The value of K_p increases as temperature increases.

85.

$$\ln\left(\frac{K_2}{K_1}\right) = \frac{-\Delta H^\circ}{R}\left(\frac{1}{T_2} - \frac{1}{T_1}\right)$$

$$\ln\left(\frac{K_{298}}{4.10 \times 10^{-4}}\right) = \frac{-1.806 \times 10^5\ \text{J/mol}}{8.314\ \text{JK}^{-1}\ \text{mol}^{-1}} \times \left(\frac{1}{298\ \text{K}} - \frac{1}{2273\ \text{K}}\right)$$

$$\ln\left(\frac{K_{298}}{4.10 \times 10^{-4}}\right) = -63.34$$

$$\left(\frac{K_{298}}{4.10 \times 10^{-4}}\right) = e^{-63.34} = 3.11 \times 10^{-28}$$

$$K_{298} = 1.28 \times 10^{-31}$$

87.

$$\ln\left(\frac{K_2}{K_1}\right) = \frac{-\Delta H^\circ}{R}\left(\frac{1}{T_2} - \frac{1}{T_1}\right)$$

$$\ln\left(\frac{1.5 \times 10^5}{23}\right) = \frac{-\Delta H^\circ}{8.314\ \text{JK}^{-1}\ \text{mol}^{-1}} \times \left(\frac{1}{703\ \text{K}} - \frac{1}{1273\ \text{K}}\right)$$

$$\Delta H^\circ = -1.1 \times 10^5\ \text{J/mol}$$

89. $$K_p = \frac{1}{(P_{N_2})(P_{H_2})^3}$$

90. $$K_p = \frac{1}{(P_{SO_3})(P_{H_2O})}$$

91. They are not included because the concentration of a solid is considered to be constant.

92. $CaCO_3(s) \rightleftharpoons CaO(s) + CO_2(g)$

$$\Delta H^o_{Rxn} = \sum n\Delta H_{f^o\,Products} - \sum n\Delta H_{f^o\,Reactants}$$

$$\Delta H^o_{Rxn} = [-634.9\text{ KJ} + -393.5\text{ KJ}] - [-1206.9\text{ KJ}]$$

$$\Delta H^o_{Rxn} = +178.5\text{ KJ}$$

$$\Delta S^o_{Rxn} = \sum nS^o_{Products} - \sum nS^o_{Reactants}$$

$$\Delta S^o_{Rxn} = [38.1\text{ J/K} + 213.6\text{ J/K}] - [92.9\text{ J/K}]$$

$$\Delta S^o_{Rxn} = 158.8\text{ J/K}$$

This reaction has a positive value for ΔH^o and a positive value for ΔS^o.

$$\Delta G = \Delta H - \Delta S$$

By examining the equation relating ΔG, ΔH, and ΔS we can see that the value of ΔG has a positive value at 0 K and decreases, at some point becoming negative, as the temperature is increased.

$$\Delta G^o = -RT \ln K$$

By examing the equation above we can see that when ΔG is positive, K is relatively small; when ΔG is negative, K is relatively large; and as ΔG becomes more negative, the value of K increases. For the given reaction, the value of K increases as the temperature for this reaction is increased since the value of ΔG begins as a positive value at 0 K, at some point becomes negative, and becomes more negative as the temperature is increased from that point.

93. a. $H_2O(g) + C(s) \rightleftharpoons CO(g) + H_2(g)$

$K_c = 3.0 \times 10^{-2}$ at 1273 K

I	0.442	0	0	$K_p = 3.1$ at 1273 K
C	$-x$	$+x$	$+x$	
E	$0.442 - x$	x	x	

$$K_p = K_c(RT)^{\Delta n}$$

$$K_p = (3.0 \times 10^{-2})[(0.0821\text{ L}\cdot\text{atm/K}\cdot\text{mol})(1273\text{ K})]^1$$

$$K_p = 3.1$$

$$K_p = \frac{[H_2O]}{[CO][H_2]}$$

$$\frac{(x)(x)}{(0.442 - x)} = 3.1$$

Solving the quadratic equation, we obtain $x = 0.393$ and $x = -3.52$.

The only useful root for our problem is $x = 0.393$.

At equilibrium, the partial pressures are as follows:

$P_{H_2O} = 0.049$ atm

$P_{CO} = P_{H_2} = 0.393$ atm

b. $H_2O(g) + C(s) \rightleftharpoons CO(g) + H_2(g)$

	$H_2O(g)$	$CO(g)$	$H_2(g)$
I	0.049	0.468 (0.393 + 0.075)	0.468 (0.393 + 0.075)
C	$+x$	$-x$	$-x$
E	$0.049 + x$	$0.468 - x$	$0.468 - x$

$$K_p = \frac{[CO][H_2]}{[H_2O]}$$

$$\frac{(0.468 - x)(0.468 - x)}{(0.049 + x)} = 3.1$$

Solving the quadratic equation, we obtain $x = 0.0162$ and $x = 4.05$.

For our problem, $x = 0.0162$ is the only useful root.

At equilibrium, the partial pressures are as follows:

$P_{H_2O} = 0.065$ atm

$P_{CO} = P_{H_2} = 0.452$ atm

95. $NH_4SH(s) \rightleftharpoons NH_3(g) + H_2S(g)$ $K_p = 0.126$ at 297 K

We were told that $P_{H_2S} = 0.355$

$$K_p = (P_{NH_3})(P_{H_2S})$$

$$0.126 = (P_{NH_3})(0.355)$$

$$P_{NH_3} = 0.355\text{ atm}$$

$$P_{NH_3} = 0.355$$

You could also solve for P_{NH_3} using the K_p expression and you will obtain the same answer.

CHAPTER 16

Equilibrium in the Aqueous Phase and Acid Rain

1. The percent dissociation of a weak acid increases as concentration decreases. Try sample calculations to prove it to yourself.

2. Since we are comparing four monoprotic acids whose concentrations are equal (1 M solutions), we can look at their respective K_a values to determine which is the strongest acid. HNO_2 has the largest K_a of the four acids; hence it is the strongest acid and will produce the highest concentration of $[H_3O^+]$. If these acids were not all monoprotic and at equal concentrations, we would likely need to perform calculations to determine which had the highest $[H_3O^+]$.

3. Since we are comparing four monoprotic acids whose concentrations are equal (0.100 M solutions), we can look at their respective percent ionization values to determine which has the smaller K_a. $HC_2H_3O_2$ has the smallest percent ionization of the four acids; hence it is the weakest acid and will have the smallest K_a. If these acids were not all monoprotic and at equal concentrations, we would likely need to perform calculations to determine which had the smallest K_a.

4. Proline is the strongest acid in formaldehyde solution. Of the three solutions, the formaldehyde solution of proline has the largest K_a and hence will have the largest $[H_3O^+]$.

5. Alanine has a higher degree of ionization in water since the K_a for alanine in water is greater than the K_a for alanine in ethanol.

6. Alanine ionizes more readily in water than in ethanol. Alanine is a stronger acid in water than it is in ethanol because water is a better base than ethanol.

7.
$$H_2A \rightleftharpoons H^+ + HA^- \quad K_{a1}$$
$$HA^- \rightleftharpoons H^+ + A^{2-} \quad K_{a2}$$

The difference in the K_a values for diprotic acids is due to Coulomb's law. K_{a2} values are smaller than K_{a1} values since the removal of the second proton is more difficult. For the second ionization, the proton must be separated to form an ion with a charge of -2. The first ionization is easier since it requires that the H^+ must be removed to form a -1 ion.

8.
$$B + H^+ \rightleftharpoons BH^+ \quad K_{b1}$$
$$BH^+ + H^+ \rightleftharpoons BH_2^{2+} \quad K_{b2}$$

K_{b1} will be larger than K_{b2}. K_{b2} values are smaller than K_{b1} values since the addition of the second proton is more difficult. The second proton must be added to an ion with a positive charge. Accepting the first is easier since the H^+ is added to a neutral species.

9.
$$HA(aq) + H_2O(l) \rightleftharpoons H_3O^+(aq) + A^-(aq)$$
$$K_a = 1.76 \times 10^{-5}$$

	HA	H_3O^+	A^-
I	0.500	0	0
C	$-x$	$+x$	$+x$
E	$0.500 - x$	x	x

$$K_a = \frac{[H_3O^+][A^-]}{[HA]}$$

$$\frac{(x)(x)}{(0.500 - x)} = 1.76 \times 10^{-5}$$

Solving for x, we obtain $x = 2.96 \times 10^{-3} M$.

$$[H_3O^+] = 2.96 \times 10^{-3} M$$

11. $$HA(aq) + H_2O(l) \rightleftharpoons H_3O^+(aq) + A^-(aq)$$

I	1.00	0	0
C	$-x$	$+x$	$+x$
E	$1.00 - x$	x	x

We were told that the acid was 2.94% ionized. Since 2.94% of 1.00 M is 0.0294 M, $x = 0.0294M$.

$$K_a = \frac{[H_3O^+][A^-]}{[HA]}$$

$$= \frac{(0.0294)(0.0294)}{(1.00 - 0.0294)}$$

Solving for K_a we obtain $K_a = 8.91 \times 10^{-4}$.

13. $$HA(aq) + H_2O(l) \rightleftharpoons H_3O^+(aq) + A^-(aq)$$
$K_a = 3.0 \times 10^{-4}$

I	0.100	0	0
C	$-x$	$+x$	$+x$
E	$0.100 - x$	x	x

$$K_a = \frac{[H_3O^+][A^-]}{[HA]}$$

$$3.0 \times 10^{-4} = \frac{(x)(x)}{(0.100 - x)}$$

Solving for x we obtain $x = 5.3 \times 10^{-3}M$.

$$\text{Solving for percent ionization} = \frac{[A^-]}{[HA]_{\text{Initial}}} \times 100$$

$$\text{Percent ionization} = \frac{5.3 \times 10^{-3}}{0.100} \times 100$$

$$= 5.3\%$$

15. $$HA(aq) + H_2O(l) \rightleftharpoons H_3O^+(aq) + A^-(aq)$$

I	0.250	0	0
C	$-x$	$+x$	$+x$
E	$0.250 - x$	x	x

We were told that $[H_3O^+] = 4.07 \times 10^{-3}$ M at equilibrium, so $x = 4.07 \times 10^{-3}$ M.

$$K_a = \frac{[H_3O^+][A^-]}{[HA]}$$

$$K_a = \frac{(x)(x)}{(0.250 - x)}$$

Solving for K_a by substituting for x, we obtain $K_a = 6.7 \times 10^{-5}$.

$$\text{Solving for percent ionization} = \frac{[A^-]}{[HA]_{\text{Initial}}} \times 100$$

$$\text{Percent ionization} = \frac{4.07 \times 10^{-3}}{0.250} \times 100$$

$$= 1.63\%$$

17. Since the HCl is a strong acid, we will assume that it will completely ionize. The $[H_3O^+] = 0.100$ M from the HCl alone. We will set up an ICE table to determine if the HF will contribute to the ionization.

$$HF(aq) + H_2O(l) \rightleftharpoons H_3O^+(aq) + F^-(aq)$$
$K_a = 6.8 \times 10^{-4}$

I	0.100	0.100	0
C	$-x$	$+x$	$+x$
E	$0.100 - x$	$0.100 + x$	x

$$K_a = \frac{[H_3O^+][F^-]}{[HF]}$$

$$\frac{(0.100 + x)(x)}{(0.100 - x)} = 6.8 \times 10^{-4}$$

Solving for x, we obtain $x = 6.75 \times 10^{-5}M$.

$$[H_3O^+] = 0.100\,M + 6.75 \times 10^{-5}M = 0.100\,M$$

The weak acid, HF, made essentially no contribution to the acidity of the solution.

19. For the first ionization

$$H_2A(aq) + H_2O(l) \rightleftharpoons H_3O^+(aq) + HA^-(aq)$$
$K_{a1} = 7.94 \times 10^{-5}$

I	0.250	0	0
C	$-x$	$+x$	$+x$
E	$0.250 - x$	x	x

$$K_a = \frac{[H_3O^+][HA^-]}{[H_2A]}$$

$$\frac{(x)(x)}{(0.250 - x)} = 7.94 \times 10^{-5}$$

Solving for x, we obtain $x = 4.41 \times 10^{-3}M$.

For the second ionization

$$HA^-(aq) + H_2O(l) \rightleftharpoons H_3O^+(aq) + A^{2-}(aq)$$
$K_{a2} = 1.62 \times 10^{-12}$

I	4.41×10^{-3}	4.41×10^{-3}	0
C	$-x$	$+x$	$+x$
E	$4.41 \times 10^{-3} - x$	$4.41 \times 10^{-3} + x$	x

$$K_a = \frac{[H_3O^+][A^{2-}]}{[HA^-]}$$

$$\frac{(4.41 \times 10^{-3} + x)(x)}{(4.41 \times 10^{-3} - x)} = 1.62 \times 10^{-12}$$

Solving for x, we obtain $x = 1.62 \times 10^{-12}$ M.

$$[H_3O^+] = 4.41 \times 10^{-3}M + 1.62 \times 10^{-12}\,M$$
$$= 4.41 \times 10^{-3}M.$$

Percent ionization based on the first ionization

$$(K_{a1}) = \frac{[HA^-]}{[H_2A]_{Initial}} \times 100$$

$$\text{Percent ionization} = \frac{(4.41 \times 10^{-3})}{(0.250)} \times 100$$

$$= 1.76\%$$

Percent ionization based on the second ionization

$$(K_{a2}) = \frac{[HA^-]}{[H_2A]_{Initial}} \times 100$$

$$\text{Percent ionization} = \frac{(4.41 \times 10^{-3})}{(0.250)} \times 100$$

$$= 1.76\%$$

The second ionization does not contribute significantly to the overall percent ionization.

21.
$$M_1V_1 = M_2V_2$$
$$(15.0M)(10.0 \text{ mL}) = (M_2)(1.00 \times 10^3 \text{ mL})$$
$$M_2 = 1.50 \times 10^{-1} M$$

Since nitric acid is a strong acid that completely ionizes, $[H_3O^+] = 1.50 \times 10^{-1}\ M$.

23. $$\frac{0.0800 \text{ mol Sr(OH)}_2}{\text{L}} \times \frac{2 \text{ mol OH}^-}{1 \text{ mol Sr(OH)}_2} = 0.160\ M\text{OH}^-$$

25.
$$2.50 \text{ L} \times \frac{0.70 \text{ mol OH}^-}{\text{L}} \times \frac{1 \text{ mol NaOH}}{1 \text{ mol OH}^-} \times \frac{40.00 \text{ g NaOH}}{1 \text{ mol NaOH}} = 70.0 \text{ g NaOH}$$

To prepare 2.50 L of 0.70 M NaOH solution, slowly combine 70.0 g of NaOH with enough water to make 2.50 L of solution.

27. $NaNO_2$ is a stronger electrolyte than HNO_2. Since HNO_2 is a weak acid, it will not completely ionize, while a solution of the same molarity of $NaNO_2$ can be assumed to ionize completely. Since there are more ions in the $NaNO_2$ solution, it is a more electrically conductive medium.

28. The HCl and the H_2O react to form H_3O^+ and Cl^-. Having these ions in solution facilitates the conduction of electricity.

29. CH_3NH_2 and H_2O are attracted together by dipole–dipole forces. The lone pair on the nitrogen in CH_2NH_2 forms a coordinate covalent bond with one hydrogen from water. This results in the formation of $CH_3NH_3^+$ and OH^-. The production of OH^-ions yields a basic solution.

30. Sulfur reacts with oxygen to form sulfur trioxide. Sulfur trioxide reacts with water to form sulfuric acid. Sulfuric acid reacts with calcium carbonate to form calcium sulfate, water, and carbon dioxide.

31. Brønsted-Lowry acids are proton donors. BF_3 does not contain hydrogen and hence it cannot be considered a Brønsted-Lowry acid. NH_3 does contain hydrogen, but NH_3 does not tend to donate protons (H^+ ions); it tends to accept protons which results in the formation of NH_4^+. Both HCl and HNO_3 contain hydrogen and tend to donate hydrogen ions; therefore they are the only two compounds of the four that are considered Brønsted-Lowry acids.

32. BF_3 can accept a pair of electrons and can be a Lewis acid. BF_3 is not capable of being a Brønsted-Lowry acid since it has no hydrogens to donate.

33. S is more electronegative than Se. Both of the oxoacids contain the same number of oxygens attached to the central atom (S or Se). The strength of the O—H bond in these two acids will be different because of the different electronegativities of their central atoms. The S will more effectively draw electron density away from the O in the O—H bond. This weakens the O—H bond and allows it to ionize more readily in H_2SO_4 relative to H_2SeO_4.

34. In order to understand why the K_{a1} of H_2SO_4 is much greater than the K_{a1} of H_2SO_3 we must look at the differences in compounds and how these differences affect the acidity of the compounds. The O—H bond of H_2SO_4 is weaker than that of H_2SO_3 because of the decreased electron density of the O—H bond. This decreased electron density is due to more oxygen atoms being attached to sulfur in H_2SO_4. Oxygen, being more electronegative than sulfur, withdraws electron density from the sulfur. The more electron deficient a central atom is, the weaker the O—H bond, and the weaker the O—H bond is, the stronger the acid. The general trend is that in oxoacids that differ only in the number of oxygens, the acid containing more oxygen atoms will be more acidic.

35. a. Acid: HNO_3; base: NaOH
b. Acid: HCl; base: $CaCO_3$
c. Acid: HCN; base: NH_3

37. $B(aq) + H_2O(l) \rightleftharpoons BH^+(aq) + OH^-(aq)$

$$K_b = 5.9 \times 10^{-4}$$

I	1.20×10^{-3}	0	0
C	$-x$	$+x$	$+x$
E	$1.20 \times 10^{-3} - x$	x	x

$$K_b = \frac{[BH^+][OH^-]}{[B]}$$

$$5.9 \times 10^{-4} = \frac{(x)(x)}{(1.20 \times 10^{-3} - x)}$$

Solving for x, we obtain $x = 5.97 \times 10^{-4} M$.
Since $x = [OH^-]$, then $[OH^-] = 6.0 \times 10^{-4} M$.

39.

Acid	Conjugate Base
HNO_2	NO_2^-
HOCl	OCl^-
H_3PO_4	$H_2PO_4^-$
NH_3	NH_2^-

41. a. Since both oxoacids contain the same number of oxygens attached to the central atom, the compound containing the more electronegative central atom (H_2SO_3) is the stronger acid.
b. Since both oxoacids contain the same central atom, the compound containing more oxygens attached to the central atom (H_2SeO_4) is the stronger acid.

43. pH values decrease as $[H_3O^+]$ increases because pH is the *negative* log of $[H_3O^+]$.

44. The pH value of solution A is two pH units lower than the pH value of solution B (–log 100 = 2).

45. When $[H_3O^+]$ is greater than 1 *M* the calculated pH of the solution will be negative.

46. In order for water or ethanol to autoionize, it must be able to act as an acid and a base. This means that a water molecule must be able donate a proton to another water molecule and that the second water molecule must be able to accept the proton. The same is true for the autoionization of ethanol. Since the equilibrium constant for the autoionization of ethanol is much less than the autoionization constant of water, water must be a better base and acid unto itself than ethanol is unto itself.

47. The following relationships can be used to calculate pH and pOH:

$$\mathrm{pH} = -\log[H_3O^+] \quad \mathrm{pOH} = -\log[OH^-]$$

$$\mathrm{pH} + \mathrm{pOH} = 14$$

a. pH = 7.462	pOH = 6.538	basic
b. pH = 4.70	pOH = 9.30	acidic
c. pH = 7.15	pOH = 6.85	basic
d. pH = 10.932	pOH = 3.068	basic

49.

$$\mathrm{pH} = -\log[H_3O^+]$$

$$\mathrm{pH} = -\log 0.155$$

$$\mathrm{pH} = 0.810$$

51.

$$\mathrm{pOH} = -\log[OH^-]$$

$$\mathrm{pOH} = -\log 0.0450$$

$$\mathrm{pOH} = 1.347$$

$$\mathrm{pH} + \mathrm{pOH} = 14$$

$$\mathrm{pH} = 14 - 1.347$$

$$\mathrm{pH} = 12.653$$

53.

$$\mathrm{pH} = -\log[H_3O^+]$$

$$\mathrm{pH} = -\log(1.33)$$

$$\mathrm{pH} = -0.124$$

55. $HA(aq) + H_2O(l) \rightleftharpoons H_3O^+(aq) + A^-(aq)$
$K_a = 1.8 \times 10^{-4}$

I	0.100	0	0
C	$-x$	$+x$	$+x$
E	$0.100 - x$	x	x

$$K_a = \frac{[H_3O^+][A^-]}{[HA]}$$

$$\frac{(x)(x)}{(0.100 - x)} = 1.8 \times 10^{-4}$$

Solving for x, we obtain $x = 4.2 \times 10^{-3}\ M$.

$$[H_3O^+] = 4.2 \times 10^{-3}\ M$$

$$\mathrm{pH} = 2.38$$

57. $[H_3O^+] = 10^{-\mathrm{pH}}$

$$[H_3O^+]_{\mathrm{Maine}} = 3 \times 10^{-5}\ M$$

$$[H_3O^+]_{\mathrm{Minnesota}} = 6 \times 10^{-6}\ M$$

59. For the first ionization:

$H_2A(aq) + H_2O(l) \rightleftharpoons H_3O^+(aq) + HA^-(aq)$
$K_{a1} = 3.71 \times 10^{-5}$

I	0.175	0	0
C	$-x$	$+x$	$+x$
E	$0.175 - x$	x	x

$$K_a = \frac{[H_3O^+][HA^-]}{[H_2A]}$$

$$3.71 \times 10^{-5} = \frac{(x)(x)}{(0.175 - x)}$$

Solving for x, we obtain $x = 2.53 \times 10^{-3}\ M$.
For the second ionization:

$HA^-(aq) + H_2O(l) \rightleftharpoons H_3O^+(aq) + A^{2-}(aq)$
$K_{a2} = 3.87 \times 10^{-6}$

I	2.53×10^{-3}	2.53×10^{-3}	0
C	$-x$	$+x$	$+x$
E	$2.53 \times 10^{-3} - x$	$2.53 \times 10^{-3} + x$	x

$$K_a = \frac{[H_3O^+][A^{2-}]}{[HA^-]}$$

$$3.87 \times 10^{-6} = \frac{(2.53 \times 10^{-3} + x)(x)}{(2.53 \times 10^{-3} - x)}$$

Solving for x, we obtain $x = 3.86 \times 10^{-6}\ M$.

Solving for $[H_3O^+] = 2.53 \times 10^{-3} + x = 2.53 \times 10^{-3} + 3.86 \times 10^{-6} = 2.53 \times 10^{-3}\ M$.

$$pH = -\log [H_3O^+] = -\log (2.53 \times 10^{-3})$$

$$pH = 2.60$$

61. Base strength increases as the number of methyl groups increases. Since pK_a increases as the number of methyl groups increases, the corresponding pK_b values decrease. This comes from the relationship $pK_a + pK_b = 14$.
Since $K_b = 10^{-pK_b}$, mathematically K_b increases as pK_b decreases.

62. K_b values are easy to calculate by using the relationship $K_aK_b = K_w$

63. Of the three ionic compounds given, only ammonium nitrate (NH_4NO_3) will yield an acidic solution. NH_4NO_3 dissolves in water to produce its component ions:

$$NH_4NO_3(s) \rightarrow NH_4^+(aq) + NO_3^-(aq)$$

Of these two ions, only NH_4^+ hydrolyzes:

$$NH_4^+(aq) + H_2O(l) \rightleftharpoons H_3O^+(aq) + NH_3(aq)$$

NO_3^-, the conjugate base of a strong acid, has no tendency to act as a base.

Sodium formate would produce a basic solution as the formate ion hydrolyzes. Note that the sodium ion does not hydrolyze.

Both of the ions in ammonium acetate hydrolyze, but the K_a of NH_4^+ is equal to the K_b of $C_2H_3O_2^-$, so a neutral solution is produced.

64. Of the three ionic compounds given, only NaF will yield a basic solution.

When NaF dissolves in water, Na^+ has no tendency to act as an acid or a base (i.e., to hydrolyze), but the fluoride ion hydrolyzes as follows:

$$F^-(aq) + H_2O(l) \rightleftharpoons HF(aq) + OH^-(aq)$$

to produce a basic solution.

When KCl dissolves in water, neither K^+ or Cl^- hydrolyzes, so the solution is neutral.

When NH_4Cl dissolves in water, NH_4^+ hydrolyzes to produce hydrogen ion, as follows:

$$NH_4^+(aq) + H_2O(l) \rightleftharpoons H_3O^+(aq) + NH_3(aq)$$

Cl^- does not hydrolyze, so the solution is acidic.

65. $B(aq) + H_2O(l) \rightleftharpoons BH^+(aq) + OH^-(aq)$

$pK_b = 5.79$

$K_b = 10^{-pK_b} = 10^{-5.79}$

$K_b = 1.6 \times 10^{-6}$

I	0.115	0	0
C	$-x$	$+x$	$+x$
E	$0.115 - x$	x	x

$$K_b = \frac{[BH^+][OH^-]}{[B]}$$

$$\frac{(x)(x)}{(0.115 - x)} = 1.6 \times 10^{-6}$$

Solving for x, we obtain $x = 4.3 \times 10^{-4}\ M$; so $[OH^-] = 4.3 \times 10^{-4}\ M$.

$$pOH = -\log [OH^-]$$

$$pOH = -\log 4.3 \times 10^{-4}$$

$$pOH = 3.37$$

$$pH = 10.63$$

67. Hydrolysis of the first (most basic) nitrogen:

$B(aq) + H_2O(l) \rightleftharpoons BH^+(aq) + OH^-(aq)$

$K_{b1} = 3.31 \times 10^{-6}$

I	0.01050	0	0
C	$-x$	$+x$	$+x$
E	$0.01050 - x$	x	x

$$K_{b1} = \frac{[BH^+][OH^-]}{[B]}$$

$$\frac{(x)(x)}{(0.01050 - x)} = 3.31 \times 10^{-6}$$

Solving for x, we obtain $x = 1.85 \times 10^{-4}\ M$.

Hydrolysis of the second (less basic) nitrogen:

$BH^+(aq) + H_2O(l) \rightleftharpoons BH_2^{2+}(aq) + OH^-(aq)$

$K_{b2} = 1.35 \times 10^{-9}$

I	1.85×10^{-4}	0	1.85×10^{-4}
C	$-x$	$+x$	$+x$
E	$1.85 \times 10^{-4} - x$	x	$1.85 \times 10^{-4} + x$

$$K_{b2} = \frac{[B_2H^{2+}][OH^-]}{[BH^+]}$$

$$\frac{(x)(1.85 \times 10^{-4} + x)}{(1.85 \times 10^{-4} - x]} = 1.35 \times 10^{-9}$$

Solving for x, we obtain $x = 1.35 \times 10^{-9}\ M$.

$$[OH^-] = 1.85 \times 10^{-4}\ M + 1.35 \times 10^{-9}\ M$$

$$= 1.85 \times 10^{-4}\ M$$

$$pOH = -\log [OH^-] = -\log 1.85 \times 10^{-4}$$

$$pOH = 3.73$$

$$pH = 10.27$$

69. $$B(aq) + H_2O(l) \rightleftharpoons BH^+(aq) + OH^-(aq)$$
$K_{b1} = 1.05 \times 10^{-6}$

I	0.00100	0	0
C	$-x$	$+x$	$+x$
E	$0.00100 - x$	x	x

$$K_{b1} = \frac{[BH^+][OH^-]}{[B]}$$

$$1.05 \times 10^{-6} = \frac{(x)(x)}{(0.00100 - x)}$$

Solving for x we obtain $x = 3.19 \times 10^{-5} = [OH^-]$

pOH = 4.496
pH = 9.504

Since K_{b2} is 10^5 smaller than K_{b1}, we do not expect it to contribute significantly to the pH of the nicotine solution. The calculation below justifies this expectation.

For the second ionization:

$$BH^+(aq) + H_2O(l) \rightleftharpoons BH_2^{2+}(aq) + OH^-(aq)$$
$K_{b2} = 1.32 \times 10^{-11}$

I	3.19×10^{-5}	0	3.19×10^{-5}
C	$-x$	$+x$	$+x$
E	$3.19 \times 10^{-5} - x$	x	$3.19 \times 10^{-5} + x$

$$K_{b2} = \frac{[BH^+][OH^-]}{[B]}$$

$$1.32 \times 10^{-11} = \frac{(x)(3.19 \times 10^{-5} + x)}{(3.19 \times 10^{-5} - x)}$$

Solving for x we obtain $x = 1.32 \times 10^{-11}$

$[OH^-] = 3.19 \times 10^{-5} + x$
$= 3.19 \times 10^{-5} + 1.32 \times 10^{-11} = 3.19 \times 10^{-5}$

pOH = 4.496
pH = 9.504

K_{b2} does not contribute significantly to the pH of the given solution.

71. $$(K_a)(K_b) = K_w$$

$$(2.1 \times 10^{-11})(K_b) = 1.0 \times 10^{-14}$$

$$K_b = 4.8 \times 10^{-4}$$

$$pK_b = 3.32$$

73. $$F^-(aq) + H_2O(l) \rightleftharpoons HF(aq) + OH^-(aq)$$
$K_b = K_w/K_a$
$K_b = 1.5 \times 10^{-11}$

I	0.00339	0	0
C	$-x$	$+x$	$+x$
E	$0.00339 - x$	x	x

$$K_b = \frac{[HF][OH^-]}{[F^-]}$$

$$\frac{(x)(x)}{(0.00339 - x)} = 1.5 \times 10^{-11}$$

Solving for x, we obtain $x = 2.3 \times 10^{-7} M$; so $[OH^-] = 2.3 \times 10^{-7}\ M$

$$pOH = -\log [OH^-] = -\log (1.2 \times 10^{-2})$$

$$pOH = 6.64$$

$$pH = 7.36$$

75. Buffers need the ability to neutralize both acids and bases. A solution that contains NaCl and HCl can neutralize added bases since the HCl can react with added bases, but the solution has no ability to neutralize acids. Therefore, the solution of NaCl and HCl cannot be considered a buffer. A solution that contains sodium acetate and acetic acid can act as a buffer since the sodium acetate can react with added acid and the acetic acid can react with added base.

76. Buffers have the ability to neutralize both acids and bases. A solution containing only a weak base will not be able to neutralize bases. A solution containing a weak base and its conjugate acid will be able to neutralize added acids and bases. The weak base component of this solution will react with added acid and the conjugate acid (a weak acid) component of this solution will react with added base.

77. $$pH = pK_a + \log \frac{[base]}{[acid]} \quad (16.20)$$

At 25°C $pH = 4.74 + \log \frac{(0.122)}{(0.244)}$ $pH = 4.44$

At 0°C $pH = 4.79 + \log \frac{(0.122)}{(0.244)}$ $pH = 4.48$

79. $$HPO_4^{2-}(aq) + H_2O(l) \rightleftharpoons PO_4^{3-}(aq) + H_3O^+(aq)$$

$$pH = pK_a + \log \frac{[base]}{[acid]}$$

$$pH = pK_a + \log \frac{[PO_4^{3-}]}{[HPO_4^{2-}]}$$

$$pH = -\log(4.2 \times 10^{-13}) + \log \frac{[0.225M]}{[0.225M]} = 12.38$$

$$pH + pOH = 14$$

$$12.38 + pOH = 14$$

$$pOH = 1.62$$

81. $$pH = pK_a + \log \frac{[base]}{[acid]}$$

$$3.56 = 4.74 + \log \frac{[acetate\ ion]}{[acetic\ acid]}$$

$$-1.18 = \log \frac{[\text{acetate ion}]}{[\text{acetic acid}]}$$

$$10^{-1.18} = \frac{[\text{acetate ion}]}{[\text{acetic acid}]}$$

$$6.6 \times 10^{-2} = \frac{[\text{acetate ion}]}{[\text{acetic acid}]}$$

83. We have been asked to lower the pH of the given buffer to 5.00 by adding a strong acid, HNO_3. By adding HNO_3 to the given buffer, acetate ions will react to form acetic acid. The net effect will be to decrease the moles of acetate ions present in the solution and to increase the amount of acetic acid in the solution.

$$HNO_3(aq) + C_2H_3O_2^- \rightarrow HC_2H_3O_2(aq) + NO_3^-(aq)$$

Using the Henderson-Hasselbalch equation we can obtain the ratio of base:acid needed to obtain pH = 5.00.

$$\text{pH} = \text{p}K_a + \log \frac{[\text{base}]}{[\text{acid}]}$$

$$\text{pH} = -\log(1.8 \times 10^{-5}) + \log \frac{[\text{acetate ion}]}{[\text{acetic acid}]}$$

$$5.00 = 4.74 + \log \frac{[\text{acetate ion}]}{[\text{acetic acid}]}$$

$$0.26 = \log \frac{[\text{acetate ion}]}{[\text{acetic acid}]}$$

$$1.82 = \frac{[\text{acetate ion}]}{[\text{acetic acid}]}$$

We know that the Henderson-Hasselbalch equation can use molarity or moles. We can set up a table that will help us determine how many moles of HNO_3 need to be added in order to obtain the desired base:acid ratio. In the table below we are using moles of HNO_3, $HC_2H_3O_2$, and $C_2H_3O_2^-$ and assuming the reaction of the strong acid with a weak base goes to completion. After the reaction has gone to completion, we have the moles of components left, and these can be plugged into the Henderson-Hasselbalch equation—or we can convert the moles to molarities and plug them into an ICE table or the Henderson-Hasselbalch equation. Since we have been asked to calculate the amount of HNO_3 needed, the moles of HNO_3 have been assigned a variable, X.

$$1.00\,\text{L}\left(\frac{0.010 \text{ moles } HC_2H_3O_2}{\text{L}}\right)$$
$$= 0.010 \text{ moles } HC_2H_3O_2$$

$$1.00\,\text{L}\left(\frac{0.10 \text{ moles } C_2H_3O_2^-}{\text{L}}\right)$$
$$= 0.10 \text{ moles } HC_2H_3O_2^-$$

	HNO_3 +	$C_2H_3O_2^-$ →	$HC_2H_3O_2 + NO_3^-$
Before	X mol	0.10 mol	0.010 mol
Change	$-X$	$-X$	$+X$
After	0	$0.10 - X$	$0.010 + X$

Note that the HNO_3 must be completely used up in order for a buffer to exist in our solution. Also note that after the HNO_3 reacted, the amount of $C_2H_3O_2^-$ decreased and the amount of $HC_2H_3O_2$ increased as we expected.

Using the ratio we determined earlier and the information from the table:

$$1.82 = \frac{[\text{acetate ion}]}{[\text{acetic acid}]}$$

$$1.82 = \frac{[0.10 - X]}{[0.010 + X]}$$

Solving for X we find $X = 2.9 \times 10^{-2}$ moles.

$$2.9 \times 10^{-2} \text{ moles } HNO_3 \left(\frac{1\text{ L}}{10 \text{ moles } HNO_3}\right) \times \left(\frac{1000\text{ mL}}{1\text{ L}}\right)$$
$$= 2.9\text{ mL}$$

2.9 mL of 10 *M* HNO_3 are needed to lower the pH of the buffer to 5.00.

85. $$HCl(aq) + NH_3(aq) \rightarrow NH_4^+ + Cl^-(aq)$$

The equation above illustrates how NH_3 and HCl react to form ammonium ion and chloride ion. We can set up a table that illustrates how many moles of each component are left after the equal volumes of 0.050 *M* NH_3 is reacted with 0.025 *M* HCl. In the table below we are using moles of HCl, NH_3, and NH_4^+, and assuming the reaction of the strong acid with the weak base goes to completion. After the reaction has gone to completion, we have the moles of components left and these can be plugged into the Henderson-Hasselbalch equation—or convert the moles to molarities and plug these into an ICE table. The table below assumes 1.0 L of each solution.

$$1.0\,\text{L}\left(\frac{0.050 \text{ moles } NH_3}{\text{L}}\right) = 0.050 \text{ moles } NH_3$$

$$1.0\,\text{L}\left(\frac{0.025 \text{ moles HCl}}{\text{L}}\right) = 0.025 \text{ moles HCl}$$

	HCl +	NH_3 →	$NH_4^+ + Cl^-$
Before	0.025 mol	0.050 mol	0 mol
Change	−0.025	−0.025	+0.025
After	0	0.025 mol	0.025 mol

We know that the Henderson-Hasselbalch equation can use molarities of acid and base or moles of acid and base.

$$pH = pK_a + \log \frac{[\text{base}]}{[\text{acid}]}$$

$$pH = -\log(5.6 \times 10^{-10}) + \log \frac{[NH_3]}{[NH_4^+]}$$

$$pH = 9.26 + \log \frac{[0.025 \text{ moles}]}{[0.025 \text{ moles}]} = 9.26$$

87. Before the strong acid is added:

$$pH = pK_a + \log \frac{[\text{base}]}{[\text{acid}]}$$

$$pH = -\log(4.5 \times 10^{-4}) + \log \frac{[0.150]}{[0.120]} = 3.44$$

After the strong acid is added:

$$HCl(aq) + NO_2^-(aq) \rightarrow HNO_2(aq)$$

The equation above illustrates how the buffer responds to added HCl. In this buffer, NO_2^- and HCl react to form HNO_2. We can set up a table that will help us determine how many moles of each component are left after the HCl is added to the buffer. In the table below we are using moles of HCl, HNO_2, and NO_2^-, and assuming the reaction of the strong acid with the weak base goes to completion. After the reaction has gone to completion, we have the moles of components left and these can be plugged into the Henderson-Hasselbalch equation or we can convert the moles to molarities and use them in and ICE table or the Henderson-Hasselbalch equation.

$$1.00 \text{ L} \left(\frac{0.120 \text{ moles } HNO_2}{\text{L}}\right) = 0.120 \text{ moles } HNO_2$$

$$1.00 \text{ L} \left(\frac{0.150 \text{ moles } NO_2^-}{\text{L}}\right) = 0.150 \text{ moles } NO_2^-$$

$$1.00 \text{ mL} \left(\frac{\text{L}}{1000 \text{ mL}}\right)\left(\frac{12.0 \text{ moles HCl}}{\text{L}}\right)$$

$$= 0.0120 \text{ moles HCl}$$

	HCl +	NO_2^- →	HNO_2
Before	0.0120 mol	0.150 mol	0.120 mol
Change	−0.0120	−0.0120	+0.0120
After	0	0.138	0.132

$$pH = 3.35 + \log \frac{[0.138 \text{ moles}]}{[0.132 \text{ moles}]} = 3.37$$

pH = 3.44 before addition of acid

pH = 3.37 after addition of acid

89. Strong acid–strong base titration curves have more gradual pH changes before the equivalence point compared to weak acid–strong base titration curves. Strong acid–strong base titration curves have more drastic pH changes in the region surrounding the equivalence point compared to weak acid–strong base titration curves. The text contains figures for both of these general types of titrations along with weak base–strong acid titrations (see Figures 16.18 and 16.19).

90. Yes. When strong acids (e.g., HCl) are reacted with strong bases (e.g., NaOH) the pH will be 7 at the equivalence point.

91. No, the pH at the equivalence point will depend upon the K_a of the acid.

92. A pH indicator should be one color over one pH range and another color over an adjoining pH range. The difference in appearance of the indicator within the two pH regions should be easily distinguished and color change should also be easy to visually observe.

93.
$$NaOH(aq) + HC_2H_3O_2(aq) \rightarrow NaC_2H_3O_2(aq) + H_2O(l)$$

Removing the spectator ions:

$$OH^-(aq) + HC_2H_3O_2(aq) \rightarrow C_2H_3O_2^-(aq) + H_2O(l)$$

The equation above illustrates how NaOH reacts with $HC_2H_3O_2$. We can set up a table that will help us determine how many moles of each component are left after the added NaOH has reacted. In the table below we are using moles of OH^-, $HC_2H_3O_2$, and $C_2H_3O_2^-$, and assuming the reaction of the strong base with the weak acid goes to completion. After the reaction has gone to completion, we have the moles of components left and these can be plugged into the Henderson-Hasselbalch equation, or the moles can be converted into molarities and these can be plugged into an ICE table or the Henderson-Hasselbalch equation. $K_a = 1.8 \times 10^{-5}$ for $HC_2H_3O_2$.

When 10.0 mL of 0.125 *M* NaOH has been added:

$$25.0 \text{ mL} \left(\frac{\text{L}}{1000 \text{ mL}}\right)\left(\frac{0.100 \text{ moles } HC_2H_3O_2}{\text{L}}\right)$$

$$= 0.00250 \text{ moles } HC_2H_3O_2$$

$$10.0 \text{ mL} \left(\frac{\text{L}}{1000 \text{ mL}}\right)\left(\frac{0.125 \text{ moles NaOH}}{\text{L}}\right)$$

$$= 0.00125 \text{ moles NaOH}$$

	OH^- +	$HC_2H_3O_2$ →	$C_2H_3O_2^-$ + H_2O
Before	0.00125 mol	0.00250 mol	0 mol
Change	−0.00125	−0.00125	+0.00125
After	0	0.00125	0.00125

$$pH = 4.74 + \log \frac{[0.00125 \text{ moles}]}{[0.00125 \text{ moles}]} = 4.74$$

The point in a titration where half of the moles have been titrated is often called the halfway point. For the titration of a weak acid with a strong base, pH = pK_a at the halfway point. We could have also set up an ICE table and used it to obtain the same pH.

When 20.0 mL of 0.125 *M* NaOH has been added:

$$25.0\text{ mL}\left(\frac{\text{L}}{1000\text{ mL}}\right)\left(\frac{0.100\text{ moles }HC_2H_3O_2}{\text{L}}\right)$$

$$= 0.00250\text{ moles }HC_2H_3O_2$$

$$20.0\text{ mL}\left(\frac{\text{L}}{1000\text{ mL}}\right)\left(\frac{0.125\text{ moles NaOH}}{\text{L}}\right)$$

$$= 0.00250\text{ moles NaOH}$$

	OH^- +	$HC_2H_3O_2$ →	$C_2H_3O_2^-$ + H_2O
Before	0.00250 mol	0.00250 mol	0 mol
Change	−0.00250	−0.00250	+0.00250
After	0	0	0.00250 mol

All of the weak acid has reacted, the solution is at the equivalence point, and we are left with a solution of the weak base. $C_2H_3O_2^-$. We cannot use the Henderson-Hasselbalch equation since this solution is not a buffer.

$$[C_2H_3O_2^-] = 0.00250\text{ mol}/0.0450\text{ L} = 0.0556\ M$$

$$C_2H_3O_2^-(aq) + H_2O(l) \rightleftharpoons HC_2H_3O_2(aq) + OH^-(aq)$$

$$K_b = 5.6 \times 10^{-10}$$

	$C_2H_3O_2^-$	$HC_2H_3O_2$	OH^-
I	0.0556	0	0
C	$-x$	$+x$	$+x$
E	$0.0556 - x$	x	x

$$K_b = \frac{[HC_2H_3O_2][OH^-]}{[C_2H_3O_2^-]}$$

$$5.6 \times 10^{-10} = \frac{(x)(x)}{(0.0556 - x)}$$

Solving for x, we obtain $x = 5.6 \times 10^{-6} \rightarrow [OH^-] = 5.6 \times 10^{-6}\ M$.

$$pOH = -\log[OH^-] = -\log(5.6 \times 10^{-6})$$

$$pOH = 5.25$$

$$pH = 8.75$$

When 30.0 mL 0.125 *M* NaOH has been added:

$$25.0\text{ mL}\left(\frac{\text{L}}{1000\text{ mL}}\right)\left(\frac{0.100\text{ moles }HC_2H_3O_2}{\text{L}}\right)$$

$$= 0.00250\text{ moles }HC_2H_3O_2$$

$$30.0\text{ mL}\left(\frac{\text{L}}{1000\text{ mL}}\right)\left(\frac{0.125\text{ moles NaOH}}{\text{L}}\right)$$

$$= 0.00375\text{ moles NaOH}$$

	OH^- +	$HC_2H_3O_2$ →	$C_2H_3O_2^-$ + H_2O
Before	0.00375 mol	0.00250 mol	0 mol
Change	−0.00250	−0.00250	+0.00250
After	0.00125	0	0.00250

At this point in the titration we are past the equivalence point. In order to calculate the pH, just use the concentration of the remaining strong titrant. Any weak acids or bases left in solution generally do not make a significant contribution to pH in such situations.

$$[OH^-] = 0.00125\text{ mol}/0.0550\text{ L} = 0.0227\ M$$

$$pOH = -\log(0.0227)$$

$$pOH = 1.64$$

$$pH = 12.36$$

95. $NaOH(aq) + HNO_2(aq) \rightarrow NO_2^-(aq) + H_2O(l)$

The equation above illustrates how NaOH and HNO_2 react.

$$50.0\text{ mL}\left(\frac{\text{L}}{1000\text{ mL}}\right)\left(\frac{0.250\text{ moles }HNO_2}{\text{L}}\right) \times \left(\frac{1\text{ mole NaOH}}{1\text{ mole }HNO_2}\right)\left(\frac{1000\text{ mL}}{1.00\text{ mole NaOH}}\right) = 12.5\text{ mL}$$

12.5 mL 1.00 *M* NaOH are needed to reach the equivalence point.

	OH^- +	HNO_2 →	NO_2^- + H_2O
Before	0.0125 mol	0.0125 mol	0 mol
Change	−0.0125	−0.0125	+0.0125
After	0	0	0.0125 mol

All of the weak acid and strong base have reacted and we are left with a solution of the weak base, NO_2^-. We cannot use the Henderson-Hasselbalch equation since this solution is not a buffer.

$$[NO_2^-] = 0.0125\text{ mol}/0.0625\text{ L} = 0.200\ M$$

$$NO_2^-(aq) + H_2O(l) \rightleftharpoons HNO_2(aq) + OH^-(aq)$$

	NO_2^-	HNO_2	OH^-
I	0.200	0	0
C	$-x$	$+x$	$+x$
E	$0.200 - x$	x	x

$$K_a = 4.5 \times 10^{-4}$$

$$K_b = 2.2 \times 10^{-11}$$

$$K_b = \frac{[HNO_2][OH^-]}{[NO_2^-]}$$

$$2.2 \times 10^{-11} = \frac{(x)(x)}{(0.200 - x)}$$

Solving for x we obtain $x = 2.1 \times 10^{-6}\ M = [OH^-]$

$$pOH = -\log[OH^-] = -\log(2.1 \times 10^{-6})$$

$$pOH = 5.68$$

$$pH = 8.32$$

Sketch of the titration curve for the titration of 50.0 mL of 0.0250 M HNO_2 with 0.100 M NaOH:

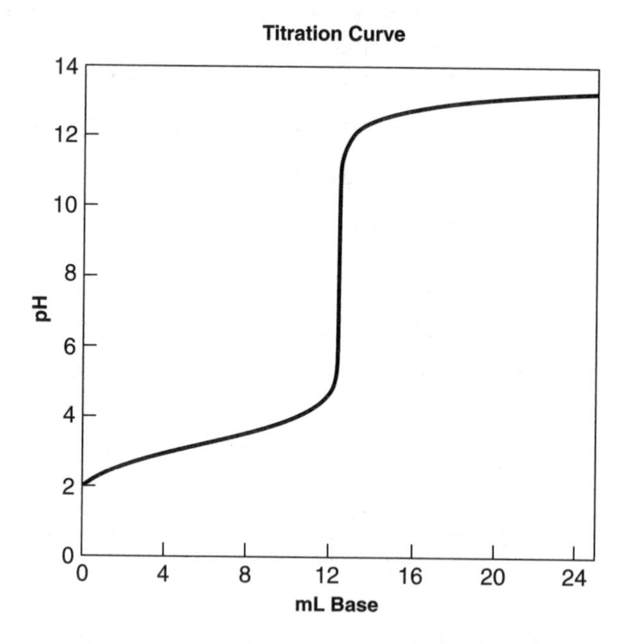

The pH values at three important points are

0 mL NaOH	pH = 1.98	Starting point
6.25 mL NaOH	pH = 3.35	Halfway point
12.5 mL NaOH	pH = 8.32	Equivalence point

97. A sketch of the titration curve for the titration of 40.0 mL of 0.100 M quinine with 0.100 M HCl:

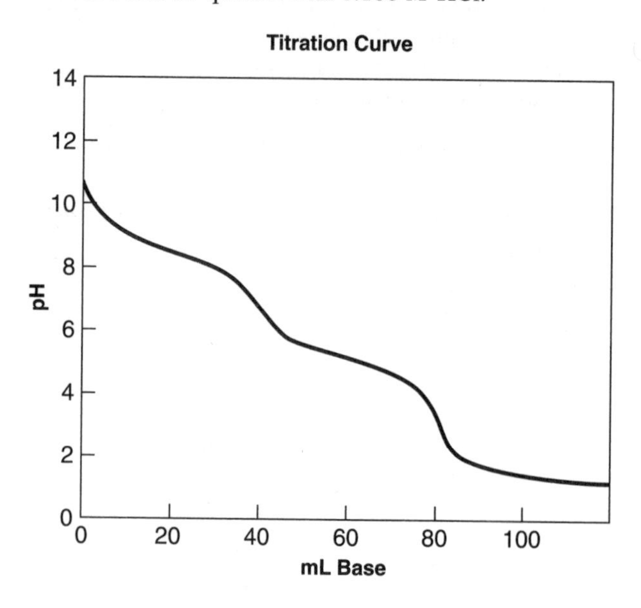

$K_{b1} = 3.31 \times 10^{-6}$ $K_{b2} = 1.35 \times 10^{-9}$

The pH values at the following important points are

0 mL pH = 10.76	Starting point
20 mL pH = 8.52	First halfway point
40 mL pH ≈ 6.8	First equivalence point
60 mL pH = 5.13	Second halfway point
80 mL pH ≈ 3.3	Second equivalence point

99. $$NH_3(aq) + HCl(aq) \rightarrow NH_4Cl(aq)$$

According to the above equation, NH_3 and HCl react on a 1 mole to 1 mole basis.

$$\left(\frac{22.35 \text{ mL HCl}}{100.0 \text{ mL NH}_3}\right)\left(\frac{\text{L}}{1000 \text{ mL}}\right)\left(\frac{0.1145 \text{ moles HCl}}{\text{L}}\right)$$
$$\times\left(\frac{1 \text{ mole NH}_3}{1 \text{ mole HCl}}\right)\left(\frac{1000 \text{ mL NH}_3}{\text{L}}\right)$$
$$= \frac{0.02559 \text{ mole NH}_3}{\text{L}}$$

101. Solubility is the concentration of a compound that dissolves in solution. Solubility product, better known as solubility product constant, is a mass action expression specifically for a solid compound dissolving to form a solution.

For the general equation:

$$M_mX_x(s) \rightleftharpoons mM(aq) + xX(aq)$$
$$K_{sp} = [M]^m[X]^x$$

102. The common ion effect is a consequence of Le Châtelier's principle. As a common ion (an ion included in the mass action expression) is added, the equilibrium shifts so as to reduce the amount of the added ion.

103. $SrCO_3$ will precipitate first (see K_{sp} values in Appendix 4). Of the three compounds, $SrCO_3$ has the smallest K_{sp}. Direct comparison of K_{sp} values is only useful if the compounds dissociate in similar fashions. In this case, all of the compounds dissociate into one cation and one anion.

104. As the solubility of a compound increases, the value of K_{sp} increases as well.

105. In order for the solubility of a compound to increase, the K_{sp} of the compound must also increase. Since K_{sp} increases with increasing temperature, solubility increases with increasing temperature and the reaction must be endothermic. In terms of Le Châtelier's principle, increasing the temperature must shift the reaction toward products, dissolved ions.

$$\text{HEAT} + M_mX_x(s) \rightleftharpoons mM(aq) + xX(aq)$$

Heat must then be considered a reactant; hence the reaction must be endothermic.

106. To answer this question we must assume that the anion of the aluminum salt will not react with the added NaOH. If this is the case, then we expect the solubility

of Al^{3+} to initially decrease as NaOH is added, since $Al(OH)_3$ is fairly insoluble with a K_{sp} of 1.9×10^{-33}. As additional OH^- is added and the pH increases, we expect the solubility of Al^{3+} to increase as $Al(OH)_4^-(aq)$ becomes the dominant form of aluminum in solution.

107. Hydroxyapatite, $Ca_5(PO_4)_3OH$, is more susceptible to erosion by acids than fluoroapatite, $Ca_5(PO_4)_3F$, because the OH^- in $Ca_5(PO_4)_4\boldsymbol{OH}$ is a stronger base compared to the F^- in $Ca_5(PO_4)_3\boldsymbol{F}$.

108. Acidic solutions react with the hydroxide ion, OH^-, that is, in hydroxyapatite, $Ca_5(PO_4)_3\boldsymbol{OH}$. This reaction begins the process of enamel erosion or tooth decay.

109. $BaSO_4(s) \rightleftharpoons Ba^{2+}(aq) + SO_4^{2-}(aq)$

$$K_{sp} = [Ba^{2+}][SO_4^{2-}]$$
$$= (1.04 \times 10^{-5})(1.04 \times 10^{-5})$$
$$= 1.08 \times 10^{-10}$$

111.

	$CuCl(s)$	$\rightleftharpoons$	$Cu^+(aq)$	+	$Cl^-(aq)$
I			0		0
C			$+x$		$+x$
E			x		x

$$K_{sp} = [Cu^+][Cl^-]$$
$$(x)(x) = 1.02 \times 10^{-6}$$
$$x = 1.01 \times 10^{-3}M$$

$$[Cu^+] = [Cl^-] = 1.01 \times 10^{-3}M$$

113.

	$CaCO_3(s)$	$\rightleftharpoons$	$Ca^{2+}(aq)$	+	$CO_3^{2-}(aq)$
I			0		0
C			$+x$		$+x$
E			x		x

$$K_{sp} = [Ca^{2+}][CO_3^{2-}]$$
$$(x)(x) = 9.9 \times 10^{-9}$$
$$x = 9.9 \times 10^{-5}$$

$$\frac{9.9 \times 10^{-5} \text{ mole } CaCO_3}{L}\left(\frac{100.09 \text{ g } CaCO_3}{\text{mole } CaCO_3}\right)$$
$$\times\left(\frac{L}{1000 \text{ mL}}\right) = \frac{9.9 \times 10^{-6} \text{ g}}{\text{mL}}$$

115.

	$AgOH(s)$	$\rightleftharpoons$	$Ag^+(aq)$	+	$OH^-(aq)$
I			0		0
C			$+x$		$+x$
E			x		x

$$K_{sp} = [Ag^+][OH^-]$$
$$(x)(x) = 1.52 \times 10^{-8}$$
$$x = 1.23 \times 10^{-4}M$$

We calculate the pH as follows:

$$[OH^-] = 1.23 \times 10^{-8}M$$
$$pOH = -\log 1.23 \times 10^{-8}$$
$$pOH = 3.91$$
$$pH = 10.09$$

117. The solubility of $CaCO_3$ is governed by the following equilibrium:

$$CaCO_3(s) \rightleftharpoons Ca^{2+}(aq) + CO_3^{2-}(aq)$$
$$CO_3^{2-}(aq) + H_2O \rightleftharpoons HCO_3^- + OH^-$$

The presence of a common ion will decrease the solubility and the ability to remove Ca^{2+} or CO_3^{2-} from the solution will increase the solubility.

a. 0.1 *M* NaCl contains no common ions.
b. 0.1 *M* Na_2CO_3 contains $Ca^{2+}(aq)$. $CaCO_3$ will have a lower solubility in solution *b* than in solution *a*.
c. 0.1 *M* NaOH contains no common ions. It does contain OH^- ions, which are produced by the hydrolysis of CO_3^{2-}. The increased concentration of OH^- will increase the concentration of CO_3^{2-}, so $CaCO_3$ will have a lower solubility in solution *c* than in solution *a*.
d. 0.1 *M* HCl contains no common ions. It is a strong acid and the H_3O^+ produced will react with the CO_3^{2-} to produce $CO_2(g)$ and $H_2O(l)$. The removal of CO_3^{2-} will cause more $CaCO_3$ to dissolve so $CaCO_3$ will have greater solubility in solution *d* than in *a*, *b*, or *c*.

119.

	$CaCO_3(s)$	$\rightleftharpoons$	$Ca^{2+}(aq)$	+	$CO_3^{2-}(aq)$
I			0		0.00100
C			$+x$		$+x$
E			x		$0.00100 + x$

$$K_{sp} = [Ca^+][CO_3^{2-}]$$
$$(x)(0.00100 + x) = 9.9 \times 10^{-9}$$
$$x = 9.8 \times 10^{-6}$$

The solubility of $CaCo_3$ in 0.00100 *M* Na_2CO_3 is 9.8×10^{-6} *M*.

121. $Ag^+(aq) + 2\ S_2O_3^{2-}(aq) \rightleftharpoons Ag(S_2O_3)_2^{3-}(aq)$

$$K_f = \frac{[Ag(S_2O_3)_2^{3-}]}{[Ag^+][S_2O_3^{2-}]^2}$$
$$5 \times 10^{13} = \frac{[Ag(S_2O_3)_2^{3-}]}{[Ag^+](0.233)^2}$$
$$2.7 \times 10^{13} = \frac{[Ag(S_2O_3)_2^{3-}]}{[Ag^+]}$$
$$4 \times 10^{-13} = \frac{[Ag^+]}{[Ag(S_2O_3)_2^{3-}]}$$

123. In order to calculate the pH of the saturated solution of lithium carbonate we will first calculate the concentration of carbonate ion supplied by the dissolving salt, Li_2CO_3. The carbonate ion, CO_3^{2-}, then hydrolyzes to produce OH^-. The pH will be obtained by calculating the $[OH^-]$ in solution.

	$Li_2CO_3(s)$	$\rightleftharpoons$	$2\,Li^+(aq)$	+	$CO_3^{2-}(aq)$
I			0		0
C			$+2x$		$+x$
E			$2x$		x

$$K_{sp} = [Li^+]^2[CO_3^{2-}]$$

$$(2x)^2(x) = 1.7 \times 10^{-3}$$

$$x = 7.5 \times 10^{-2}$$

$$x = [CO_3^{2-}]\ 7.5 \times 10^{-2} M$$

In order to calculate the pH, we will determine $[OH^-]$ by taking into account the hydrolysis of CO_3^{2-} and HCO_3^-.

Carbonate hydrolyzes according to the following equation:

	$CO_3^{2-}(aq) + H_2O(l)$	$\rightleftharpoons HCO_3^-(aq)$	+ $OH^-(aq)$
I	7.5×10^{-2}	0	0
C	$-x$	$+x$	$+x$
E	$7.5 \times 10^{-2} - x$	x	x

$$K_{b1} = \frac{[HCO_3^-][OH^-]}{[CO_3^{2-}]}$$

$$\frac{(x)(x)}{(7.5 \times 10^{-2} - x)} = 1.8 \times 10^{-4}$$

$$x = 3.7 \times 10^{-3} = [OH^-] = [HCO_3^-]$$

Bicarbonate hydrolyzes according to the following equation:

	$HCO_3^-(aq) + H_2O(l)$	$\rightleftharpoons H_2CO_3(aq)$	$+OH^-(aq)$
I	3.7×10^{-3}	0	3.6×10^{-3}
C	$-x$	$+x$	$+x$
E	$3.7 \times 10^{-3} - x$	x	$3.6 \times 10^{-3} + x$

$$K_{b2} = \frac{[H_2CO_3][OH^-]}{[HCO_3^-]}$$

$$\frac{(x)(3.7 \times 10^{-3} + x)}{(3.7 \times 10^{-3} - x)} = 2.3 \times 10^{-8}$$

$$x = 2.3 \times 10^{-8}$$

$$[OH^-] = 3.7 \times 10^{-3} + x = 3.7 \times 10^{-3} + 2.3 \times 10^{-8}$$
$$= 3.7 \times 10^{-3} M$$

K_{b2} does not make a significant contribution to the pH.

pOH = 2.43

pH = 11.57

CHAPTER 17 Electrochemistry and Electrical Energy

1. To keep the charge balanced between the two half-cells and to complete the circuit.

2. A wire would not be able to transfer ions without being part of the electrochemical reaction itself.

3. Oxidation occurs at the anode and reduction occurs at the cathode. Since electrons are one of the products in an oxidation half-reaction, the anode is assigned a minus sign. Similarly, the cathode is assigned a positive charge because electrons are consumed in a reduction half-reaction.

4. Ions can pass freely between the two aqueous solutions.

5. a. Anode: $Pb(s) \rightarrow Pb^{2+}(aq) + 2\,e^-$
   Cathode: $Zn^{2+}(aq) + 2\,e^- \rightarrow Zn(s)$
   b. $Pb(s) + Zn^{2+}(aq) \rightarrow Pb^{2+}(aq) + Zn(s)$
   c. Sketch of the cell for this reaction.

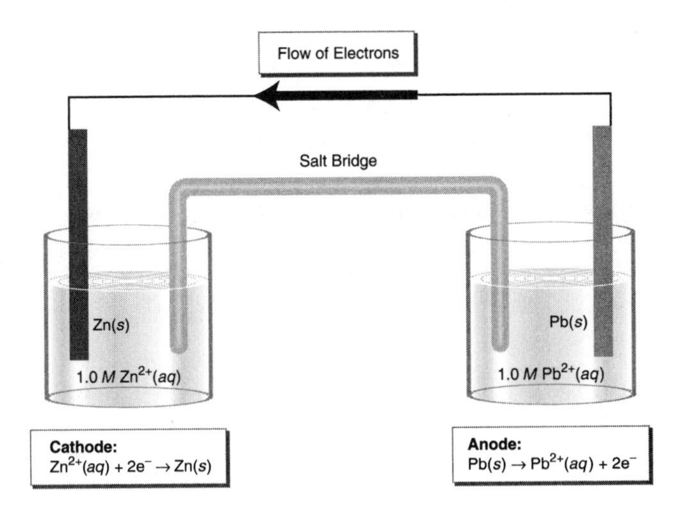

7. a. Anode: $Cd(s) + 2\,OH^-(aq) \rightarrow Cd(OH)_2(s) + 2\,e^-$
   Cathode: $2\,H_2O(l) + MnO_4^- + 3\,e^- \rightarrow MnO_2(s) + 4\,OH^-$
   b. $3\,Cd(s) + 4\,H_2O(l) + 2\,MnO_4^-(aq) \rightarrow 2\,MnO_2(s) + 2\,OH^- + 3\,Cd(OH)_2(s)$
   c. Sketch of the cell for this reaction:

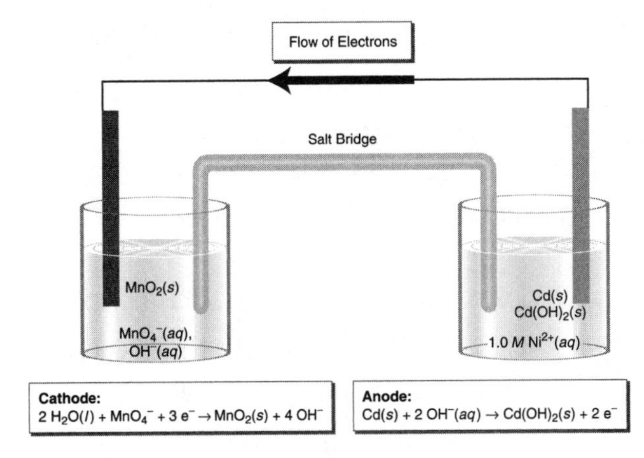

9. a. Six electrons are transferred; $n = 6$ for this reaction.
   b. In K_2FeO_4, the iron is in the +6 oxidation state.
   In Fe_2O_3, the iron is in the +3 oxidation state.
   In Zn, the zinc is in the zero-oxidation state.
   In ZnO, the zinc is in the +2 oxidation state.
   In K_2ZnO_2, the zinc is in the +2 oxidation state.
   c. The anode is made of $Zn(s)$.

11. a. Bromine is being oxidized (−1 to 0) and chlorine is being reduced (+7 to −1).
   b. Chlorine is both oxidized (0 to +1) and reduced (0 to −1).
   c. This is not a redox reaction.

13. We know that $\Delta G^\circ = -nFE^\circ_{cell}$. In this equation, ΔG° is directly proportional to E°_{cell} when comparing two reactions with equal values of n. Assuming that the two voltaic cells have the same value of n, the value of ΔG° for cell with the larger potential will be double that of the cell with the smaller potential because of the 2:1 ratio of cell potentials.

14. In electrochemical work, electric charge is being moved against Coulombic forces through an external circuit.

15. a.
$$\Delta G^\circ = \sum n\Delta G^\circ_{f,products} - \sum m\Delta G^\circ_{f,reactants}$$
$$= (+65.5\text{ kJ } + 0\text{ kJ}) - [2(50.0\text{ kJ})]$$
$$= -34.5\text{ kJ}$$

$$\Delta G^\circ = -nFE^\circ_{cell}$$
$$-3.45 \times 10^{-4}\text{ J/mol} = -(1)(96485\text{ C/mol})E^\circ_{cell}$$
$$E^\circ_{cell} = +0.358\text{ V}$$

b.
$$\Delta G^\circ = \sum n\Delta G^\circ_{f,products} - \sum m\Delta G^\circ_{f,reactants}$$
$$= (+77.1\text{ kJ} + -78.9\text{ kJ}) - (0\text{ kJ} + -4.7\text{ kJ})$$
$$= +2.9\text{ kJ}$$

$$\Delta G^\circ = -nFE^\circ_{cell}$$
$$+2.9 \times 10^{3}\text{ J/mol} = -(1)(96485\text{ C/mol})E^\circ_{cell}$$
$$E^\circ_{cell} = -0.030\text{ V}$$

17. Anode: $Al + 3\,OH^- \rightarrow Al(OH)_3 + 3\,e^-$
Cathode: $O_2 + 2\,H_2O + 4\,e^- \rightarrow 4\,OH^-$

Overall: $4\,Al(s) + 3\,O_2(g) + 6\,H_2O(l) \rightarrow 4\,Al(OH)_3(s)$

19. a.
$$6\,H^+(aq) + 2\,MnO_4^-(aq) + 5\,H_2O_2(aq) \rightarrow 2\,Mn^{2+}(aq) + 8\,H_2O(l) + 5\,O_2(g)$$

b.
$$4\,H^+(aq) + MnO_2(s) + 2\,Cl^-(aq) \rightarrow Mn^{2+}(aq) + 2\,H_2O(l) + Cl_2(g)$$

c.
$$4\,H^+(aq) + 3\,MnO_4^{2-}(aq) \rightarrow 2\,MnO_4^-(aq) + 2\,H_2O(l) + MnO_2(s)$$

21. The platinum serves only as an inert conductor.

22. Yes. It is possible to build a battery in which the anode chemistry involves two soluble species if an inert electrode is used.

23. The cell potential should not be affected by the increase in the thickness of the zinc anode.

24. The potential will be smaller for the Ni/Zn cell.
For the Cu/Zn cell:

$$E^\circ_{cell} = E^\circ_{cathode} - E^\circ_{anode}$$
$$= +0.3419\text{ V} - (-0.7618\text{ V})$$
$$= +1.1037\text{ V}$$

For the Ni/Zn cell:

$$E^\circ_{cell} = E^\circ_{cathode} - E^\circ_{anode}$$
$$= -0.257\text{ V} - (-0.7618\text{ V})$$
$$= +0.505\text{ V}$$

25. The reaction at the anode is an oxidation and we must change the sign of the value listed to get the anode potential since the table contains only reduction potentials. This change of sign us is taken care of by subtracting E°_{anode} from $E^\circ_{cathode}$. This is analogous to switching the sign of ΔH when we reverse a thermochemical equation.

26.
$$2\,H_2O(l) + 2\,e^- \rightarrow H_2(g) + 2\,OH^-(aq)$$
$$E^\circ_{red} = -0.8277\text{ V}$$

If the above reaction were assigned a standard reduction potential of 0 V, this would shift all other reduction potentials by +0.8277 V.

a. $E^\circ_{H^+} = 0\text{ V} + 0.8277\text{ V} = +0.8277\text{ V}$

b. $E^\circ_{Fe^{3+}} = 0.777\text{ V} + 0.8277\text{ V} = +1.605\text{ V}$

c.
$$E^\circ_{cell} = E^\circ_{cathode} - E^\circ_{anode}$$
$$= (1.6913\text{ V} + 0.8277\text{ V}) - (-0.126\text{ V} + 0.8277\text{ V}) = +1.817\text{ V}$$

Note that this is the same potential as when $E^\circ_{H^+} = 0.00$ V, since E°_{cell} is based on the difference in standard reduction potentials. Shifting both individual potentials by the same amount will have no effect on their difference.

27. No. The reaction will not proceed since E°_{cell} is negative.

$$E^\circ_{cell} = E^\circ_{cathode} - E^\circ_{anode}$$
$$= +0.3419\text{ V} - (+0.7996\text{ V})$$
$$= -0.4577\text{ V}$$

28. Yes, The reaction will proceed since E°_{cell} is positive.

$$E^\circ_{cell} = E^\circ_{cathode} - E^\circ_{anode}$$
$$= -0.136\text{ V} - (-0.403\text{ V})$$
$$= +0.267\text{ V}$$

29. a.
$$E^\circ_{cell} = E^\circ_{cathode} - E^\circ_{anode}$$
$$= -0.136\text{ V} - (+0.3419\text{ V})$$
$$= -0.478\text{ V}$$

$$\Delta G^\circ = -nFE^\circ_{\text{cell}}$$
$$= -(2)(96485 \text{ C/mol})(-0.478 \text{ V})$$
$$= +92200 \text{ J} = +92.2 \text{ kJ}$$

b.
$$E^\circ_{\text{cell}} = E^\circ_{\text{cathode}} - E^\circ_{\text{anode}}$$
$$= -0.257 \text{ V} - (-0.7618 \text{ V})$$
$$= +0.505 \text{ V}$$

$$\Delta G^\circ = -nFE^\circ_{\text{cell}}$$
$$= -(2)(96485 \text{ C/mol})(+0.505 \text{ V})$$
$$= -97400 \text{ J} = -97.4 \text{ kJ}$$

31. a. $Zn(s) + MnO_2(s) + H_2O(l) \rightarrow Mn(OH)_2(s) + ZnO(s)$

Manganese is reduced at the cathode and zinc is oxidized at the anode.

b. $S_8(s) + 16\ Na(s) \rightarrow 16\ Na^+(aq) + 8\ S^{2-}(aq)$

S_8 is reduced at the cathode and Na is oxidized at the anode.

c.
$$4\ Al(s) + 4\ OH^-(aq) + 3\ O_2(g) + 6\ H_2O(l)$$
$$\rightarrow 4\ Al(OH)_4{}^-(aq)$$

O_2 is reduced at the cathode and Al is oxidized at the anode.

33. From Appendix 6:

$$Cu^{2+} + 2\,e^- \rightarrow Cu \qquad E^\circ_{\text{red}} = +0.3419 \text{ V}$$
$$Cd^{2+} + 2\,e^- \rightarrow Cd \qquad E^\circ_{\text{red}} = -0.403 \text{ V}$$

In order to obtain a positive standard cell potential, Cu^{2+} must be reduced and Cd must be oxidized. Electron flow is from anode to cathode. $Cu^{2+}(aq)$ will be reduced at the cathode to form $Cu(s)$, and $Cd(s)$ will be oxidized at the anode to form $Cd^{2+}(aq)$.

35. a.
$$NiOOH(s) + 2\ H_2O(l) + TiZr_2(s)$$
$$\rightarrow Ni(OH)_2(s) + 2\,OH^-(aq) + TiZr_2H(s)$$

b.
$$E^\circ_{\text{cell}} = E^\circ_{\text{cathode}} - E^\circ_{\text{anode}}$$
$$= 1.32 \text{ V} - 0.00 \text{ V}$$
$$= +1.32 \text{ V}$$

37. No. The equal molar concentrations of the oxidizing and reducing agents make the concentration-dependent portion of the Nernst equation go to zero.

$$E_{\text{cell}} = E^\circ_{\text{cell}} - \frac{0.0592}{n} \log \frac{[Zn^{2+}]}{[Cu^{2+}]}$$
$$= +1.10 \text{ V} - \frac{0.0592}{2} \log \frac{(0.25)}{(0.25)} = +1.10 \text{ V}$$

38. Yes, cell potentials are concentration dependent. We can check this by using the Nernst equation. $[Zn^{2+}]$ will increase and $[Cu^{2+}]$ will decrease as the reaction occurs. This will cause a decrease in cell potential (see sample calculation that follows). E_{cell} will continue to decrease until it equals zero. At that point the reaction will be at equilibrium.

$$E_{\text{cell}} = E^\circ_{\text{cell}} - \frac{0.0592}{n} \log \frac{[Zn^{2+}]}{[Cu^{2+}]}$$
$$= +1.10 \text{ V} - \frac{0.0592}{2} \log \frac{(1.30)}{(0.70)} = +1.09 \text{ V}$$

39.
$$MH(s) + NiO(OH)(s) \rightarrow M(s) + Ni(OH)_2(s)$$

The potential will not change. The concentration of the aqueous electrolyte is not included in the reaction quotient (Q) portion of the Nernst equation since it is not found in the balanced redox equation. Therefore a change in [KOH] or $[OH^-]$ will not affect E_{cell}.

40.
$$Cd(s) + 2\ NiO(OH)(s) + 2\ H_2O(l)$$
$$\rightarrow 2\ Cd(OH)_2(s) + 2\ Ni(OH)_2(s)$$

As the cell discharges, the potential does not change because all species in the balanced redox equation are solids or pure liquids. (The concentrations of pure solids and liquids are considered constant.) As the electrochemical reaction proceeds, the amount of reactants and products will change, but their concentrations will remain constant. Therefore we expect E_{cell} to remain fairly constant as the reaction proceeds.

41.
$$E_{\text{cell}} = E^\circ_{\text{cell}} - \frac{0.0592}{n} \log \frac{[Fe^{2+}][Cr^{3+}]}{[Fe^{3+}][Cr^{2+}]}$$
$$= [+0.770 \text{ V} - (-0.41 \text{ V})]$$
$$- \frac{0.0592}{1} \log \frac{(2.5 \times 10^{-4})(2.5 \times 10^{-4})}{(1.50 \times 10^{-3})(1.50 \times 10^{-3})}$$
$$= +1.27 \text{ V}$$

43.
$$\log K = \frac{nE^\circ_{\text{cell}}}{0.0592} = \frac{(1)[0.770 - (-0.41)]}{0.0592} = 19.9$$
$$K = 8.56 \times 10^{19}$$
$$= 9 \times 10^{19}$$

45.
$$E = E^\circ - \frac{0.0592}{n} \log \frac{P_{H_2}}{[H^+]^2}$$
$$E = 0.000 \text{ V} - \frac{0.0592}{2} \log \frac{P_{H_2}}{[1 \times 10^{-7}]^2}$$
$$E_{\text{red}} = -0.414 \text{ V}$$

47.

$$E_{\text{cell}} = E^\circ_{\text{cell}} - \frac{0.0592}{n} \log \frac{[SO_4^{2-}]^3[OH^-]^2}{[MnO_4^-]^2[SO_3^{2-}]^3}$$

$$E_{\text{cell}} = +0.59 - (-0.92\text{ V}) - \frac{0.0592}{6}$$

$$\times \log \frac{[0.178]^3[0.0100]^2}{[0.150]^2[0.256]^3}$$

$$E_{\text{cell}} = +1.54\text{ V}$$

If the reaction proceeds spontaneously as written, the initial value of E_{cell} will be greatest at the instant the reaction begins and will decrease as reactants are turned into products. Prove this for yourself by performing sample calculations with the Nernst equation.

49. a.

$$E^\circ_{\text{cell}} = E^\circ_{\text{cathode}} - E^\circ_{\text{anode}}$$

$$= 0.96\text{ V} - 0.3419\text{ V}$$

$$= +0.62\text{ V}$$

b. $$E_{\text{cell}} = E^\circ_{\text{cell}} - \frac{0.0592}{n} \log \frac{(P_{NO})^2[Cu^{2+}]^3}{[NO_3^-]^2[H^+]^8}$$

$$= 0.62\text{ V} - \frac{0.0592}{6} \log \frac{(0.00150)^2(0.0375)^3}{(0.0250)^2(0.100)^8}$$

$$= +0.61\text{ V}$$

51. a.

$$E^\circ_{\text{cell}} = E^\circ_{\text{cathode}} - E^\circ_{\text{anode}}$$

$$= 1.50\text{ V} - 1.23\text{ V}$$

$$= +0.27\text{ V}$$

b. $$E_{\text{cell}} = E^\circ_{\text{cell}} - \frac{0.0592}{n} \log \frac{[NO_3^-][H^+]^2}{[NH_4^+](P_{O_2})^2}$$

At equilibrium, $E_{\text{cell}} = 0$, so

$$0 = 0.27\text{ V} - \frac{0.0592}{8} \log \frac{[NO_3^-][2.5 \times 10^{-6}]^2}{[NH_4^+](0.21)^2}$$

$$36.5 = \log \frac{[NO_3^-][2.5 \times 10^{-6}]^2}{[NH_4^+](0.21)^2}$$

$$3.07 \times 10^{36} = \frac{[NO_3^-][2.5 \times 10^{-6}]^2}{[NH_4^+](0.21)^2}$$

$$\frac{[NO_3^-]}{[NH_4^+]} = 2.2 \times 10^{46}$$

53. a. Both batteries will likely have the same E°_{cell} and E_{cell}.
b. Both batteries will likely utilize the same anode half-reaction.
c. The total masses of electrode material are likely to be different since the CCA rating is directly related to the amounts of Pb and PbO_2 in the battery.
d. The number of cells is likely to be the same since both are 12 V batteries and under standard conditions each cell has a potential of about 2 V. Connecting the six cells in parallel yields a 12 V battery.
e. Since both are lead-acid batteries, their electrolytes will be composed of the same materials.
f. The combined surface areas of the electrodes are likely to be different since an increase in surface area of the electrodes increases the capacity of the cell to deliver current.

54. a. F Their cell potentials are different, 1.10 V for Zn/Cu and 0.745 V for Cd/Cu.
b. F The masses of the anodes will be the same since both are made of 1 mole of Cu; however, the masses the cathodes will be different since Cd has a larger molar mass than Zn.
c. T The quantities of electrical charge they can produce are equal, since both reactions have $n = 1$ and the amount of reactants and products present are equal.
d. F Since $\Delta G^\circ = -nFE^\circ_{\text{cell}}$, the reaction with the higher potential, Zn/Cu, can produce more energy.

55. The reaction involving Al can produce the greatest quantity of charge per gram of anode material. Three moles of electrons can be transferred from 27 g of an aluminium anode, whereas the reaction involving Cd can only produce 2 moles of electrons per 112 g of anode material.

57. The units of power break down to CV/s (J/s). Whichever reaction produces more CV will be able to produce the most power per gram at a given amperage. The voltage values used in this problem were either given in the problem or obtained from the proper appendix table.

For the reaction having the Cd anode:

$$E^\circ_{\text{cell}} = E^\circ_{\text{cathode}} - E^\circ_{\text{anode}}$$

$$E^\circ_{\text{cell}} = +1.34V$$

$$1\text{ g Cd}(s)\left(\frac{\text{mol Cd}}{112.4\text{ g Cd}}\right)\left(\frac{2\text{ mol e}^-}{1\text{ mol Cd}}\right)\left(\frac{96{,}485\text{ C}}{\text{mol e}^-}\right)$$

$$\times (+1.34\text{ V}) = 2.3 \times 10^3\text{ CV}$$

For the reaction having the Li anode:

$$1\text{ g Li}(s)\left(\frac{\text{mol Li}}{6.94\text{ g Li}}\right)\left(\frac{1\text{ mol e}^-}{1\text{ mol Li}}\right)\left(\frac{96{,}485\text{ C}}{\text{mol e}^-}\right)$$

$$\times (+2.69\text{ V}) = 3.7 \times 10^4\text{ CV}$$

The reaction with the Li(s) anode can produce more power per gram of anode material.

59. No, Mg^{2+}($E^\circ_{\text{red}} = -2.37$ V) is easier to reduce than Na^+($E^\circ_{\text{red}} = -2.71$ V) and so we do not expect the presence of NaCl to interfere with the electrolysis of $MgCl_2$.

60. Br_2 will be formed first in the electrolysis since Br^- ($E^\circ_{\text{red}} = -2.37$ V) is easier to oxidize than Cl^- ($E^\circ_{\text{red}} = +1.066$ V).

61. Strongly negative cathode potentials (< −2.0 V vs. SHE) are needed to reduce the ions of the alkaline earth metals (including Mg^{2+}) to their free metals. However, water is reduced at less negative cathode potentials:

$$2\,H_2O(l) + 2\,e^- \quad H_2(g) + 2\,OH^-(aq) \quad E^\circ = -0.8277\text{ V}$$

and so the water in seawater would be reduced instead of Mg^{2+}.

62. A strong electrolyte, such as Na_2CO_3, is added to increase the conductivity of the water. Greater conductivity leads to greater current flow and more rapid electrolysis.

63. If NaCl were added, Cl^- ions would be oxidized at the anode to Cl_2 instead of water being oxidized to O_2. The two half-reactions have similar E° values (1.358 V for the reduction of Cl_2; 1.229 V for the reduction of O_2) but standard conditions means pH = 0.00 for the reduction of O_2 (or the oxidation of water). At neutral pH Cl^- ions are much more easily oxidized than molecules of H_2O.

64. A cathode potential $\frac{0.0591}{2}\log\left(\frac{1}{1\times10^{-4}}\right)$V more negative than 0.34 V vs. SHE is necessary to reduce the concentration of Cu^{2+} to $1\times10^{-4}\,M$.

65. $$1.7\text{ amp}\cdot\text{hr}\left(\frac{3600\text{ s}}{\text{hr}}\right)\left(\frac{1\text{ mol e}^-}{96{,}485\text{ A}\cdot\text{s}}\right)\left(\frac{1\text{ mol Ag}}{1\text{ mol e}^-}\right)$$

$$\times\left(\frac{107.9\text{ g Ag}}{\text{mol Ag}}\right) = 6.8\text{ g Ag}$$

67. 5.00 g Cd is equivalent to 0.0445 moles of Cd.
4.10 g NiO(OH) is equivalent to 0.0447 moles of NiO(OH).

$$Cd(s) + 2\,OH^-(aq) \rightarrow Cd(OH)_2(s) + 2\,e^-$$

$$NiO(OH)(s) + H_2O(l) + e^- \rightarrow NiO(OH)_2(s) + OH^-(aq)$$

Since the Cd half-reaction requires 2 electrons and the NiO(OH) half-reaction only requires 1 electron, we should base our calculation on 50% of the NiO(OH) being reacted.

50% of 0.0447 moles of NiO(OH) is 0.0224 moles of NiO(OH).

$$0.0224\text{ moles NiO(OH)}\left(\frac{1\text{ mol e}^-}{\text{mol NiO(OH)}}\right)\left(\frac{96{,}485\text{C}}{\text{mol e}^-}\right)$$

$$= 2160\text{ C} = 2160\text{ A}\cdot\text{s}$$

$$\frac{2160\text{ A}\cdot\text{s}}{2.00\text{ A}} = 1080\text{ s or about 18 minutes}$$

69. a. $$2\,H_2O \rightarrow O_2 + 4\,H^+ + 4\,e^-$$

$$(1\text{ hr})(0.025\text{ A})\left(\frac{3600\text{ s}}{\text{hr}}\right)\left(\frac{\text{mol e}^-}{96{,}485\text{ A}\cdot\text{s}}\right)\left(\frac{1\text{ mol }O_2}{4\text{ mol e}^-}\right)$$

$$= 2.3\times10^{-4}\text{ mol }O_2$$

$$V = \frac{nRT}{P}$$

$$= \frac{(2.3\times10^{-4}\text{ mol})(0.0821\text{ L}\cdot\text{atm/K}\cdot\text{mol})(273\text{ K})}{1.00\text{ atm}}$$

$$= 0.0052\text{ L }O_2$$

b. No, because in pH 8 seawater Cl^- ions are more easily oxidized to Cl_2 gas than molecules of H_2O are oxidized to O_2 gas (see the solution to Problem 63).

71. $$E = E^\circ - \frac{0.0592}{n}\log\frac{1}{[Ni^{2+}]}$$

$$E = -0.257\text{ V} - \frac{0.0592}{2}\log\frac{1}{0.35} = -0.27\text{ V}$$

The cathode potential would have to be −0.270 V or a more negative value.

73. Advantages of hybrid systems over all-electric systems include the ability to continue to utilize widely available gasoline when necessary, increased driving range, more efficient propulsion system, lower cost because fewer batteries are needed—and many hybrids have regenerative braking systems that harness some of the energy normally wasted in the deceleration process. Disadvantages include higher emissions.

74. Advantages of fuel cell systems over internal combustion engines include the fact that water is the only emission and less energy is wasted in converting fuels into mechanical energy.

75. a.

Species	***Oxidation States***
CH_4	C, +4; H, +1
H_2O	H, +1
CO	C, +2
H_2	H, 0
CO_2	C, +4

b. First step:

$$\Delta G^\circ = \sum n\Delta G^\circ_{\text{f,products}} - \sum m\Delta G^\circ_{\text{f,reactants}}$$
$$= [-110.5\text{ kJ} + 3(0)\text{ kJ}]$$
$$- [-74.8\text{ kJ} + (-241.8\text{ kJ})]$$
$$= +206.1\text{ kJ}$$

Second step:

$$\Delta G^\circ = \sum n\Delta G^\circ_{\text{f,products}} - \sum m\Delta G^\circ_{\text{f,reactants}}$$
$$= [0\text{ kJ} + (-393.5\text{ kJ})]$$
$$- [-110.5\text{ kJ} + (-241.8\text{ kJ})]$$
$$= -41.2\text{ kJ}$$

Overall:

$$\Delta G^\circ = +206.1\text{ kJ} + (-41.2\text{ kJ}) = 164.9\text{ kJ}$$

CHAPTER 18 Materials Chemistry: Past, Present, and Future

1. The coppers-containing mineral is completely converted by oxidation into copper(II) oxide. The copper(II) oxide is then reacted with CO which produces copper metal. Overall, an ionic form of copper is reduced to elemental copper.

2. Cu^{+} is oxidized to Cu^{2+}.
   O_2 is reduced to O^{2-}.
   S^{2-} is oxidized to S^{+4}.

3. Higher temperatures promotes melting because the $C(s)$ can be converted to $CO(g)$. The gaseous CO reacts with the solid mineral more readily than does the solid carbon.

4. $AlO_2^{-} + 2\,H_2O \rightarrow Al(OH)_4^{-}$

5. Magnetite, Fe_3O_4, contains the highest percent iron by mass (72.4%) of the four given ores. $FeCO_3$, Fe_2O_3, and FeS_2 contain 48.2%, 69.9%, and 46.3% iron by mass, respectively.

7. $2\,FeCuS_2(s) + 3\,O_2(g)$
   $\rightarrow 2\,FeO(s) + 2\,SO_2(g) + 2\,CuS(s)$
   $FeO(s) + SiO_2(s) \rightarrow FeSiO_3(s)$
   $FeCuS_2$: Fe has an oxidation state of +3.
   FeO : Fe has an oxidation state of +2.
   $FeSiO_3$: Fe has an oxidation state of +2.

9. $2\,MgCaO_2(s) + FeSi(s) \rightarrow 2\,Mg(s) + Ca_2SiO_4(s) + Fe(s)$
   In the given reaction, FeSi is the reducing agent and $MgCaO_2$ is the oxidizing agent. The silicon in FeSi has an oxidation state of zero and is oxidized to the +4 oxidation state in the compound Ca_2SiO_4. The magnesium in $MgCaO_2$ has an oxidation state of +2 and is reduced to an oxidation state of zero in elemental Mg.

11. $2\,Ga_2O_3(s) + 2\,Mg(s) + O_2(g) \rightarrow 2\,MgGa_2O_4(s)$

    Ga_2O_3: Ga, +3
    Mg: Mg, 0
    $MgGa_2O_4$: Mg, +2; Ga, +3

13. $Cu_2O(s) + CO(g) \rightarrow CO_2(g) + 2\,Cu(s)$

$$\Delta H^\circ_{rxn} = \sum n\Delta H^\circ_{f,\ products} - \sum m\Delta H^\circ_{f,\ reactants}$$
$$= [(-393.5\,kJ) + 2(0\,kJ)] - [(-168.6\,kJ) + (-110.5\,kJ)]$$
$$= -114.4\,kJ$$

$$CuCO_3 \cdot Cu(OH)_2(s) + 2\,CO(g) \rightarrow 2\,Cu(s) + 3\,CO_2(g) + H_2O(g)$$

$$\Delta H^\circ_{rxn} = \sum n\Delta H^\circ_{f,\ products} - \sum m\Delta H^\circ_{f,\ reactants}$$
$$= [2(0\,kJ) + 3(-393.5\,kJ) + (-136.3\,kJ)] - [(-1051.4\,kJ) + 2(-110.5\,kJ)]$$
$$= -44.4\,kJ$$

$$(CuCO_3)_2 \cdot Cu(OH)_2(s) + 3\,CO(g) \rightarrow 3\,Cu(s) + 5\,CO_2(g) + H_2O(g)$$

$$\Delta H^\circ_{rxn} = \sum n\Delta H^\circ_{f,\ products} - \sum m\Delta H^\circ_{f,\ reactants}$$
$$= [3(0\,kJ) + 5(-393.5\,kJ) + (-136.3\,kJ)] - [(-163.6\,kJ) + 3(-110.5\,kJ)]$$
$$= -1608.7\,kJ$$

15. Pure gold is more malleable than white gold. The mixture of different size atoms in the solid phase of white gold makes it more difficult for the atoms to slide past one another in the solid state. The result is that pure gold is more malleable than white gold.

16. There is no difference between a solid solution and a homogeneous alloy. *An alloy is a solution* of two or more metals and solutions are homogeneous mixtures.

17. Zinc occurs only as zinc compounds and not as free Zn because it is more readily oxidized than Cu. Copper ions are more readily reduced to copper metal than zinc ions are to zinc metal.

18. The alloy melts at a lower temperature than either component because the mixture of different size atoms in the solid phase weakens the metallic bonds that hold the lattice together.

19. In a lattice, there is one octahedral hole per atom that defines the lattice and there are two tetrahedral holes per atom that defines the lattice. A face-centered unit cell contains four atoms and hence contains four octahedral holes and eight tetrahedral holes.
    a. If the face-centered unit cell is defined by atom of A and atoms of B occupy all of the octahedral holes in the unit cell, then the unit cell contains two atoms of A and four atoms of B and the formula of the alloy would be AB.
    b. If the face-centered unit cell is defined by atom of A and atoms of B occupy half of the octahedral holes in the unit cell, then the unit cell contains four atoms of A and two atoms of B and the formula of the alloy would be A_2B.
    c. If the face-centered unit cell is defined by atom of A and atoms of B occupy one fourth of the octahedral holes in the unit cell, then the unit cell contains four atoms of A and four atom of B and the formula of the alloy would be A_4B.

21. In a lattice, there is one octahedral hole per atom that defines the lattice. There are two tetrahedral holes per atom that defines the lattice. A face-centered unit cell contains four atoms and hence contains four octahedral holes. Since there are five octahedral holes for every five lattice atoms, and there is one atom of B for every five atoms of A one fifth of the octahedral holes are occupied.

23. A face-centered unit cell contains four atoms, three X atoms from the six face atoms ($6 \times \frac{1}{2}$) and one Y atom from the eight corner atoms ($8 \times \frac{1}{8}$). The formula of this alloy would be X_3Y.

25. The radius ratio of the two elements is 77pm/135pm = 0.57. Since this value is between 0.41 and 0.71, we expect the carbon to occupy octahedral holes.

27. a.
$$4.00\,\text{hr}\left(\frac{3600\,\text{s}}{1\,\text{hr}}\right)\left(\frac{1.00 \times 10^5\,\text{C}}{\text{s}}\right) \times \left(\frac{1\,\text{mol e}^-}{96{,}485\,\text{C}}\right)\left(\frac{24.31\,\text{g Mg}}{2\,\text{mol e}^-}\right) = 1.81 \times 10^5\,\text{g Mg}$$

    b.
$$4.00\,\text{hr}\left(\frac{3600\,\text{s}}{1\,\text{hr}}\right)\left(\frac{1.00 \times 10^5\,\text{C}}{\text{s}}\right) \times \left(\frac{1\,\text{mol e}^-}{96{,}485\,\text{C}}\right)\left(\frac{1\,\text{mol Cl}_2}{2\,\text{mol e}^-}\right)$$
$$= 7.46 \times 10^3\,\text{mol Cl}_2$$

    The volume of Cl_2 is calculated as follows:

$$PV = nRT$$
$$(1.00\,\text{atm})V = (7.46 \times 10^3\,\text{mol}) \times (8.314\,\text{L} \cdot \text{atm/K} \cdot \text{mol})(1073\,\text{K})$$
$$V = 6.66 \times 10^7\,\text{L Cl}_2(g)$$

29. When kaolinite is fired, the water that was adsorbed onto the clay particles is first driven off. Next, some hydroxide ions are decomposed to water and oxide ions. The water formed in this process is driven off and the oxides remain. Structural changes then occur and lastly some silica is formed. The process transforms $Al_2Si_2O_5(OH)_4$ into $Al_2O_3 \cdot 2\,SiO_2$.

30. When fired, kaolinite does not shrink as much as most other clays and it can be molded into a shape that is retained after firing.

31. Metal oxides have higher melting points than free metals because there are greater attractive forces between the metal ions and the oxide ions in these compounds than between the metal atoms of the free metals. The metal oxides have ion-ion attractive forces while the free metals only have London dispersion forces.

32. a. ZrO_2 will have the higher melting point than CaF_2 since the ions that make it up have charges of +4 and −2, while CaF_2 is made up of ions with charges of +2 and −1. Based on Coulomb's law, we expect a compound with larger ionic charges to have greater electrostatic attractive forces and therefore a higher melting point.
    b. CeO_2 will have a higher melting point than UO_2. These compounds are made up of ions with charges of the same magnitude (cations +4, anions −2). For ions with charges of the same magnitude, smaller ions will have greater electrostatic attraction because of Coulomb's law, so should have higher melting points.
    c. BeO will have a higher melting point than MgO. The cations and anions in these species have charges of the same magnitude; therefore the electrostatic attraction will depend on the size of the ions. Be^{2+} is smaller than Mg^{2+}, so BeO will have stronger ion–ion attractive forces and will have the higher melting point.

33. Since both of these compounds are metal oxides, the aluminum of Al_2O_3 is the dopant or impurity. The

substitution of Al for Zn would create an n-type semiconductor. Each atom of the Al impurity has three valence electrons while each atom of the replaced Zn contains two valence electrons. Aluminum is considered an electron rich additive relative to Zn. Semiconductors that contain electron rich additives are called n-type semiconductors.

34. Since both of these compounds are metal oxides, the lithium of Li_2O is the dopant or impurity. The substitution of Li for Zn would create an p-type semiconductor. Each atom of the Li impurity has one valence electron while each atom of the replaced Zn contains two valence electrons. Lithium is considered an electron deficient additive relative to Zn. Semiconductors that contain electron deficient additives are called p-type semiconductors.

35. $$Al_2O_3 + 3\,C + N_2 \rightarrow AlN + 3\,CO$$

In the above reaction carbon is oxidized (0 to +2) and nitrogen is reduced ($0 \rightarrow -3$).

$$3\,SiO_2 + 6\,C + 2\,N_2 \rightarrow Si_3N_4 + 6\,CO$$

In the above reaction carbon is oxidized (0 to +2) and nitrogen is reduced (0 to −3).

37. The larger spacing corresponds to the compound containing potassium. The only difference in the formulas of the two compounds is that muscovite contains a single potassium ion where paragonite contains a single sodium ion. Potassium ions are larger than sodium ions, so we expect a larger spacing in the muscovite.

39. a. Al^{3+} will be smaller than Ca^{2+}.
    b. Al^{3+} will likely fill tetrahedral holes and Ca^{2+} will likely fill octahedral holes.

41. $$3\,Al_2Si_2O_5(OH)_4 \rightarrow Al_6Si_2O_{13} + 4\,SiO_2 + 6\,H_2O$$

44.99% of the mass is lost in the conversion of kaolinite to mullite.

43. The band gap for InN corresponds to radiation with wavelength 620 nm, which is in the visible region.

For InN:

$$\lambda = \frac{hc}{E} = \frac{(6.626 \times 10^{-34}\ \text{J} \cdot \text{s})(2.998 \times 10^{8}\ \text{m/s})}{(1.929 \times 10^{5}\ \text{J/mol})/(6.022 \times 10^{23}\ \text{mol}^{-1})}$$

$$= 6.20 \times 10^{-7}\ \text{m} \times \frac{10^{9}\ \text{nm}}{\text{m}}$$

$$= 620\ \text{nm}$$

Similar calculations show that AlN emits radiation at 206 nm and GaN emits radiation at 371 nm.

45. $$\Delta G^\circ = -RT \ln K$$

$$183\ \text{kJ/mol} = -RT \ln K$$

$$1.83 \times 10^{5}\ \text{J/mol} = -(8.314\ \text{J/K} \cdot \text{mol})(1400) \ln K$$

$$-15.722 = \ln K$$

$$K = 1.5 \times 10^{-7}$$

$$K = (P_{Zn})^2(P_{O_2})$$

$$1.5 \times 10^{-7} = (P_{Zn})^2(0.21\ \text{atm})$$

$$P_{Zn} = 8.4 \times 10^{-4}\ \text{atm}$$

47. Since the reaction involves many individual species combining into large polymeric units we expect $\Delta S < 0$. This leads us to expect $\Delta H < 0$. The reaction must be exothermic if $\Delta S < 0$. If not, the reaction would be nonspontaneous at all temperatures as are all reactions where $\Delta S < 0$ and $\Delta H > 0$.

48. Two purposes of the catalyst are to promote the reaction by lowering the activation energy of the reaction and to help control or manipulate the properties of the resultant polymer.

49. The carbons in $(CH_2)_n$ are sp^3.

50. Primary structure tells us only the order of amino acids in the protein.

    Secondary structure locates the amino acids that form β-sheets, β-turns, and so on, in the protein.

    Tertiary structure tells us how the various areas of the protein fold over on each other.

51. The primary structure of silk gives us the amino acid sequence. Ultimately it is the ordering of the amino acids that is responsible for the properties of silk, but we can be more specific by looking beyond the primary structure. Because of the pattern of the primary sequence, the secondary structure of silk contains areas of β-strands. When several regions of β-strands on different proteins interact, they form β-pleated sheets. These areas of β-pleated sheets impart strength to the silk protein. β-pleated sheets are an example of tertiary structure. There are also areas seemingly without order, which are appropriately called random coils. The random coils impart elasticity to the protein because they have little interchain hydrogen bonding, so these areas can stretch and contract as force is applied and removed from the silk fibers.

52. Hydrogen bonding is responsible for the helical secondary structure in proteins, while London dispersion forces are responsible for helical secondary structure in polypropylene.

53. No, in Kevlar, the π-electrons are not likely to be delocalized over the polymer chain. Delocalization along the entire polymer chain would be likely if there were alternating C—C and C=C bonds along the entire backbone of the chain.

54. No, the π-electrons in Nomex are not likely to be delocalized over the polymer chain. Delocalization along the entire polymer chain would be likely if there were alternating C—C and C=C bonds along the entire backbone of the chain.

55. Tyrosine, tryptophan, serine, threonine, lysine, arginine, histidine, aspartate, glutamate, asparagine, and glutamine can all form more than two hydrogen bonds.

56. Serine, in general, forms more interstrand hydrogen bonds than either glycine or alanine. It is likely that a higher percentage of serine will yield a less elastic polymer.

57. The polymerization of propene requires the breaking of one C=C bond requiring 614 kJ. The polymerization then requires the making of two C—C bonds per propene molecule releasing 348 kJ each. The diagram on the next page is included to help visualize the bonds broken and made and does not represent how the polymerization actually occurs. Also note that this enthalpy value obtained is for an infinitely long polymer without terminal units on the two ends of the polymer chain.

$$\Delta H = [614\,\text{kJ}] - [2(348\,\text{kJ})]$$
$$\Delta H = -82\,\text{kJ}$$

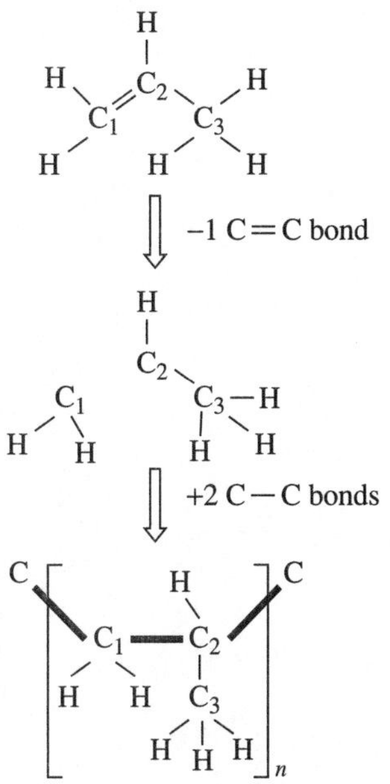

−1 C=C bond

+2 C−C bonds

59. $[-O-Si(CH_3)_2-]_n$

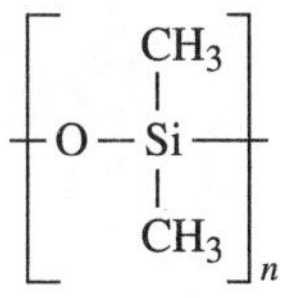

61. Silicones with $-CH_3$ or $-C_6H_5$ side chains are water repellent since their side chains are hydrophobic, or "water hating."

63. The polymer with more sp^3 hybridized atoms in its backbone will be more elastic. The presence of sp or sp^2 bonded atoms tends to make polymer backbones more rigid and/or less able to be stretched or compressed.

65. Both *a* and *b* could produce isotactic polymers. Isotactic polymers have a nonrandom order of side groups along the backbone; they are all found on one "side" or with the same orientation. A polymer made with monomer *c* would not be considered to have tacticity because it only has one possible arrangement of side groups.

67. Poly(thiophene) and poly(pyrrole) are both addition polymers. Their respective starting materials are thiophene and pyrrole.

Thiophene   Pyrrole

69. Poly(methylmethacrylate) is an addition polymer made from the following monomer:

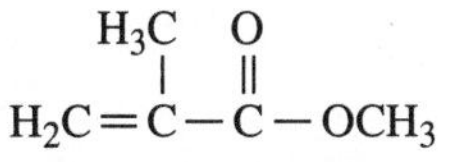

$$H_2C=C(CH_3)-C(=O)-OCH_3$$

Methyl methacrylate

The polymer on the right could be formed by addition polymerization of the following monomer:

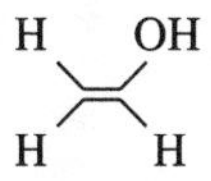

Vinyl alcohol

71. Poly(vinyl acetate)

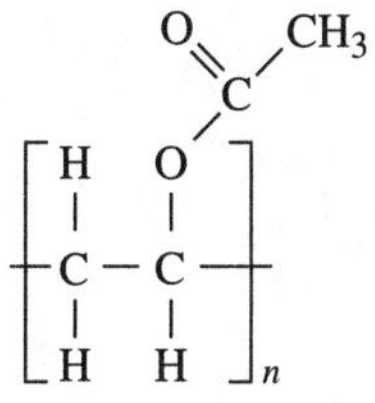

$[-CH_2-CH(OC(=O)CH_3)-]_n$